Springer Series in Advanced Manufacturing

Series editor

Duc Truong Pham, Cardiff, UK

For further volumes:
http://www.springer.com/series/7113

Elsa Henriques · Paulo Peças
Arlindo Silva
Editors

Technology and Manufacturing Process Selection

The Product Life Cycle Perspective

Springer

Editors
Elsa Henriques
Paulo Peças
Arlindo Silva
IDMEC, Instituto Superior Técnico
Universidade de Lisboa
Lisbon
Portugal

ISSN 1860-5168 ISSN 2196-1735 (electronic)
ISBN 978-1-4471-5543-0 ISBN 978-1-4471-5544-7 (eBook)
DOI 10.1007/978-1-4471-5544-7
Springer London Heidelberg New York Dordrecht

Library of Congress Control Number: 2013953217

Printed on acid-free paper

Springer is part of Springer Science+Business Media (www.springer.com)

Editorial Board

Preface

In a global market, competitive advantage lies not only on the mastering of existing processes and methodologies, but most of all on the ability to pursue different avenues, with an increased value. This can only be achieved with an up-to-date technological knowledge and scientific principles materialized in the design and manufacturing of new products, with the goal of protecting the environment and conserving resources, while encouraging economic progress, keeping in mind the need for sustainability. Design and process engineering problems are frequently of an ill-defined nature, demanding for the analysis and evaluation of complex alternative solutions, in which environmental, economic, and functional performance criteria interact in a complex net of influences, with an emergent behavior. Moreover, even when decisions are made in a well-defined and narrow timeframe, their effects are normally felt over a larger time sphere and scope domain, shaping the future further than anticipated and in eventually unsought ways.

Technology and manufacturing process selection is essential in the continuous improvement of existing products and processes as a key factor to competitiveness and sustainability. Technology-based innovation relies on the combination of design and manufacturing areas, bringing together a multidisciplinary team with different expertise and perspectives. The complexity of the decision-making process under such a widespread knowledge framework implies the use of efficient and reliable approaches. The analysis and synthesis mechanisms to support this decision-making process must also be effective in the early design phases and integrate all the aspects related with the life cycle stages of both product and technologies.

To deploy a technology evaluation and selection process under a life cycle scope, it is essential to capture all the evolutions and impacts of the selected alternatives, frequently supported on vague information and uncertain data. In fact, nowadays product developers need to address not only the production costs, but also all the costs incurred throughout the entire product life cycle (Life Cycle Cost -LCC). The estimation of all the costs associated with a product in a "cradle to grave" perspective—or, even in a broader way, from "cradle to cradle"—integrates the analysis of the impact of design for cost, design for maintainability, design for assembly, design for recycling, etc. With the aim of providing drivers and indicators for a sustainable engineering practice, it is also important to design and evaluate the technological alternatives on a life cycle environmental basis, namely involving Life Cycle Assessment (LCA) methods. Accordingly, the use of methodologies like LCA to estimate the environmental performance supports the disciplines of design for the environment, design for recycling, design for standards, etc.

The main reason for including a life cycle perspective in the early stages of product and process development is that decisions taken at the front end of the development largely influence the production of competitive products with high quality standards in regards to functional performance, cost and environmental impact for their entire life. Therefore, to better design for the entire life, Design-for-X strategies, supported by the corresponding tools, have been increasingly and successfully applied. These strategies drive the design team in the creation of products, processes, and services that achieve a specific target or that maximize the performance in a wide range of engineering fields (cost, environment, assembly, etc.). The problem then becomes one of striking a balance between different "optimizations," as optimizing for recycling will necessarily lead to a different outcome than optimizing for manufacturing and assembly, which further enhances the need to better understand the way in which these dispersed approaches/tools need to be used in a coherent and comprehensive way.

The consideration of all life cycle stages of a product in the early design phase allows a more complete perception of the product's value in the market and in society. This way of designing and developing a product can be called Design for the Life Cycle. To differentiate it from the regular DfX strategies, several authors prefer to denominate it as Life Cycle Engineering, understood as a decision-making methodology that considers functional performance, environmental, and cost dimensions throughout the duration of a product or, in a narrower sense, throughout the time horizon affected by an engineering decision, guiding design engineers toward informed decisions.

The research in Life Cycle Engineering challenges the academic world because it endorses a multidisciplinary approach on a problem solving framework. In fact the development of Life Cycle Engineering tools and its implementation in product design and development requires the collaboration of different areas of expertise during several phases of such a project. Therefore, the incorporation of concurrent engineering practices is recommended, if not mandatory.

In conclusion, the development of decision-making methodologies based on Life Cycle approaches is extremely important to support informed and reliable assessment and selection of technological solutions. Based only on singular types of performance or integrating several types of performance, these methodologies are under development by several research groups worldwide.

This book provides specific topics intending to contribute to an improved knowledge on Technology Evaluation and Selection in a Life Cycle Perspective. Although each chapter will present possible approaches and solutions, there are no recipes for success. Each reader will find his/her balance in applying the different topics to his/her own specific situation. Case studies presented throughout will help in deciding what fits best to each situation, but most of all any ultimate success will come out of the interplay between the available solutions and the specific problem or opportunity the reader is faced with. Contributions were accepted from 47 authors in seven countries from around the world: China, France, Germany, Italy, Portugal, Sweden, and the United States of America.

Editing a book embodies team work and represents considerable work from the authors, editors, and editorial advisory board. This collaborative teamwork involves keeping track of contacts of authors and their contributions, exchanging information and ideas, managing the review process, feeding back review to the authors, managing conflicting perspectives, and integrating contents into a reasonable structure, with the ultimate goal of developing a product that adds value to the readers' body of knowledge.

As team leaders we, the editors, have to thank our team members for the effort involved in this initiative. This book is primarily supported by the team of professionals from Springer. We thank them for the opportunity and constant support in editing the book, timely suggestions, prompt feedback, and friendly reminders about deadlines. To the Members of the Editorial Board, our gratitude for sharing with us their knowledge and experience in the support of the decision-making processes inherent to the project, for assisting in the review process, and for their help in shaping the book. We acknowledge all the authors, without whom there would be no book in the first place! Many contributions were not considered, despite their merit, either because they were out of the scope for this book, of time limitations, or other constraints. A special word to our home institution, the Instituto Superior Técnico of the Technical University of Lisbon, for providing the infrastructure, material resources, and logistics required for our work.

We hope the book will enlighten the reader in the same way it enlightened us during the editing process, and that its contents will help foster new and innovative research worldwide.

Elsa Henriques
Paulo Peças
Arlindo Silva

Contents

Product Architecture Decision Under Lifecycle Uncertainty Consideration: A Case Study in Providing Real-time Support to Automotive Battery System Architecture Design 1
Qi D. Van Eikema Hommes and Matthew J. Renzi

Consideration of Legacy Structures Enabling a Double Helix Development of Production Systems and Products 21
Magnus Wiktorsson

Six Sigma Life Cycle 33
Pedro A. Marques, Pedro M. Saraiva, José G. Requeijo and Francisco Frazão Guerreiro

On the Influence of Material Selection Decisions on Second Order Cost Factors 59
Marco Leite, Arlindo Silva and Elsa Henriques

Aircraft Industrialization Process: A Systematic and Holistic Approach to Ensuring Integrated Management of the Engineering Process 81
José Manuel Lourenço da Saúde and José Miguel Silva

Material Flow Cost Accounting: A Tool for Designing Economically and Ecologically Sustainable Production Processes 105
Ronny Sygulla, Uwe Götze and Annett Bierer

Life Cycle Based Evaluation and Interpretation of Technology Chains in Manufacturing 131
F. Klocke, B. Döbbeler, M. Binder, R. Schlosser and D. Lung

Selecting Manufacturing Process Chains in the Early Stage of the Product Engineering Process with Focus on Energy Consumption 153
Martin Swat, Horst Brünnet and Dirk Bähre

Manufacturing with Minimal Energy Consumption: A Product Perspective 175
Alexandra Pehlken, Alexandra Kirchner and Klaus-Dieter Thoben

Integrated Framework for Life Cycle-Oriented Evaluation of Product and Process Technologies: Conceptual Design and Case Study 193
Uwe Götze, Andrea Hertel, Anja Schmidt, Erik Päßler and Jörg Kaufmann

Life Cycle Engineering Framework for Technology and Manufacturing Processes Evaluation 217
Inês Ribeiro, Paulo Peças and Elsa Henriques

Proposal for an Architectural Solution for Economic and Environmental Global Eco-Cost Assessment: Model Combination Analysis 239
Nicolas Perry, Alain Bernard, Magali Bosch-Mauchand, Julien Le Duigou and Yang Xu

The Ecodesign of Complex Electromechanical Systems: Prioritizing and Balancing Performance Fields, Contributors and Solutions 257
S. Esteves, M. Oliveira, F. Almeida, A. Reis and J. Pereira

Composite Fiber Recovery: Integration into a Design for Recycling Approach 281
Nicolas Perry, Stéphane Pompidou, Olivier Mantaux and Arnaud Gillet

Design for Disassembly Approach to Analyze and Manage End-of-Life Options for Industrial Products in the Early Design Phase 297
Claudio Favi and Michele Germani

Index 323

Product Architecture Decision Under Lifecycle Uncertainty Consideration: A Case Study in Providing Real-time Support to Automotive Battery System Architecture Design

Qi D. Van Eikema Hommes and Matthew J. Renzi

Abstract Flexibility is valuable when the future market and customer needs are uncertain, especially if the product development process is long. This chapter focuses on what the firm can do to increase their flexibility before a product is produced and sold. The flexibility is built into the product architecture, which then enables the firm to take a staged decision process. Flexibility-in-the-Project approach was developed by de Neufville and Sholtes (2011), and has been successfully applied to large infrastructure projects. Real options analysis has only been utilized in high-level product planning decisions. The case study described in this chapter is the first successful application of the Flexibility-in-the-Project framework, providing real-time engineering design decision support to Ford Motor Company engineering efforts in future vehicle electrification. In hybrid and electric vehicle applications, the high voltage battery pack hardware and control system architecture will experience multiple engineering development cycles in the next 20 years. Flexibility in design could mitigate risk due to uncertainty in both engineering and consumer preferences. Core engineering team decisions on battery pack voltage monitoring, thermal control, and support software systems will iterate as technology evolves. The research team valued key items within the technology subsystems and developed flexible strategies to allow Ford to capture upside potential while protecting against downside risk, with little-to-no extra cost at this early stage of development. The methodology used to evaluate the uncertainty, identify flexibility, and provide the real options value of flexibility is presented.

Q. D. V. E. Hommes (✉) · M. J. Renzi
Engineering Systems Division, Massachusetts Institute of Technology, 77 Massachusetts Ave, Cambridge, Massachusetts, USA
e-mail: qhommes@mit.edu

M. J. Renzi
e-mail: Matthew.rezi@sloan.mit.edu

E. Henriques et al. (eds.), *Technology and Manufacturing Process Selection*, Springer Series in Advanced Manufacturing, DOI: 10.1007/978-1-4471-5544-7_1,

1 Introduction

Much data have shown that the most important decisions about a product are made in the early phase of the design process, when the design is still fluid, and changes are relatively inexpensive (Fig. 1). However, making decisions in this phase of the design can be very challenging, because the prediction about the future markets and operations demand has high uncertainty, especially when the product development cycle is long.

The historical gasoline price data is a good example to illustrate the challenges in forecasting (Fig. 2). The United States Energy Information Administration (EIA) provides a concise explanation of the factors influencing the gasoline prices (EIA 2012), many of which are attributed to global social, political, and economical dynamics that are impossible to accurately predict. Therefore, the large fluctuation of gasoline prices often surprises and frustrates industries and individuals, and sends the equity market on a roller coaster ride.

The inability to accurately forecast gasoline price has a strong impact on the US automotive sales in various segments such as small car, SUV, etc., as demonstrated during the 2009 financial crisis period. Typically, new automobile models take 3–5 years to design, engineer, and manufacture. Forecast based on the 2003 gasoline price made the truck and SUV segment seem highly profitable. The sales volume assumptions were based on consumer purchase decisions at the low gasoline prices. After developing these new models of the SUVs and trucks for several years and bringing them to market, many automotive companies found themselves stuck with a large inventory of SUVs and trucks as consumers quickly switched to buying small cars, reacting to the soaring gasoline price in 2008. The automotive companies weren't able to quickly change to making small cars. Years of engineering efforts seemed to have been set in the wrong direction.

The main reason for which the automotive companies weren't able to quickly react to market changes is that their entire cost structure were optimized to making SUV and trucks, based on the point forecast made in earlier years. Thirty years before the 2009 financial crisis and the struggle of the American automotive

Fig. 1 Committed lifecycle cost against time

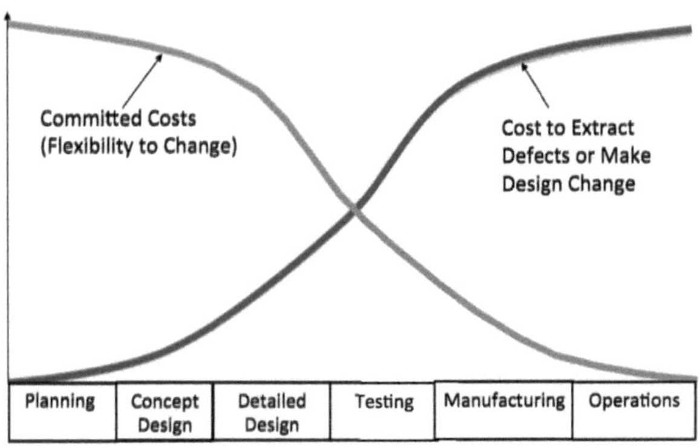

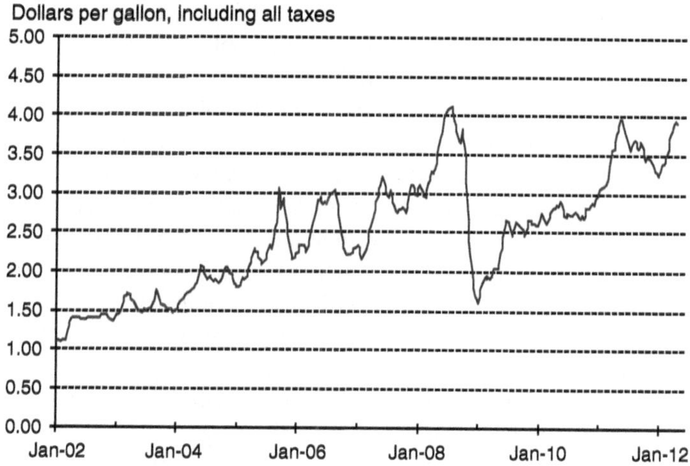

Fig. 2 United States motor gasoline price data (*Source* U.S. Department of Energy, Energy Information Administration, Weekly Retail Gasoline Prices, available at http://eia.doe.gov/ as of April 2012)

companies, Abernathy (1978) argued that automobiles running on gasoline internal combustion engines had arrived at a dominant architecture. The focus of the manufacturers turned to process innovation—optimizing the productivity of the production process for a few mature architectures. Abernathy gave an example on why it could not be profitable for an automotive company to manufacture small vehicles in manufacturing plants optimized for making large vehicles. He pointed out that in order to stay competitive, firms should be careful not to let productivity kill the flexibility to innovate. Unfortunately, history repeated itself 30 years later, due to precisely the same cause that Abernathy had identified—lack of flexibility to react to the market when the market isn't what is forecasted years ago.

Remaining flexible is important because forecasting is inherently uncertain, as no one has been able to predict the future accurately. Many assumptions enter forecasting models so that mathematical calculations can be performed (Stock and Watson 2007, Train 2003). The data collection methods for market and consumer information used to feed the forecasting models are also not perfect (Aaker et al. 2010). Questionnaire design can strongly affect the responses, depending on how questions are worded, and how they are interpreted (Brace 2004; and Harkness et al. 2003). Consumers' actual purchase decision may be very different from what they say in a market clinic or when they answer a survey (Kahneman and Tversky 1979; Tversky and Kahneman 1981; Kahneman et al. 1990; Gladwell 2005).

Although the forecasted values are often uncertain, the customary practice is to use the average forecasted value in planning (Ulrich and Eppinger 2008 (Chap. 15)). Many of the optimization and trade-off studies are done based on average forecasted values. Yet, average values are highly flawed (Savage 2009). The Iridium fleet of communication satellites was a good example on decisions made based on average forecasted demand, which was so far off the reality that the company went into bankruptcy (de Weck et al. 2004). In their 2011 book, de Neufville and Scholtes provide many additional real examples to illustrate this point. To make things worse, many large capital-intensive products, such as automobiles, take years to develop and manufacture. Even if the forecasted average value was close to reality at the time, things change overtime, and the future can be very uncertain (Fig. 3).

Literature does advise conducting what-if scenario sensitivity analysis, after assessing the most-likely case using average numbers (Ulrich and Eppinger 2008 (Chap. 15)). This additional step is much better than basing the decision only on an average forecast. However, as de Neufville and Scholtes (2011) point out, this approach is a "bunker mentality:" Will we be able to survive adverse futures? Will we be able to sustain risks? It is an afterthought of having optimized the design following the averaged forecast. It does not design with the uncertainties in mind so that uncertainties can be leveraged to our advantage.

The discussions in this chapter is about gaining the ability for a product design to remain flexible for long term future uncertainty, and even taking advantage of the uncertainty when possible. As Fig. 1 has illustrated, the best place to incorporate such thinking is in the early phase of the product design process. Specifically, this chapter focuses on how to assess the value of flexibility embedded in product architecture, during the concept design phase of the product development process (Ulrich and Eppinger 2008). The concept of architecture used in this chapter follows the definition in Ulrich and Eppinger (2008):

> The architecture of a product is the scheme by which the functional elements of the product are arranged into physical chunks and by which the chunks interact.

The methodology presented in this chapter is a support framework for product architecture selection in real time. This framework focuses on three questions: Why do we need flexibility, when will we need it, and how much will it cost? The framework contains four steps (de Neufville and Scholtes 2011): (1) establish the key uncertainties, (2) determine points of flexibility, (3) provide a financial model incorporating the key uncertainties, and (4) establish the value of flexibility. The framework was proven successful when applied as real-time support for the Ford Motor Company's decision process on core technology for the thermal control of an electrified vehicle battery system.

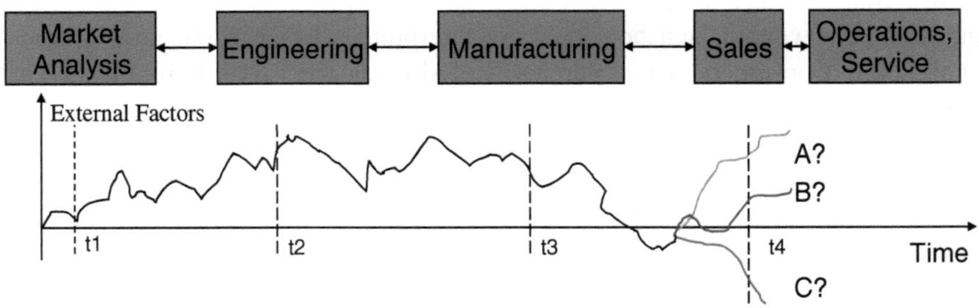

Fig. 3 Lifecycle view of the early product decisions

2 Literature Review

2.1 Uncertainty Consideration in Product Design and Manufacturing

There exists rich literature in addressing uncertainties in product design and manufacturing. This section organizes the literature around two questions:

1. What are the uncertainties being considered?
2. What are the strategies developed to address these uncertainties?

2.1.1 Types of Uncertainties

The first type of uncertainty is the recognition that customer requirements are usually not set at fixed points. Products are usually designed for a market segment, in which the customer requirement is not uniform, but rather a distribution. For an individual user, the utility for a certain performance metric varies within a range of acceptable values. Requirements can be balanced to maximize the utility of the overall product. Work in the area of Multidisciplinary Design Optimization (MDO) represents concern of this type of uncertainty (Donndelinger et al. 2003; Papalambros and Wilde 2000; Ferguson and Siddiqui 2007; Chen and Yuan 1999; Ross et al. 2008).

The second type of uncertainty is that the product usage may change after the product is deployed (Ferguson and Siddiqui 2007; Olewnik et. al. 2004; Olewnik and Lewis 2006; Saleh et al. 2003; Skiles et al. 2006; Haulbelt et al. 2002, Frick and Shulz 2005; Ross et al. 2008; Shah et al. 2008; Matin and Ishii 2002; Lieke et al. 2008). Customers may face new usage situations. The operating environment may be unpredictable. The product may degrade over time.

The third type of uncertainty involve customer and market needs change over time (Saleh et al. 2003; Keese et al. 2006 and 2007; Clarkson et al. 2004; Eckert et al. 2004; Fricke and Shulz 1999; 2005; Ross et al. 2008; Shah et al. 2008; Martin and Ishii 2002; Allada and Jiang 2001 and 2002; Sethi and Sethi 1990; Gustavsson 1984; Gerwin 1982, and Kapoor and Kazmer 1997). Customers may want new functionalities or higher quality. Government regulatory requirements may change. Industry standards can change. Technology competition may change the requirements. Societal and economical trends may also change what consumers want. The market demand (quantity) may change over time (Pandey and Thurston 2008).

Additional uncertainties mentioned in many literature include the introduction of new technology (Keese 2006; Fricke and Shulz 1999; 2005; Ross et. al 2008; Shah et. al 2008; Martin and Ishii 2002; Sethi and Sethi 1990; Gustavsson 1984; Gerwin 1982, and Kapoor and Kazmer 1997), manufacturing piece to piece variation and other operating noise factors (Phadke 1989; Kazmer and Roser 1999; Parkinson and Chase 2000), and system integration issues and emergent behaviors (Ekert et al. 2004).

2.1.2 Strategies to Address Uncertainties

Many strategies have been developed to address the aforementioned uncertainties. They include utility theory, MDO, and Pareto frontier (Papalambros and Wilde 2000; Donndelinger et al. 2003; Olewinik and Lewis 2006; Chen and Yuan 1999), Robust Design (Phadeke 1989; Kazmer and Roser 1999; Gustavsson 1984), Design for manufacturing variation and tolerance stack up (Kazmer and Roser 1999; Whitney 2004), systems engineering methods include FMEA, P diagram, etc. (Sage and Rouse 1999), product architecture and platform design such as the architecture guidelines, modularity decisions, interface management, product family design (Haubelt et al. 2002; Fricke and Schulz 1999 and 2005; Allada and Jiang 2001; 2002; Lieke et al. 2008; Meyer and Lehnerd 1997; Simpson et. al 2006; Rechtin 1991; Maier and Rechtin 2002; Baldwin and Clark 2000; Suh et al. 2007), adaptive and reconfigurable system design (Skiles et al. 2006; Ferguson and Lewis 2004; Ferguson et al. 2007), principles and guidelines for flexible system design such as those summarized from patent searches, product reverse engineering, and design rules (Keese et al. 2007; Skiles et al. 2006; Qureshi et al. 2006; Baldwin and Clark 2000; Tilstra et al. 2008). Additionally, manufacturing process flexibility and production volume plant allocation flexibility have also been extensively studied (Sethi and Sethi 1990; Gustavsson 1984; Gerwin 1982; Kapoor and Kazmer 1997; Balakrishnan and Geunes 2003).

To better understand the structure of the literature concerning various methodologies to address uncertainties, two additional dimensions are employed. First is the phase in the product life cycle when the methodology can be applied. Saleh et al. (2003) suggest that we can look at these strategies based on whether they can be applied before or after the system deployment. The system designer/manufacturer can exercise flexibility options before deployment, while users/operators or the products themselves are the ones that can exercise flexibility after deployment. The second dimension is the degree of change in the product. Sethi and Sethi (1990) suggest that flexibility embedded in a system can be state flexibility or action flexibility. State flexibility is the capacity to continue functioning effectively despite changes in the environment. Action flexibility is the capacity for taking new action to meet new circumstances. It can include changing design variable level, changing design variable set, or changing the architecture of the product completely.

The two tables in Fig. 4 summarize the literature relative to the types of uncertainties they address, the time that they are useful (before or after deployment), the level of changes they require, and what existing methods may be useful.

Specifically, the focus of this chapter is on the uncertainty of "Consumer requirements on the same type of product can change in the long term," and what

Uncertainties	Before Deployment			
	State Flexibility	Action Flexibility		
	No Change	Change Design Variable Level	Change Design Variable Set	Change Architecture
Predefined customer requirements usually are not at fixed points	Multidisciplinary Optimization			
Product usage may change after the product is deployed	Robust Design Systems Engineering Methods		Product Architecture and Platform Strategy	Product Architecture and Platform Strategy
Customer requirements on the same type of product can change long term		Product Architecture	Product Architecture and Platform Strategy	Total Architecture Redesign
Market demand (quantity/volume) change over time	Manufacturing Flexibility		Modular Architecture Helps Manufacturing Flexibility	
New technology			Modular Architecture	Architecture Change, Principles of Flexibility
Manufacturing piece-to-piece variation, other noise factors	Robust Design			
System integration issues and emergent changes	System Engineering Methods	System Engineering Methods	Modular Architecture Design, Axiomatic Design	Principles of Flexible Product Design

Uncertainties	After Deployment			
	State Flexibility	Action Flexibility		
	No Change	Change Design Variable Level	Change Design Variable Set	Change Architecture
Predefined customer requirements usually are not at fixed points				
Product usage may change after the product is deployed		Adaptive and Reconfigurable System		
Customer requirements on the same type of product can change long term		Adaptive and Reconfigurable System	Upgradable Modular Design Based on Architecture	
Market demand (quantity/volume) change over time				
New technology			Upgradable Modular Design Based on Architecture, Principles of Flexible Design	
Noise factors during usage	Robust Design			
System integration issues and emergent changes		Adaptive and Reconfigurable System	Upgradable Modular Design Based on Architecture	

Fig. 4 Literature survey summaries

the producers can do before deployment. Based on literature, the method to be examined is system architecture decision embedding flexibility options.

2.1.3 Valuation of Flexibility

People have recognized since long ago that flexibility is valuable. The concept of options formally appeared first in the financial industry. Investing in a financial option is investing in the right but not the obligation to purchase investment equity. Black and Sholes (1973) developed the famous formula to put a fair value on options as a financial instrument. Buying option is cheaper than investing in the stock itself, and hence allowing hedging against future uncertainties. This concept was later extended to capital investment projects, and became Real Options (Trigeorgis 1996; Copeland and Antikarov 2003).

De Neufville pioneered the work to extend the real options thinking to engineering decisions (de Neufville and Sholtes 2011). He coined the phrase "Flexibility in a Project," where flexibility is built into the system and designed with long-term uncertainty in mind (Wang and de Neufville 2005). Some researchers try to apply the Black and Sholes formula or other options valuation techniques such as binomial lattice to engineering design decisions (Engel and Browning 2008; Mathews et al. 2007). However, the formulas used for financial investments are only valid under a set of assumptions about markets and available investment choices. These assumptions do not necessarily hold true for engineering design projects. De Neufville and Sholtes (2011) propose instead to employ Monte-Carlo simulations, whose computation does not require assumptions on the market and investment choices. Their approach is briefly described below.

The first step of the valuation process is to start with the traditional business case model. Typically, a business case model using point forecasts is created to determine the Net Present Value (NPV) of the project. NPV assumes that future cash flows are worth less than current cash flows, as are future capital expenditures (CAPEX). The equation for NPV is defined as:

$$NPV = \sum_{t=0}^{n} \frac{CF_t}{(1+r)^t}$$

where n is the period over which the analysis is calculated (often in "years"), r is the discount rate, or the competing rate of interest as a benchmark, and CF is the cash flow in year t. For a very safe investment, the discount rate might be the rate on a similar US Treasury bill. For a riskier investment, such as a startup venture, a substantially higher interest rate may be used. A short-term CAPEX would be a negative cash flow at year 0, while sales in any year would provide positive cash flows.

Building this model entails a full understanding of market possibilities and technical details. The unit cost of a technology is built from an understanding of the underlying components. A market forecast is created from sophisticated demand modeling techniques. This point forecast model is the starting point for our flexibility valuation model development and begins the process of identifying flexibility.

The future is impossible to forecast with certainty. Demand and technology conditions are uncertain. A traditional NPV assumes that conditions are static, not uncertain. The static NPV model does not capture the potential for decision-making based upon incoming data. An econometrics forecast could be made with historical data. Even with historical data, forecasts are always wrong. Forecasted sales of 10,000, with realized sales of 9,999 is still wrong. A new product entails even higher levels of uncertainty. For a new product, the NPV is unrealistic, as demand in year 10 for the product is impossible to predict with certainty.

By acknowledging uncertainty, the second step of the process is to run NPV simulation over a distribution of input variables using a Monte-Carlo simulation.

By understanding market drivers and technology risks, a decision maker can look to mitigate downside risk and capture upside gain in the system. In order to limit downside losses and capture upside gain, flexibility in the system may provide value via insurance against negative outcomes or for positive situations (de Neufville and Scholtes 2011). The deterministic NPV model fails to capture upside potential and downside risk.

However, merely running the Monte-Carlo simulation over the same design is only as good as the traditional sensitivity analysis, leading to only after-thought risk mitigation strategies. What de Neufville and Sholtes (2011) further suggest is to look for places to embed flexibility in the design itself, hence the "Flexibility in the Project" term, similar to buying stock options instead of stocks. The Monte-Carlo simulation can then be applied to assess the profitability of the embedded flexibility in the design itself. The search for where to embed flexibility in a design should come before the economic assessment using Mont-Carlo simulation.

A good example of this process can be found in the "Garage Case," which is a study on building a parking garage structure at a new shopping mall (de Neufville et al. 2006). The number of parking spaces required is highly uncertain. Build too many and there will be losses. Build too few and additional revenue will be lost.

Recognizing the uncertainties in demand, the flexibility can be built into the garage design by employing a stronger base and prefabricated connectors on the upper level for future expansion. When compared with the garage built using static forecasting and assuming pessimistic demand scenario, this design will lead to additional per floor Capital Expenditures (CAPEX) for the initial structure due to required strengthening. However, this garage design will have fewer floors and lower CAPEX than the one built with optimistic forecast using the static NPV calculation. Nonetheless, this design will enable the garage owner to quickly capture the upswing of demand if it materializes, but also be protected from the scenario in which the actual demand for parking space does not grow while the forecast says it will. The value of the flexibility can be quantified using Monte-Carlo simulation, applied to the baseline static NPV business case model.

2.1.4 System Architecture

The garage design concept incorporating flexibility requires domain expertise about how to architect the system. Architecture plays a very important role in the lifecycle of the product and product family (Ulrich and Eppinger 2008; Rechtin 1991; Maier and Rechtin 2002). It affects how a product can be changed. Integral architecture is usually optimized for its predefined requirements, but can be challenging to change if requirements change. Modular architecture makes it easy and economical to upgrade, add-on, adapt, maintain, reuse, etc. (Henderson and Clark 1990; Baldwin and Clark 2000). Modular architecture can also create a product platform that enables product variety (Simpson et al. 2006) and the use of standard components.

Much of the modular architecture and product platform discussions concern the productivity of the company. More specifically, they try to find solutions that are least costly facing varieties in customer needs. This is very important for private industries where cost is important to competitiveness. However, to stay competitive, low cost is only one part of the equation. The revenue side of the equation must also be considered. In other words, the ability to capture the upswing of the market and continue to create revenue if future demand changes are just as important as cost saving.

Engineers traditionally have been trained to generate design solutions based on predefined requirements. The architecture concepts are often a few distinctive ones. With cost pressure, and seeing architecture concepts as either-or choices, many of the companies fall into the trap of selecting an architecture that may not be flexible to future uncertainties, leading to architecture lock-in. The Iridium satellite is a good example of architecture lock-in (de Weck et al. 2004).

The idea of embedding flexibilities in the architecture has been studied by many researchers. de Weck et al. (2004) proposes a staged launch and investment process to mitigate downside risk should demand not materialize. Ford and Durward (2005) suggest that instead of following traditional wisdom to select a single design concept early on, we should leave the design space open and demonstrate how platform consideration could be made with flexibility in mind. However, to the authors' best knowledge, none of the existing literature on flexibility in architecture flexibility in architecture decision has actually applied this method in a real engineering design setting. All of the examples successfully applied in real-world case studies are on large infrastructure projects (de Neufville and Scholtes 2011).

The contribution of the case study discussed in this chapter is to test wether the "Flexibility in the Architecture" concept following the de Neufville and Sholtes (2011) framework can actually be used in product architecture decisions. If so, how effective is this framework? As embedded members of the decision making team for Ford Motor Companies advanced R&D team on battery packs, the MIT team actually had the opportunity to influence design decisions for flexibility in the system, assisting in the generation of concepts for added value and reduced risk in early stage decisions.

3 Case Study: Automotive Battery Pack Control

3.1 Uncertainties in Automotive Battery Technology

Technology and market uncertainty are high in industries undergoing product innovation. Product innovation is characterized by a sharp rise in new entrants, followed by consolidation over time as efficiency, reliability, standardization, and cost reduction reduce the field of profitable companies (Abernathy 1978; Utterback 1996). The automotive industry has been in an era of process innovation, developing vehicles of high reliability built on standardized platforms. Recent developments in battery technology and vehicle electrification are driving the automotive industry into a new disruptive innovation phase.

Product innovation in electrified vehicles creates opportunities and uncertainties for automobile manufacturers. In the scope of this case study, product innovation includes various means of electrification of traditional gasoline internal combustion engine driven automobiles, including full hybrid (FHEV), plug-in hybrid (PHEV), and battery electric vehicles (BEV). The market demand of various types of electrified vehicles is still uncertain, and largely affected by gasoline prices as well as global social and economic trends. In addition, traditional uncertainties influencing consumer preference confound with uncertainty in battery technology.

Three clear technology based uncertainties exist in the future of HEV/EV design, with direct market connections (Wang 2011; Westbrook 2001):

- Battery chemistry (and/or electricity sources) may change.
- Voltage/current may be different per vehicle depending upon the cell design (e.g. higher current chemistries and power density).
- Voltage/current may change due to vehicle class/customer power preferences (e.g. SUV, truck).

Predicting battery technology over the next 10 years is unlikely to yield accurate results. Batteries and hybrid drive systems continue to improve, both incrementally (process improvements) and disruptively (Nickel Metal Hydride batteries transitioning to Lithium-ion batteries). Lithium-ion is the state-of-the-art. They have begun to appear in electrified vehicles (e.g. the 2012 Nissan Leaf BEV and 2013 Ford Fusion Hybrid), with new challenges in the control of the technology (Ford Battery Research Team 2011; Andrea 2010).

Batteries require a control system for temperature, voltage, and current Cold temperatures deplete the usable charge within the battery by changing the anode/cathode electro-potential and lowering cell voltage. Both cold and high temperatures create risky situations that may lead to thermo runaway, presenting the risk of explosion. Over-voltage and current could also cause explosion, and deep discharge can permanently deplete the battery. Therefore, the level of risk and control requirements for Lithium-ion differ from Nickel Metal Hydride, and future chemistries may add extra requirements, including other promising energy storage technologies on the horizon (Andrea 2010).

Battery pack architecture and battery management systems must be flexible, as the arrangement and number of cells used in a vehicle will vary by vehicle class. An SUV has higher power and energy demands than a small economy car; the thermal management requirements will change due to charging and discharging. The heat loss associated with moving the vehicle requires thermal monitoring/control of the battery pack proportional to the total electric power.

Such uncertainties in battery technology require an approach different from the well-studied platform architectures approach whose focus is on low unit cost for predefined markets and mature technology. The rapid development of battery technology and the variable power requirements of different vehicle segments present uncertainties surrounding vehicle design, technology, and consumer preference. It requires a different way of thinking about architecture—from a value maximization point of view rather than a cost minimization view. "Flexibility in the Architecture" appears to be the most appropriate approach.

3.2 *Battery Thermal Control Architectures in Current Production Vehicles*

In current product vehicles, battery cells are assembled in a battery pack. The battery pack casing can protect the battery chemistry from environmental elements. The battery cells are fully sealed with external voltage connection points. The cells are placed in an impact-resistant, environmentally sealed metal compartment. The metal compartment is electrically isolated from the vehicle chassis.

As stated in the last section, Li-ion battery packs require an aggressive thermal control strategy both for performance and safety. The current standard production architecture for battery temperature control in FHEVs is sealed-cabin-air cooling (Fig. 5). Air intake ports located above the rear seats provide a flow channel for cabin air, drawn via fans, to enter into the battery enclosure, secured in the trunk. The cabin-air is conditioned to an acceptable temperature by the vehicle occupants, which is also suitable for the battery operations. The cabin air battery cooling architecture currently meets the requirements for most full hybrid vehicles and is used across the industry in production vehicles, including the Ford Fusion Hybrid.

However, cabin air-cooling reduces the useable cabin space because it constraints both rear seat positioning and the overall trunk space. It is also not flexible for several future uncertainties. First, large size plug-in hybrid (PHEV) or full battery electric (BEV) vehicles require additional power, using more batteries than the FHEV. The physical size of the batteries limits the ability to package the battery inside the trunk, as evidenced by the choices made by Nissan and GM in designing the Nissan Leaf BEV and Chevy Volt PHEV, both placing a very large battery pack beneath the vehicle. Second, as Li-ion batteries' performance characteristics become better understood and technology advances, batteries may be operated at higher thermal loads with fewer cells in order to reduce cell cost and pack size. Similar trends have been observed in the Lead-acid and Nickel-metal Hydride batteries. Higher thermal loads will call for more stringent cooling

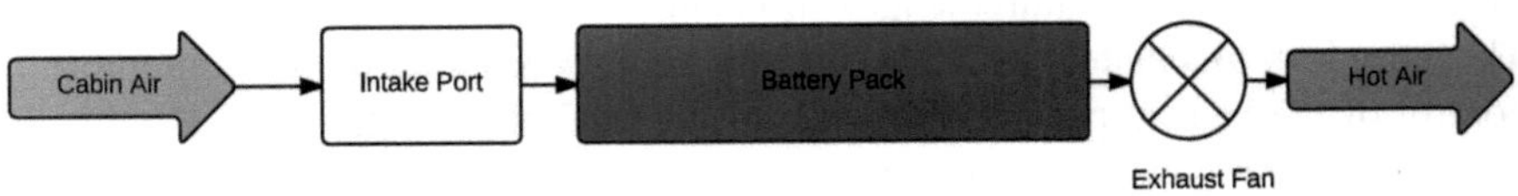

Fig. 5 Sealed-cabin-air cooling, Toyota Prius, 2nd gen (Renzi 2012)

requirements. Third, PHEVs or BEVs will have increased thermal requirements due to the higher dependence on the electric powertrain.

In fact, vehicles in the market today already started exploring alternative cooling architectures. The Chevy Volt uses underbody-mounted batteries with liquid cooling (General Motors 2011), but it suffered from safety problems during NHTSA testing (Tran 2012). The Nissan Leaf uses passive air-cooling with underbody mounting (Nissan 2012), utilizing the airflow during driving to cool the battery system. The recently announced Ford Focus BEV uses a tight temperature control technique with both active heating and cooling elements (Siri 2010; Ford 2011). It was designed for all environments, with highly regulated temperatures.

With such varieties of design alternatives and insights on future uncertainties, Ford was interested in finding out how to select a battery cooling system architecture. The "Flexibility in the Project (Architecture)" approach is thus appropriate to help Ford making the architecture decision.

3.3 Generating Alternative Battery Thermal Control Architectures

In support of Ford, the MIT team worked with the Ford engineering team to develop and analyze a group of unique Ford-specific thermal control concepts. In order to protect Ford proprietary design information, the rest of the discussion will not provide details to the specific architecture design, but instead discuss them generically using the names Architecture 1, 2, 3, and 4.

At the beginning of the project, Ford engineering experts generated architecture concepts that were stand-alone solutions. A thermal engineer is an expert in finding solutions to a given system requirement. However, these solutions were not developed with requirement uncertainties in mind. Changes in future requirements either due to consumer preferences or technology change, are likely to cause major redesign in each concept, and potentially leading to product introduction delay.

After learning from the MIT team about designing flexibility and "options" into the architecture to mitigate future uncertainties, Ford engineers generated very different architecture concepts. For the analysis, Architecture 1 is the baseline cabin air cooling concept in current production vehicles. Architecture 2, 3, and 4 are potential solutions for FHEVs, PHEVs, and BEVs, with physical embedded options that enable the design to switch among the three architectures, and hence providing flexibility to adapt the architecture to future technology and market uncertainties.

3.4 Flexibility Valuation Modeling Approach

In order to compare the value of flexibility among these alternative architectures, the Ford engineering team and the MIT research team extended the garage case modeling approach to this project (de Neufville et al. 2006). Flexibility in this project means the ability of the thermal control system to support the potential thermal loads across vehicle types, FHEV, PHEV, and BEV, reflecting uncertainty in demand.

First, a "fixed forecast" NPV was developed. A point forecast for market demand for hybrid vehicles is generated using multiple sources of forecasting information. JD Power estimates global hybrid demand at 7.3 % of total sales in 2020, or 5.2 million units. Ford sold 5.3 million vehicles globally in 2010 out of about 72 million total light vehicles sold, for about a 7.4 % market share (Stenquist 2011). Assuming that Ford will be aggressive in the hybrid market given their wide range of planned products, it is reasonable to estimate that Ford will achieve at least the 7.4 % market share of the 5.2 million units sold in 2020, or close to 400,000 vehicles.

In the second stage, sales projections were converted to a normal distribution with uncertainty based on the wide ranges in forecasts that have been seen in the past. In 2003, JD Power estimated 500,000 sales by 2008 (Hybrid Market Forecasts 2006). Realized sales were 314,000 (Hybridcars.com 2009). For 2011 sales, in Q3 2008 JD Power estimate 1,000,000 sales in 2011 and a market share of over 6 % (Omotoso 2008). Actual sales were 270,000 and a 2.1 % market share (Hybridcars.com 2012). These are not exceptions; there are many expert predictions of hybrid sales with very similar deviation (Hybrid Market Forecasts 2006). Uncertainty tends to increase with the number of years from the date of forecast. Given the uncertainty in predictions, we'll assume from the prior forecasts that Ford sells 400,000 electrified vehicles in 2020, and assume ~200,000 as a standard deviation when building the demand model. We use the median of 400,000 to provide the peak of a normal distribution. One standard deviation is assumed to be 200,000 a, 50 % value from the peak. More sophisticated median, deviation, and distribution estimates were acquired from propriety data and used by Ford, but are not included in this publication.

Finally, a flexibility rule is incorporated for any uncertainty that might impact sales. A flexibility rule is an "if" statement within the simulation that represents a decision management might make, given changing circumstances. Management "exercises the option" that flexibility represents when conditions arise. For the garage case (de Neufville 2006), management would expand based upon prior year demand, e.g. "if prior year demand >x, expand 1-level in next year". In this case study, switching between Architecture 1 and the rest incur a high cost. As noted, with flexible options in mind, Architecture 2, 3, and 4 were designed to be compatible. Switching among them incurs much lower cost.

The simulation's primary assumptions and uncertainties can be found in Renzi (2012). The model ran 2000 Monte Carlo simulations, randomly sampling consumer demand. The simulation results are shown in Fig. 6. The quantitative

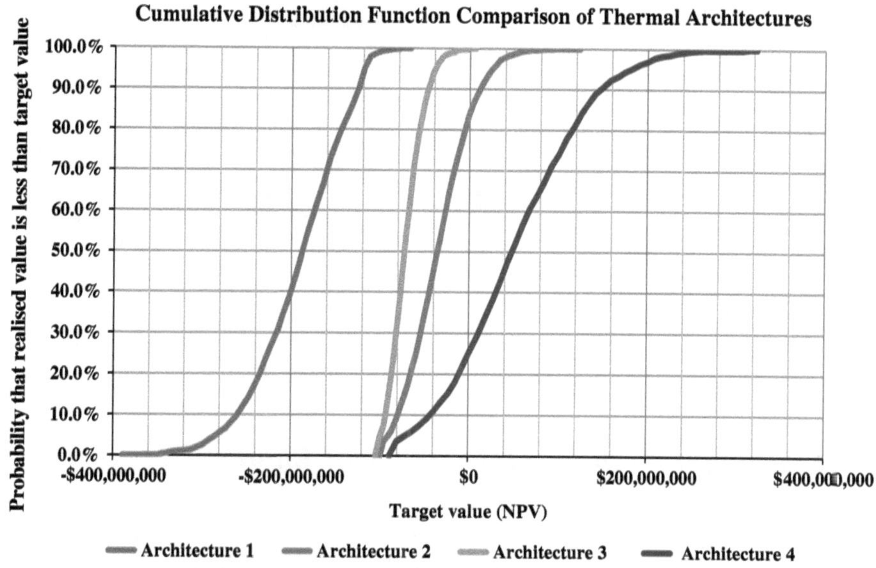

Fig. 6 NPV comparison of thermal architecture under uncertainty

assessment results indicate that the newly generated Architecture 2, 3, and 4 with uncertainty consideration in mind are better at capturing the upswing of the future markets. First, Architecture 2, 3, and 4 can be used for PHEVs and BEVs as well as FHEVs, capturing future sales without additional engineering cost that would be required for architecture 1. Second, Architecture 2, 3, and 4 are associated with positive consumer willingness-to-pay values due to improvements in consumer comfort and convenience feature of the vehicle. Additionally, architecture 4 has fewer components than 2 and 3, further driving down the unit cost, making it the leading choice among the four alternatives.

The above simulation assumes that Architecture 1, 2, 3, or 4 will actually be successful at achieving the thermal performance requirements, integrating into production vehicle programs, and be manufacturable. However, in the early architecture generation and selection phase of the product development process, there is no guarantee that any of these three downstream activities will be successful, except for the architecture that is already in production today. Therefore, additional valuation of the architecture will be necessary to make an educated decision.

A decision tree was built (Fig. 7) to further assess the viability of each of the architecture. The mean value of the Monte-Carlo simulation was used as the end points of the tree. The probability to success at each stage of the product development process was estimated. Due to the flexibility to easily switch among Architecture 2, 3, and 4 as built into the architecture concept themselves, Ford now have the flexibility to carry out a staged decision process (de Neufville and Sholtes 2011), enabling the company to be prepared not only for the uncertainties in the market, but also the uncertainties in the product integration process.

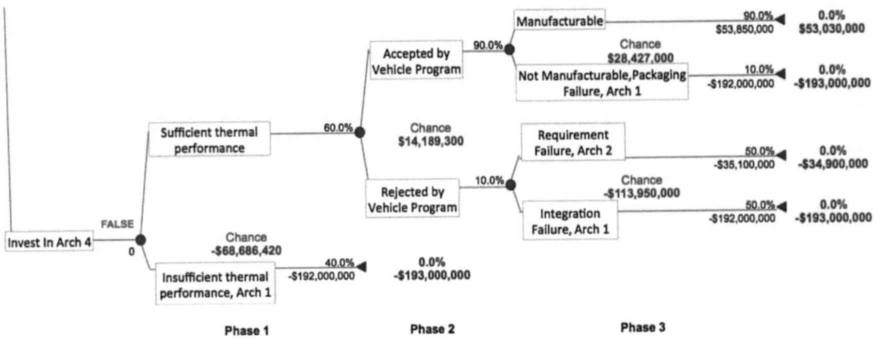

Fig. 7 Decision tree of investment in architecture 4 only (Renzi 2012)

3.5 Learning from the Real-time Support Effort

After comparing and assessing each of the four architectures, Ford engineering team was able to select the most suitable architecture with future uncertainties in mind. The flexibility valuation model helped provide justification to management for this decision. As mentioned earlier in this chapter, to the authors' best knowledge, this is the first application example using the "Flexibility in the Project" framework in real time to support the lifecycle consideration of an actual engineering architecture decision. The authors were able to successfully integrate the framework into Ford's existing engineering workflow.

In our experiences, the biggest contribution of this method was not the quantitative assessment, although it was helpful. To gather large quantity of market forecasting data in order to build up detailed econometric model for engineering architecture decisions was not practical due to time and resource constraints. What was very effective was the design philosophy change introduced by this method. Instead of looking for and optimizing design solutions based on point forecast, the engineers learned how to integrate flexibility into their designs very early on in the product development process. The role of the flexibility valuation model was providing an objective framework to carry out architecture selection discussions and offering directional indications for what-is scenario analyses. The Architecture 2, 3, and 4 generated in this case were very different from the initial design, and engineers almost immediate recognized them as better and more flexible, without having to be convinced by quantitative simulations. In another application carried out in this project but not detailed in this chapter, we had the same experience. Once learned about how to look for "options" in their architecture and prepare the concept for future uncertainties, Ford engineers generated a flexible design solution that was so new from the existing ones that Ford was able to file for a patent. In that project, engineers did not even felt the need for quantitative analysis. They just moved right on to the project described in this chapter.

4 Summary

In summary, this chapter described a real-time support application of the flexibility in engineering design framework. The project was to assist Ford battery cooling system architecture decision. By adopting a flexible architecture solution, the Ford design can transition from one architecture to another if there are changes in future vehicle requirements, with little impact to the vehicle program cost and timing. Acknowledging the risk of meeting requirements and investing in a flexible architecture, the team will easily be able to "fallback" to the most aggressive architecture solution if the lowest unit cost is unable to meet requirements. As the first application case, this study has demonstrated the effectiveness of applying this framework in real engineering applications.

References

Aaker DA, Kumar V, Day GS, Leone RP (2010) Marketing research. 10th edn. Wiley, New York

Abernathy W (1978) Productivity dilemma: roadblock to innovation in the automobile industry. The Johns Hopkins University Press, Baltimore

Allada V, Jiang L (2001) Design for robustness of modular product families for current and future markets. In: Proceedings of the DETC01, ASME design engineering and technical conference and computers and information in engineering conference, Pittsburgh, PA, 9–12 Sept 2001

Allada V, Jiang L (2002) New modules launch planning for evolving modular product families. DETC2002/DFM-34190. In: Proceedings of ASME 2002 design engineering technical conferences and computer and information in engineering conference montreal, Canada, Sept 29–Oct 2, 2002

Andrea D (2010) Battery management systems. Artech House, Norwood

Balakrishnan A, Geunes J (2003) Production planning with flexible product specifications: an application to specialty steel manufacturing. Oper Res 51(1):94–112

Baldwin CY, Clark KB (2000) Design Rules, vol 1. MIT Press, Cambridge

Black F, Scholes M (1973) The pricing of options and corporate liabilities. J Polit Econ 81(3):637–654

Brace I (2004) Questionnaire design : how to plan, structure, and write survey material for effective market research. Kogan Page, London

Chen W, Yuan C (1999) A probabilistic design model for achieving flexibility in design. ASME J Mech Des 121(1):77–83

Clarkson PJ, Simons CS, Eckert CM (2004) Predicting change propagation in complex design. ASME J Mech Des 126(5):788–797

Copland T, Antikarov V (2003) Real options: a practitioner's guide. Texere EIA (US Energy Information Administration) (2012). http://www.eia.gov/energyexplained/index.cfm?page=gasoline_factors_affecting_prices. Accessed May 2012

de Weck O, de Neufville R, Chaize M (2004) Staged deployment of communications satellite constellations in low Earth orbit. J Aerosp Comput Inf Commun 1:119–136

de Neufville R, Scholtes S, Wang T (2006) Real options by spreadsheet: parking garage case example. J Infrastruct Syst 12(3):107–111

de Neufville R, Scholtes S (2011) Flexibility in engineering design. MIT Press, Cambridge

Donndelinger J, Ferguson S, Lewis K (2003) Exploring mass trade-offs in preliminary vehicle design using pareto sets. AIAA conference 2003

Eckert CP, Clarkson J, Zanker W (2004) Change and customisation in complex engineering domains. Res Eng Des 15:1–21
Engel A, Browning TR (2008) Designing systems for adaptability by means of architecture options. Syst Eng 11(2):125–146
Ferguson S, Lewis K (2004) Effective development of flexible systems in multidisciplinary optimization, AIAA 2004
Ferguson S, Siddiqi A (2007) Flexible and reconfigurable systems: nomenclature and review. DETC 2007-35745. In: Proceedings of the ASME 2007 international design engineering technical conferences & computers and information in engineering conference, Las Vegas, NV, 4–7 September 2007
Ferguson S, Kasprzak E, Lewis K (2007) Designing a family of reconfigurable vehicles using multilevel mutidiscilinary design optimization. AIAA-2007-1880, the 48th AIAA/ASMR/ASCE/AHS/ASC structures, structural dynamics, and materials conference, 2007
Ford DN, Durward KS (2005) Adapting real options to new product development by modeling the second Toyota Paradox. IEEE Trans Eng Manage 52(2)
Ford (2011) New Ford focus electric keeps its cool when the temperature heats up. http://media.ford.com/article_display.cfm?article_id=34774. Accessed 28 May 2012
Ford Battery Research Team (2011) Personal conversations
Fricke E, Schulz AP (1999) Incorporating flexibility, agility, robustness, and adaptability within the design of integrated systems—Key to success. In: Proceedings of 18th digital avionics systems conference, 1999. IEEE. St. Louis, MO, vol. 1, pp 1.A.2-1–1.A.2-8
Fricke E, Schulz AP (2005) Design for changeability (DfC): principles to enable changes in systems throughout their entire lifecycle. Syst Eng 8(4):342–358
General Motors (2011) Cooling fins help keep chevrolet volt battery at ideal temperature. http://media.gm.com/product/public/us/en/volt/home.detail.html/content/Pages/news/us/en/2011/Feb/0214_battery.html
Gerwin D (1982) Do's and Don'ts of computerized manufacturing. Harv Bus Rev 60:107–116
Gladwell M (2005) Blink, the power of thinking without thinking. Little Brown and Company, New York
Gustavsson SO (1984) Flexibility and productivity in complex production processes. Int J Prod Res 22(5):801–808
Harkness JA, Van de Vijver FJR, Mohler P (2003) Cross-cultural survey methods. Wiley, New York
Haubelt C, Telch J, Richter K (2002) System design for flexibility. Design, automation and test in europe conference and exhibition, Paris, France
Henderson RM, Clark KB (1990) Architectural innovation. Adm Sci Q 35(1990):9–30
Hybrid Market Forecasts (2006) Hybridcars.com: http://www.hybridcars.com/hybrid-drivers/hybrid-market-forecasts.html. Accessed 20 Mar 2012
Hybridcars.com (2009) Hybrid Cars: http://www.hybridcars.com/hybrid-sales-dashboard/December-2008-dashboard-focus-production-numbers-25416.html Accessed 20 Mar 2012
Hybridcars.com (2012) Dashboard: Sales Still Climbing. http://www.hybridcars.com/news/December-2011-dashboard-sales-still-climbing-35093.html. Accessed 20 Mar 2012, from Dec 2011
Kahneman D, Tversky A (1979) Prospect theory, an analysis of decision under risk. Econometrica 47:263–291
Kahneman D, Knetsch JL, Thaler RH (1990) Experimental tests of the endowment effect and the coase theorem. J Polit Econ 98:1325–1347
Kazmer DO, Roser C (1999) Evaluation of product and process design robustness. Res Eng Des 11(1):21–30
Kapoor D, Kazmer D (1997) The definition and use of the process flexibility index. 1997 ASME Design engineering technical conference, DFM-97-110 Sacramento, California
Keese DA, Takawale NP, Seepersad C, Wood K (2006) An enhanced change modes and effects analysis (CMEA) tool for measuring product flexibility with applications to consumer

products. ASME international design engineering technical conferences and computers and information in engineering conferences, DETC2006-99478, Philadelphia, PA
Keese DA, Tilstra C, Seepersad C, Wood KL (2007) Empirically-derived principles for designing products with flexibility for future evolution, 2007 ASME International Design Engineering Technical Conference, Las Vegas, NV, Sept 2007. DETC2007-35695
Lieke A, Chmarra MK, Tomiyama T (2008) Modularization method for adaptable products. In: Proceedings of the 2008 design engineering technical conferences, New York, NY, 3–6 Aug 2008
Maier MW, Rechtin E (2002) The Art of Systems Architecting. CRC Press, Boca Renton
Martin MV, Ishii K (2002) Design for variety: developing standardized and modularized product platform architectures. Res Eng Des 13:213–235
Mathews SH, Datar VT, Johnson B (2007) A practical method for valuing real options. J Appl Corp Finan 9(2):95–104
Meyer MH, Lehnerd AP (1997) The power of product platforms. Free Press, New York 1997
Nissan (2012) Why did Nissan develop and EV battery? Nissan technology magazine. http://www.nissan-global.com/EN/TECHNOLOGY/MAGAZINE/ev_battery.html. Accessed May 2012
Olewnik A, Brauen T, Ferguson S, Lewis K (2004) A framework for flexible systems and its implementation in multiattribute decision making. ASME J Mech Des 126(3):412–419
Olewinik A, Lewis K (2006) A decision support framework for flexible system design. J Eng Des 17(1):75–97
Omotoso M (2008) Alternative powertrain sales forecast. University of Michigan: http://umtri.umich.edu/content/Mike.Omotoso.Forecast.pdf. Accessed 12 Mar 2012
Pandey V, Thurston D (2008) Copulas for demand estimation for portfolio reuse design decisions. In: Proceedings of the 2008 design engineering technical conferences, DETC2008-49406, New York, NY, Aug 3–6 2008
Papalambros P, Wilde D (2000) Principles of optimal design. Cambridge University Press, New York
Parkinson AR, Chase KW (2000) An introduction to adaptive robust design for mechanical assemblies. ASME design engineering technical conferences DETC2000/DAC-14241, Baltimore, Maryland
Phadke MS (1989) Quality engineering using robust design. Prentice-Hall, Englewood Cliffs
Qureshi A, Murphy JT, Kuchinsky B, Seepersad C, Wood K, Jensen D (2006) Principles of product flexibility. ASME international design engineering technical conferences and computers and information in engineering conference, DETC2006-99583, Philadelphia, PA
Rechtin E (1991) Systems architecting, creating and building complex systems. Prentice Hall, Englewood Cliffs
Renzi M (2012) System architecture decisions under uncertainty: a case study on automotive battery system design, MS Thesis. MIT Cambridge
Ross AM, Donna RH, Hastings DE (2008) Design for changeability (DfC): principles to enable changes in systems throughout their entire lifecycle. Syst Eng 11(4)
Sage AP, Rouse WB (1999) Handbook of systems engineering and management. Wiley, New York
Saleh JH, Hastings DE, Newman DJ (2003) Flexibility in system design and implications for aerospace systems. Acta Astronaut 53(12):927–944
Savage S (2009) The flaw of averages. Wiley, New York
Sethi AK and Sethi SP (1990) Flexibility in manufacturing: a survey. Int J Flex Manuf Syst. 2:289–328 (Kluwer Academic Publishers, Boston. Manufactured in The Netherlands)
Shah NBJ, Viscito WL, Ross AM, Hastings DE (2008) Quantifying flexibility for architecting changeable systems. 6th Conference on engineering research, Long Beach, CA, April 2008
Simpson TW, Siddique Z, Jiao J (2006) Product platform and product family design. Springer, Heidelberg
Siri D (2010) In race to market, Nissan's electric car takes shortcuts. Wired http://www.wired.com/autopia/2010/01/nissan-leaf-2/. Accessed 21 Feb 2012

Skiles SM, Singh V, Krager J, Wood K, Jensen D, Szmerekovsky A (2006) Adapted concept generation and computation techniques for the application of a transformer design theory. ASME International design engineering technical conferences and computers and information in engineering conference, DETC2006-99584, Philadelphia, PA

Stenquist P (2011) New York Times. Wheels: http://wheels.blogs.nytimes.com/2011/01/28/hyundai-kia-overtakes-ford-while-g-m-threatens-toyota-in-2010-global-sales/. Accessed 20 Feb 2012

Stock JH, Watson MW (2007) Introduction to econometrics. 2nd edn. Pearson, Boston

Suh E, de Weck O, Zhang D (2007) Determining product platform extent. Res Eng Des 18(2):67–89

Tilstra AH, Backlund PB, Seepersad CC, Wood KL (2008) Industrial case studies in product flexibility for furture evoluation: an application and evaluation of design guidelines. In: Proceedings of the ASME 2008 international design engineering technical conferences & computers and information in engineering conference, IDETC/CIE 2008, Brooklyn: ASME, pp 1–14

Tran L (2012) NHTSA statement on conclusion of chevy volt investigation. NHTSA: http://www.nhtsa.gov/About+NHTSA/Press+Releases/2012/NHTSA+Statement+on+Conclusion+of+Chevy+Volt+Investigation. Accessed 15 Feb 2012

Train K (2003) Discrete choice methods with simulation. Cambridge University Press, New York

Trigeorgis L (1996) Real options: managerial flexibility and strategy in resource allocation. The MIT Press, Cambridge

Tversky A, Kahneman D (1981) The framing of decisions and psychology of choice. Science 211(1981):453–458

Ulrich KT, Eppinger SD (2008) Product design and development. McGraw Hill, New York

Utterback JM (1996) Mastering the dynamics of innovation. Harvard Business Press, Boston

Wang CY (2011) Vehicle Battery Engineering, Summer 2011 Short Course. Dearborn, MI, USA

Wang T, de Neufville R (2005) Real options “in” projects. 9th Real options annual international conference, Paris

Westbrook M (2001) The electric and hybrid electric car. Society of Automotive Engineers, Warrendale

Whitney DE (2004) Mechanical assemblies: their design, manufacture, and role in product development. Oxford University Press, New York

Consideration of Legacy Structures Enabling a Double Helix Development of Production Systems and Products

Magnus Wiktorsson

Abstract Based on an increasing need of life cycle perspectives in product and production development, there is a call for more effective working methods for the reconfiguration, rearrangement, retro-fit and reuse of current equipment, systems and processes within production systems. This chapter discusses the need and character of such methods based on current research and industrial practice in production system design and development. A concluded development process is illustrated by a double helix development cycle for the production system and the product. The traditional life cycle illustration of product and production system design is in this case altered to a double helix where the same design phases of requirement analysis, alternative synthesis and alternative analysis reoccur for each project phase of conceptual design, detailed design, validation and industrialization/running-in, but for each development cycle on an elaborated level.

1 Introduction

The increasing consideration for products and production system life cycles and need for drastic increased resource efficiency in manufacturing—doing more by using less non-renewable resources—becomes clear as data on economic activity in manufacturing is presented as in Fig. 1. A number of countries show a tremendous growth in manufacturing activity. Also on a global scale manufacturing activities is increasing. During a recent decade (2001–2010), the global economic activity within manufacturing has increased by 34.7 % in constant prices over the period, while the global gross national product (GDP) has increased by 'only' 26.0 % (UN Stats 2012).

M. Wiktorsson (✉)
Associate professor in product and process development, Mälardalen University, Eskilstuna, Sweden
e-mail: magnus.wiktorsson@mdh.se

E. Henriques et al. (eds.), *Technology and Manufacturing Process Selection*, Springer Series in Advanced Manufacturing, DOI: 10.1007/978-1-4471-5544-7_2,

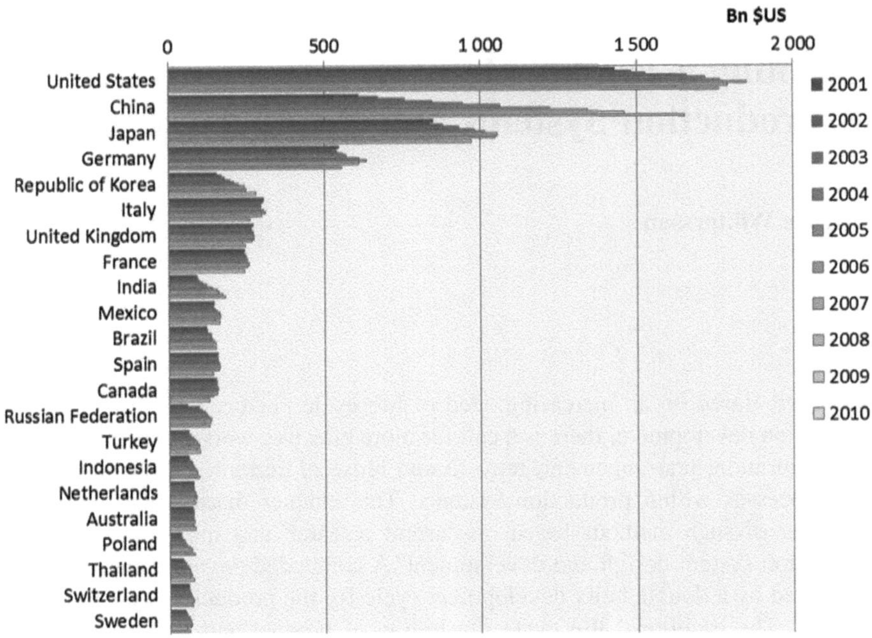

Fig. 1 Economic activity within manufacturing in constant 2005 US dollars for the top 22 countries. From year 2001 (top within each country) to year 2010 (bottom within each country). Data from UN Stats (2012)

Meanwhile, the climate change and need for absolute decrease of for instance greenhouse gases has been witnessed by numerous researchers and agreed to by governments and authorities. As industrial activities contribute to significant environmental impact, the rapid increase of manufacturing activity call for the urgent actions for resource efficiency and life cycle perspectives in product and production development.

In the light of resource efficiency and frequent product changes there is a call for effective working methods for the reconfiguration, rearrangement, retro-fit and reuse of current equipment, systems and processes. The objective of this chapter is to introduce the concept of the double helix development of production system and products, enabled by the effective consideration of legacy structures during the production system redesign and product introduction. Based on a study of industrial practice on production development processes is an elaborated production system design process presented, including redesign elements.

2 Production System Design

The research area of production system design is inherently dependent on the close interaction between academics and practitioners. State-of-the-art knowledge is created in a complex pattern of requirements, prerequisites, methodologies, empirics, implementation and deployment. The state of art can thus only be described by covering both research- as well as practitioner-based knowledge fields.

A well-established field of knowledge within production concerns implementation theories for corporate improvement initiatives with an *operational* focus, such as TPM, TQM, Six Sigma, Lean etc., represented by e.g. Ohno (1998), Liker (2004), Womack and Jones (1996). The lean production paradigm is established by theories, instruments, ontologies, values and metaphysical assumptions that are implemented in the competences, tools, methods and processes within the industry of today, as illustrated to the right in Fig. 2.

The field of knowledge concerning production system *design* have gained momentum ever since Skinner (1969) pointed out the design of the production system as a key to success by: "*what appears to be routine manufacturing decisions frequently come to limit the corporation's strategic options, binding it with facilities, equipment, personnel, basic controls and policies to a non-competitive posture, which may take years to turn around.*" Throughout the recent decades, competences, tools, methods and processes required to design and realize the lean production system have been described, represented by the left box in Fig. 2.

The two areas of knowledge illustrated in the figure are of course closely interlinked by feeding experience, knowledge and data from operations to design, and feeding decisions, plans and guidelines from design to operations.

Fig. 2 Illustrating the dual and interacting activities of production system design and production system operations

2.1 The Design Structure of Analysis and Synthesis

In order to understand the requirements on a production system design process, the specifics of a design process need to be addressed. Many are the researchers who have tried to characterise and structure design—the activity aimed *"at changing existing situations into preferred ones"*, as put by Simon (1981). The design process is often described as including both phases of intuitive creativity as well as rational phases of calculation and evaluation (e.g. Rosell 1990). These two parts of the design process—the creative and analytic viewpoints—seems to separate more and more as the technical complexity increases and specialists tend to play a more important role in the design process. This impedes the development of good design, since both the analytic and creative parts are needed in the design process and it is often in the meeting of these two competencies that a successful design is made.

A general model of the events in an engineering design process is the steps of (problem) analysis, (solution) synthesis and (solution) evaluation, presented in the general ASE model by for instance Braha and Maimon (1997) linking back to Jones (1970) and Asimow (1962). However, although the rhetorical value of such a description, there is a risk of giving the impression of the design process as a linear, predictable and rational process. Instead the design process might be described as a cyclic process where an analysis of the problem—a synthesis of solutions—an evaluation of solutions leads to a new analysis of the problem with knowledge given from the first evaluation and so on. Suh (1990) described the interaction between the solution synthesis (by Suh named synthesis capability) and the solution evaluation (by Suh named analytical ability) as a feedback control loop improving and detailing the design.

2.2 Examples on Production System Design Schemes

Different research traditions have contributed to the current state of knowledge concerning production system design. The holistic and competence perspective on production system design is pointed out by e.g. Bruch and Bellgran (2012) referring to the textbook by Love (1996). The strategic fit of the production system is discussed by e.g. Hayes and Wheelwright (1979) introducing the product-process matrix in order to choose production system lay-out according to product and process life cycle stage. From an industrial engineering perspective, the textbooks by Bennett (1986) and Bennett and Forrester (1993) are examples supporting the manufacturing engineer and management on technology selection and designing the physical system. System modelling has also been an inspiration, such as the IDEF0 based method by for example Wu (2001), but also the stage-gate model (from Robert S. Cooper) developed further by e.g. Blanchard and Fabrycky (1998), Rau and Gu (1997), and Wu (1994). The system approach is taken on the production system problem by Seliger, Viehweger et al. (1987) and Bellgran and

Säfsten (2010), as well as in the work by Wiktorsson (2000) focusing the evaluation of production systems and linking to tools for performance and behavioural validation. Also the design information in the production system design process is in focus, as by Bruch and Bellgran (2012) and requirement specification by Wiktorsson et al. (2000).

2.3 The Increasing Need of Considering Legacy Structures

The research based methods described earlier, are in most cases based on a clean-sheet design process. However, production design situation ranges from a total 'green field'/full investment situation, to a redesign; rearranging and reusing existing equipment and facilities. The most common situation encompasses both aspects; new investment as well as redesign.

The increasing need for changing and adapting the production system to the ever changing requirements drives a development towards (1) more adaptive and responsive production systems and technologies, (2) more effective working methods for the rearrangement and reuse of current equipment, systems and processes. Within the first, the integration of legacy is handled by reconfigurable platforms/modular based engineering approaches that enables the reuse of legacy structures, discussed by e.g. Bi et al. (2008), El Maraghy (2006) and Rogers and Bottaci (1997). In the second case, the integration of legacy concerns the specific production design and procurement processes, enabling the record and reuse of legacy structures.

As mentioned, the earlier works on production system design have not especially focused on the aspect of production system *re*design, where the reuse of existing equipment and facilities are of specific interest. In work by Andrew (1991) and Tobias (1991) more general key issues which determine success or failure for a redesign of a manufacturing system are discussed. The issues pointed out by Andrew (1991) include the composition of project teams, manufacturing strategy, system design, manufacturing control systems, human issues and implementation. Also the textbook 'Manufacturing Systems Redesign' by O'Sullivan (1994) discusses the subject but presents a more general structure for the design of manufacturing systems. In fact, as current academic and industrial production design processes are in most cases derivatives of product development processes, not pinpointing legacy equipment and structures, an elaborated design process is needed including redesign elements, to ensure adaptability and sustainability.

3 A Case Study Illustrating Industrial Practice

The dominant way to organize a development process from an industrial perspective is by a stage-gate model with a supporting project management infrastructure. A large number of production development processes with a stage-gate procedure are in use, often within the context of a more general product realization and development process.

The documentation and influence of legacy structures is closely linked to the acquiring, sharing and use of information during the production system design process. One example on information aspects of the production system design process is the industrial case study by Bruch and Bellgran (2012) focusing the factors facilitating the information sharing during this process. In the study it is concluded that sharing of information is promoted by formalization and the study provides strong evidence for the importance of sharing information during a design process in a more sophisticated manner.

One industrial case study more specifically contributing to the empirical basis for this chapter was conducted at a Swedish automotive manufacturer. The company's product industrialization and production procurement processes were studied, as these two processes were the ones corresponding to the production system design, as described by Netz and Wiktorsson (2009). It was concluded that the current formal processes are to a large extend based on an investment or green field situation. The management of legacy structures and information retrieval was studied in order to synthesize into an elaborated design structure for focusing the production system, but in a context of new product introduction and life cycle considerations.

This procurement process for investment projects and the gates are closely related to the gates in the formal purchasing order document currently used by the company. This process was grouped into six stages and nine phases as presented in Fig. 3.

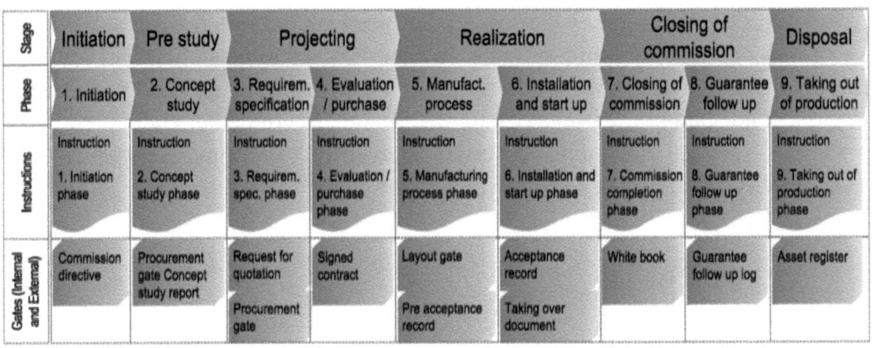

Fig. 3 The procurement process for production equipment, as used by the case company

By studying this production equipment procurement process and interviewing project members, it is concluded that the gates suitable for investment projects are not optimal for redesign projects. In a redesign there are no purchasing orders to refer to at the internal gates. Also there are important aspects in redesign project which are not emphasised in investment projects, such as in detail considering the down-time during the redesign and rearrangement. The authors conclude that there is an industrial need to formulate state-gates which could be used also for cases including legacy equipment and processes.

To conclude, by engagement and studies of product introduction projects and production system development processes, it is concluded that the observed industrial processes for production or assembly system design are realized in three forms:

- production system design considered as a sub-task in an overarching product development and industrialization process,
- production system design handled by a general project management process, not specifically addressing the challenges and characteristics of this open, complex and multidimensional system design,
- production system design handled from an equipment procurement perspective, concerning the specific elements with need of investments.

In neither of the observed industrial cases are the specific characteristics from considering legacy within a system redesign identified, and there is a potential in further formalizing the specification and information of legacy structures during the design process.

4 The Double Helix Development Cycle for Production and Product

Concluding from the brief overview of production system design schemes and industrial practice, two key characteristics of a production system development process, previously not explicitly described in production system design literature and in studied practice are:

- The design based structure of analysis and synthesis in form of three generic and iterative phases: requirement analysis, alternative synthesis and alternative analysis.
- Handling of legacy structures by a formalised specification used during the development refinement.

As these two aspects are considered, a schematically described double helix development cycle for production and product emerges, as illustrated in Fig. 4.

The traditional life cycle illustration of product and production system design is in this case altered to a double helix where the same design phases of requirement

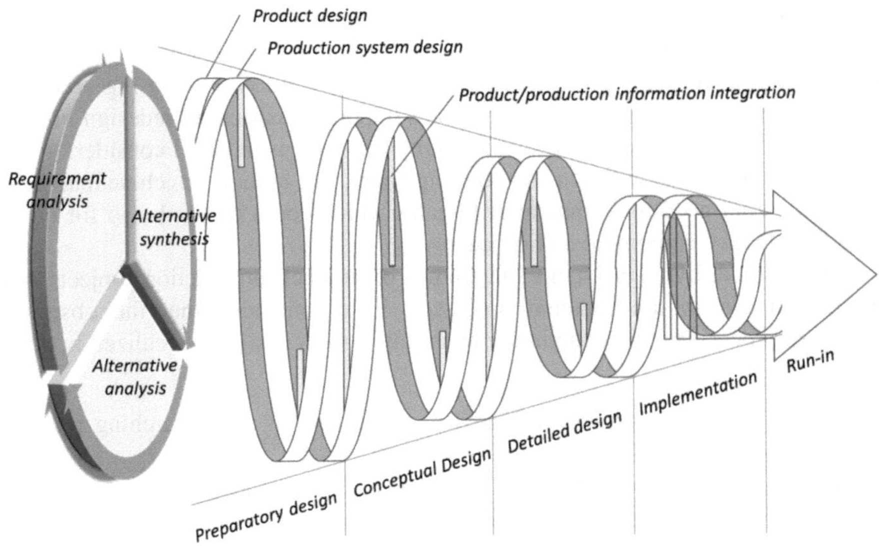

Fig. 4 The double helix development process for production system and product

analysis, alternative synthesis and alternative analysis reoccur for each project phase of conceptual design, detailed design, validation and industrialization/running-in, but for each development cycle on an elaborated level. For an efficient and goal oriented production system design helix, the synchronization with the product design helix is vital, illustrated in the double helix model by the information links between the product and production development helixes. In order to gain two harmonized, resource efficient and effective development helixes, the product and process information (the linking 'nucleobases' of this double helix) are keys, as well as the management of legacy structures.

4.1 Towards a Formal Consideration of Legacy Structures

It is concluded that from both a research perspective as well as an industrial perspective, processes and methods for production system *re*design with a focused handling of legacy in production systems are not explicitly described. The current academic and industrial production design processes are in many cases derivatives of product development processes. These process plans do not pinpoint legacy equipment and structures, since this is not in general a vital part within product development. Neither production procurement processes do for natural reasons focus on legacy infrastructure—when investing in new equipment, other aspects are more essential than considering current equipment. Production system design processes are in many cases focused on the specific details in the system that needs renewal or modification, not the entire system characteristics or architecture including legacy structures.

The consideration of legacy structures is however an established format within e.g. IT management, where a common situation is to migrate from a current situation to a new system design where current solutions are to be reused. Procedures, information formats and processes have been developed within this field. Typical solutions in this respect include discarding the legacy system and building a virtual replacement system; freezing the system and using it as a component of a new larger system; and modifying the system to give it new functionality (Lucia et al. 2008; Wang et al. 2007).

The consideration and potential reuse of legacy production structures is in the presented double helix process proposed to be solved by a formal requirement and constraint structure, to be detailed in the early design phases and used throughout the production system design helix. This phase of specification implies defining and structuring terms such as prerequisites, constraints, requirements, goals, objectives, wishes, wants, demands, musts and needs, all being internal or external. By comparison with the formulation of a traditional linear optimisation problem, the concept of constraints as a language for legacy structures is introduced by e.g. Wiktorsson et al. (2000). The requirements of the redesign are on each system level, as the iterative process proceeds, described by the four elements in the requirement analysis:

- Functional requirements: musts on performance
- Internal design constraints: musts on design solutions due to internal reasons
- External design constraints: musts on design solutions due to external reasons
- Winning criteria: wants on capabilities.

This four-element-framework is based on, encouraging and using the wide range among the criteria; from absolute musts on the system to interesting aspects to know of; from general functions to fulfil to detailed design solutions to use.

The legacy structures to be documented and considered during the development process are to be covered within the element "internal design constraints" within the framework. These constraints help the designer throughout the design by limiting the possible options.

4.2 Operationalizing the Double Helix Framework

From the conceptually described double helix development process, it is not obvious how to implement it in practice. Figure 5 illustrates an effort in clarifying the production system design process in terms of tools and phases in order to realise the concept of considering legacy structures and enabling a double helix development of production systems and products. The three design phases of analysis—synthesis—analysis, is realised by tools such as:

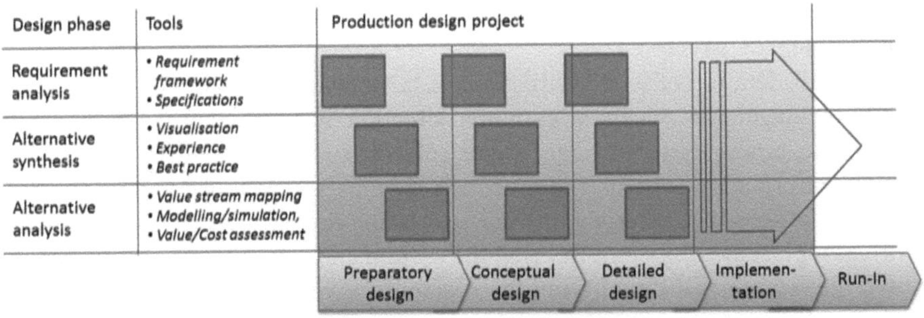

Fig. 5 Operationalizing the double helix framework into a production system design process scheme

- the requirement framework with the concept of legacy constraints for the phase of Requirement analysis,
- visualization, best practice and creativity tools for the phase of Alternative synthesis,
- evaluation, assessment and validation tools such as value stream mapping, simulation and modelling, value/cost assessment and business case models for the phase of Alternative analysis.

These three phases of design and their corresponding tools are linked to a production system design process with the classic phases within a stage-gate model from preparatory design to implementation and run-in, as illustrated in Fig. 5.

Such an implementation of the double helix model should approach the two identified weaknesses of current academic and industrial design models: the design aspects of analysis and synthesis as well as the handling of legacy structures.

By extending this operationalization to the earlier described case company, it is clear that the current development process focus on the stage-gates of the development. It describes *what* to accomplish, not *how* to accomplish it. The three design aspects could be included in a stage-gate process, similar to the one in Fig. 3 in a natural way, by specifying templates, methods and frameworks that supports the inherent design logic. By adding a helix structure of a structured iteration between requirement analysis, alternative synthesis and alternative analysis, as well as guiding instruments/tools to use during the three phases, the design logic is built into the process and the development team. In addition, the case illustrated the challenge in considering the legacy structures of manufacturing during a production development process. This is supported by adding e.g. a formal requirement and constraint structure, guidance and best practice for retro-fit of equipment, stage-gates for the redesign and rearrangement, and analysis tools for redesigned production facilities. Another example is the three industrial examples of the formal requirement and constraint structure given by Wiktorsson et al. (2000) illustrating the improvement potentials in describing objectives and legacy during development processes.

The double helix illustration of the development process emphasises the close interaction between product development and production development. However, in many situations nowadays, product development is done separately from production. From a total sustainability and efficiency perspective, there is still a need to utilize the potential in legacy manufacturing structures, even if the manufacturing is done by another company. It is here argued that the total cost, effort and environmental impact could be decreased by a closer consideration of opportunities given by the current manufacturing equipment, instead of a sequential development process where lowest offers are sought for the manufacturing of the finished product design.

5 Conclusion

To summarize, the basis for the paper is the production system life cycle and the need for more efficient reuse and adaption of the production system, both from a resource-efficiency perspective as well as a product customization perspective. From an industrial case study it is concluded that the elaborated production system design methods need to consider legacy equipment, processes and systems in an extended way that can be reused and reconfigured to suit the future need. The proposed double helix design process incorporates the inherent nature of the design process, the vital interchange of product and production information, the non-sequential nature of the design process, as well as the formal consideration of legacy structures to reuse in future production designs. Future efforts for the design process would include the formalisation of the specification structures, the incorporation of current validation and analysis tools as well as creative tools for solution synthesis.

Acknowledgments The author would like to thank the editorial team and anonymous reviewers for their valuable comments and suggestions to improve this chapter. In addition, the chapter would not have been possible without the valuable contribution from the case company, for which the author is greatly thankful.

References

Andrew MT (1991) Managing the redesign of manufacturing systems. Int J Technol Manage 6(3/4):375–384

Asimow M (1962) Introduction to design. Prentice-Hall, Englewood Cliffs

Bi ZM, Lang SYT, Shen W, Wang L (2008) Reconfigurable manufacturing systems: the state of the art. Int J Prod Res 46:967–992

Bellgran M, Säfsten K (2010) Production development: design and operation of production systems. Springer, UK

Bennett D (1986) Production systems design. Butterworth-Heinemann, London

Bennett D, Forrester PL (1993) Market-focused production systems: design and implementation. Prentice Hall, New Jersey
Blanchard B, Fabrycky W (1998) Systems engineering and analysis. Prentice-Hall, New Jersey
Braha D, Maimon O (1997) The design process: properties, paradigms and structure. IEEE Trans Syst Man Cybern 27(3):146–166
Bruch J, Bellgran M (2012) Creating a competitive edge when designing production systems: facilitating the sharing of design information. Int J Serv Sci 4(3/4):257–276
ElMaraghy H (2006) Flexible and reconfigurable manufacturing systems paradigms. Int J Flex Manuf Syst 17:261–276
Hayes RH, Wheelwright SC (1979) Link manufacturing process and product life cycle. Harvard Bus Rev 1:133–140
Jones JC (1970) Design methods: seeds of human futures. Wiley-Interscience, London
Liker JK (2004) The Toyota way: 14 management principles from the world's greatest manufacturer. McGraw-Hill, New York
Love D (1996) The design of manufacturing systems. In: Warner M (ed) International encyclopedia of business and management. Routsledge, London
Lucia AR, Francese R, Scanniello G (2008) Developing legacy system migration methods and tools for technology transfer. Methods 2008:1333–1364
Netz E, Wiktorsson M (2009) Enabling effective production system redesign: complementing investment and design processes. Swedish production symposium 2009, SPS09, Göteborg
Ohno T (1988) The Toyota production system—beyond large scale production. Productivity Press, New York
O'Sullivan D (1994) Manufacturing systems redesign: creating the integrated manufacturing environment. Prentice Hall, Englewood Cliffs, NJ
Rau HA, Gu P (1997) Design methodology and integrated approach for design of manufacturing systems. Integr Manuf Syst 8(3):159–172
Rogers G, Bottaci L (1997) Modular production systems: a new manufacturing paradigm. J Intell Manuf 8:147–156
Rosell G (1990) Notes on the design process (In Swedish). Dissertation. KTH—Royal Institute of Technology, Stockholm
Seliger GB, Viehweger, et al (1987) Decision support in design and optimization of flexible automated manufacturing and assembly. Robot Comput Integr Manuf 3(2): 221–227
Simon HA (1981) The sciences of the artificial. 2nd ed. MIT Press Cambridge, MA
Skinner W (1969) Manufacturing—missing link in corporate strategy. Harvard Business Review May–June 1969
Suh NP (1990) The principles of design. Oxford University Press, New York
Tobias AM (1991) Managing the redesign of manufacturing systems. Int J Technol Manage 6(3):375–384
UN Statistics divisions (2012) Data available online on: http://unstats.un.org/. Data adapted by the author. Accessed 20 Apr 2012
Wang X, Hu SXK, Haq E (2007) Garton H (2007) Integrating legacy systems within the service-oriented architecture. IEEE Power Eng Soc Gen Meet 95630:1–7
Wiktorsson M (2000) Performance assessment of assembly systems. Dissertation. KTH—Royal Institute of Technology, Stockholm
Wiktorsson M, Andersson T, Broman M (2000) A note on the specification of assembly systems. Int J Prod Res 38:3997–4002
Womack JP, Jones DT (1996) Lean thinking: Banish waste and create wealth in your corporation. Simon and Schuster, New York
Wu B (1994) Manufacturing systems design and analysis-context and techniques. Chapman and Hall, New York
Wu B (2001) Strategy analysis and system design within an overall framework of manufacturing system management. Int J Comput Integr Manuf 14(3):319–341

Six Sigma Life Cycle

Pedro A. Marques, Pedro M. Saraiva, José G. Requeijo and Francisco Frazão Guerreiro

Abstract This chapter presents the Six Sigma Life Cycle approach, a framework that comprises two interlinked models: the Six Sigma Project Life Cycle model and the Technical System Life Cycle model. The Six Sigma Project LC model consists of a detailed and comprehensive process intended to guide managers and team leaders along all the relevant stages of a Six Sigma project, from identification to post-project, helping them to achieve the project's goals with minimum expenditure of effort and resources. The Technical System LC model aims to assist them towards the best decision of which Six Sigma methodology, roadmap and toolbox should be used, depending on the degree and type of innovation that are involved in the project scope. Two case studies are described to demonstrate the practical application and usefulness of the framework.

1 Introduction

Since Motorola's first proposed it during the mid-80s, Six Sigma has gained increasing interest, both from the scientific and business communities. Nowadays, Six Sigma goes well beyond traditional continual improvement activities, ranging a wide set of other relevant topics, including product, service and process design, integration with management systems, strategic business planning, among others.

P. A. Marques (✉) · F. F. Guerreiro
Department of Strategy and Special Projects, ISQ–Welding and Quality Institute, Oeiras, Portugal
e-mail: pamarques@isq.pt

P. M. Saraiva
Department of Chemical Engineering, University of Coimbra, Faculty of Sciences and Technology, Coimbra, Portugal

J. G. Requeijo
Department of Mechanical and Industrial Engineering, Nova University of Lisbon, Faculty of Sciences and Technology, Lisbon, Portugal

E. Henriques et al. (eds.), *Technology and Manufacturing Process Selection*, Springer Series in Advanced Manufacturing, DOI: 10.1007/978-1-4471-5544-7_3,

Six Sigma Life Cycle (LC) provides a comprehensive and practical framework that links, in a logical manner, most of the significant issues of the Six Sigma subject. It contains two main models, which together enable professionals to efficiently manage all the phases along the path of a Six Sigma project targeting the improvement, design or redesign of any kind of technical system.

This chapter starts by reviewing, in Sect. 2, all the main concepts of Six Sigma, including the most recent issues concerning Design for Six Sigma (DFSS) and the ICRA (Innovation, Configuration, Realization, Attenuation) value creation strategy. In Sect. 3, the two models of the Six Sigma LC, herein proposed, are introduced, related and described in detail. Then, in Sect. 4, two case studies describing practical applications of the Six Sigma LC approach are presented. In Sect. 5, the practical implications and the main findings of the research are presented and discussed. Section 6 closes with a synthesis of the chapter and its main conclusions, along with the discussion of areas for future research.

2 Six Sigma

Six Sigma has been defined in a variety of ways (Schroeder et al. 2008), but it is often regarded according to the following three perspectives (McCarty et al. 2004):

1. Management system.
2. Methodology.
3. Metric.

The following subsections summarize each of these perspectives and their main concepts, based on the review of relevant literature. Many of the main recent topics are covered, including the integration of Six Sigma with other management systems, the DFSS methodological approach, and the third generation of Six Sigma.

2.1 Six Sigma as a Management System

At the highest level, Six Sigma is an enterprise wide initiative, directly linked to its strategy and business processes, which deployment relies on a well trained role structure, usually known as 'belt system', with specific levels of responsibilities (Zu et al. 2008). It is when Six Sigma is articulated with the strategic activities that organizations see its greatest impact on sustainability of the business results (McCarty et al. 2004). Creating value for the enterprise requires top-line growth (Montgomery 2005), coming primarily from actions needed to take advantage of opportunities for improvement and innovation. Only by prioritizing and selecting projects that contribute the most to organization's business performance, can management ensure significant gains from the Six Sigma initiative (Watson 2004).

One of the emerging trends of Six Sigma is its integration with a set of relevant management models (Antony et al. 2006). This happens mainly because the successful implementation of Six Sigma in an enterprise highly depends on the ability to integrate it with the already existing management initiatives (Pfeifer et al. 2004). Among this topic it is to highlight the integration of Six Sigma with the following approaches: ISO[1] 9001 quality management standard, EFQM[2] Excellence Model, Lean Management, and ITIL (Information Technology Infrastructure Library).

2.2 *Six Sigma as a Methodology*

Six Sigma makes use of systematic, highly disciplined, customer-centric and data-driven methodologies to foster business improvement or innovation, on a project by project approach (Bendell 2006). There are two main methodological approaches through which Six Sigma teams carry out their projects:

- Six Sigma projects, following the DMAIC (Define, Measure, Analyze, Improve, Control) roadmap, aimed at improving the performance of an existing technical system, by minimizing the variability regarding its critical to quality characteristics (CTQCs).
- Design for Six Sigma (DFSS) projects, carried out to design new technical systems or to redesign existing ones, aimed at ensuring high degrees of performance and customer satisfaction throughout the system's life cycle. There are many DFSS roadmaps available, but the IDOV (Identify, Design, Optimize, Validate) and the DMADV (Define, Measure, Analyze, Design, Verify) are the most often employed.

All the mentioned roadmaps are inspired in the well-known Deming/Shewhart PDCA cycle. The DMAIC roadmap is used under the context of a continual improvement process, while the mentioned DFSS roadmaps (IDOV and DMADV) are usually adopted as a design/innovation process.

The previous methodological approaches can be fitted into a value creation strategy, known by a four-step process called ICRA (Harry and Linsenmann 2006), depicted in Fig. 1, developed under the so-called third generation of Six Sigma. The ICRA process follows the same generic sequence of stages of a typical system life cycle approach, regardless of the type or tangibility of such system. Both "Innovation" and "Configuration" are steps that cover the initial stages of a technical system life cycle, where design and development activities take place, thus demanding a DFSS type of project. The remaining two steps are mainly related to efforts of optimizing the performance of a current technical system, so the Six Sigma DMAIC approach is preferable. The ICRA strategy is applicable to four distinct areas: market, business, product, and process.

---

[1] International Organization for Standardization.

[2] European Foundation for Quality Management.

Fig. 1 The relationship between the Six Sigma methodological approaches and the ICRA value creation strategy (adapted from: Harry and Linsenmann 2006)

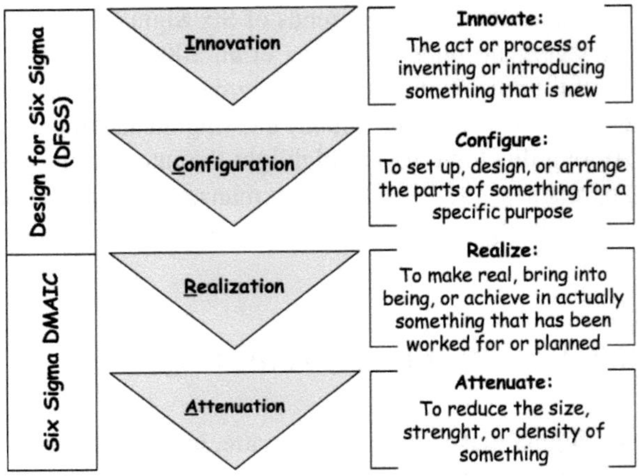

2.3 *Six Sigma as a Metric*

From the metric perspective, the sigma level is a measure of process capability. The focus of Six Sigma is reducing variability in critical to quality characteristics (CTQCs) around specified target values to the level at which failure or defects are extremely unlikely (Montgomery and Woodall 2008). The higher the value of the sigma level, also known as Z–score, the less the variability observed on its outputs, thus the better the performance of that process. When the sigma level equals six, the chance for a failure or a defect to occur is extremely unlikely. Not every process needs to operate at a Six Sigma level, since the established quality level of performance depends on its strategic importance and the cost of improvement relative to its benefit (Kumar et al. 2007).

Figure 2 shows the statistical concept behind Six Sigma for a process with two-side specification limits, under the assumption of a measurable and normally distributed CTQC. A process with a six sigma level of performance is at a distance

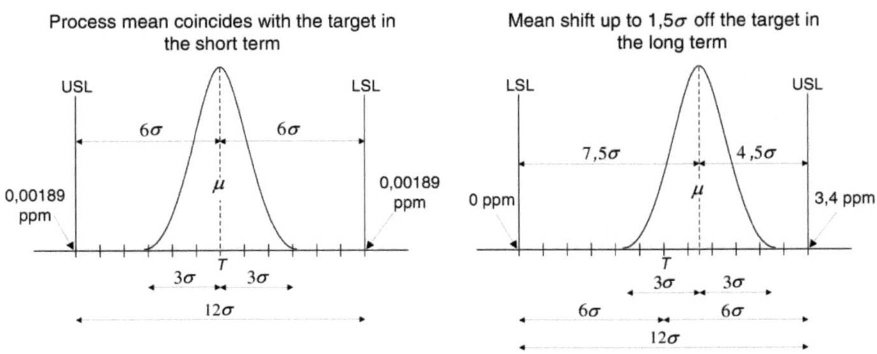

Fig. 2 Statistical concept behind Six Sigma

of six standard deviations (6σ) from both specification limits when the value of its mean equals the target value. In this case, the number of defects will be of about 0.00189 parts per million (ppm), so that it can be stated as being an almost defect free process. In the long term, such a process is robust enough so that it can also accommodate shifts of up to $\pm$ 1,5σ in its central tendency location around the target value, because the measured CTQC will not fall out of specification limits, even if that happens, more than 3.4 times over each million of opportunities.

3 The Six Sigma Life Cycle Framework

The Six Sigma LC approach incorporates two interlinked models:

1. Six Sigma Project Life Cycle model.
2. Technical System Life Cycle model.

Among other benefits, these two models are intended to be particularly useful, in guiding Six Sigma managers and team leaders to better:

- Systematize the detection of opportunities for improvement and innovation.
- Identify and properly scope potential Six Sigma projects.
- Prioritize the most promising Six Sigma projects.
- Determine the most appropriate methodological approach and roadmap to carry out a selected Six Sigma project.
- Take advantage of the synergies between the Six Sigma DMAIC and the DFSS methodological approaches.
- Target the improvement, design or redesign of any type of technical system.

Both models can be related to other common life cycle based methodologies. The Technical System LC model fits into a generic system life cycle methodology, regardless of the type or nature of such system, while the Six Sigma Project LC model can be framed in the phases of a traditional project life cycle process. Table 1 and Fig. 3 provide a comparison among the Technical System LC model and system life cycle phases suggested by the Systems Engineering (SE) approach (INCOSE 2004). Table 2 and Fig. 4 relate the Six Sigma Project LC model with the project life cycle framework proposed in the BS 6079-1:2010 standard (BSI 2010).

In the next two subsections, these models under the Six Sigma LC framework are discussed in detail.

3.1 Six Sigma Project Life Cycle model

The Six Sigma Project LC model is exhibited in Fig. 5. All Six Sigma projects need to go through the following five stages, which together comprise their life cycle:

Table 1 Relationship between the Technical System LC model and the phases of a system life cycle according to the Systems Engineering approach

System life cycle phase (SE approach)	Main activities performed in each phase	Technical System LC model
Pre-concept	Ideation (new ideas, new technology, new market needs, etc.) Technical research activities	These activities are performed during the "Identify" phase of the IDOV roadmap of the DFSS approach
	Market research activities Feasibility studies	The DMAIC roadmap is not adopted in this phase
Concept exploration	Identification of customer needs and wants Determination of design requirements Conceptual design	Conceptual design tasks only take place in radical or substantial innovation initiatives Concept exploration activities are performed during the "Design" phase of the IDOV roadmap
	Technology and risk assessment	The DMAIC roadmap is not adopted in this phase
Definition and risk reduction	High-level design review and update	These activities are performed during the "Design", Optimize" and "Validate" phases of the IDOV roadmap
	Detail design and integration of subsystems System testing and functional optimization	These activities are performed during the "Design" and "Verify" phases of the DMA(DV)C roadmap, when minor design changes are required
	Prototyping and validation of the system	The DMAIC roadmap is not adopted in this phase
Production, fielding and operations	Quality improvement efforts	The DMAIC roadmap is used in this phase for improvement initiatives
	Efficiency improvement efforts Control and monitoring activities	The DMA(DV)C roadmap is when improvement requires minor conceptual changes in the system design
	Operations management activities	The IDOV roadmap is not adopted in this phase
Disposal	System replacement Efficient disposal efforts	The DMAIC roadmap can be employed in order to improve efficiency of the disposal

1. Identification of Six Sigma projects.
2. Selection of relevant Six Sigma projects.
3. Six Sigma project planning.
4. Six Sigma project execution and completion.
5. Post Six Sigma project.

The Six Sigma Project LC model consists of a detailed and comprehensive process intended to assist professionals along all these stages, helping them to efficiently achieve the project's goals. The process under each of the project life

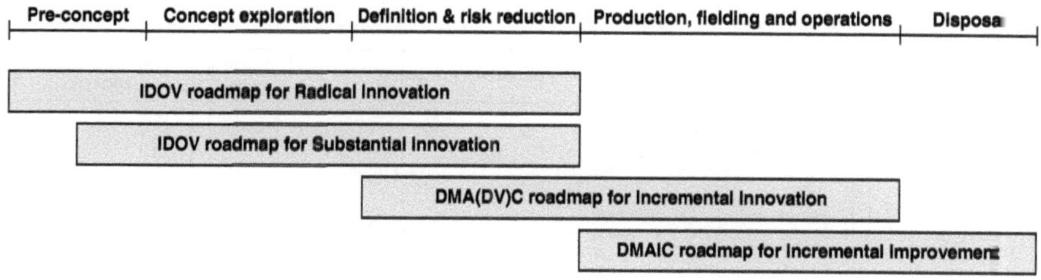

Fig. 3 The role of the Six Sigma roadmaps of the Technical System LC model within the phases of a system life cycle according to the Systems Engineering approach

Table 2 Relationship between the Six Sigma Project LC model and the components of a traditional project life cycle according to the BS 6079-1:2010 standard

Components of a project life cycle (BS 6079-1:2010)	Description of each component	Six Sigma Project LC model
Preparing for the project	Project portfolio management. Project selection Project statement, including definition of objectives and risk assessment	Identification of Six Sigma projects (Stage 1) Selection of relevant Six Sigma projects (Stage 2)
Initiating a project	Project planning Definition of roles and responsibilities Definition of milestones	Six Sigma project planning (Stage 3)
Directing and managing a project	Management of each project phase Monitoring and controlling the project, including risk management activities	Six Sigma project execution and completion (Stage 4)
Closing a project	Review the efficiency of the project and confirm that project objectives were achieved Record and communicate lessons learnt for future projects	Post Six Sigma project (Stage 5)
Reviewing a project outcome	Determine the extent to which actual benefits match those predicted in the business case	Post Six Sigma project (Stage 5)
Project gates	Check if the project is still required and viable Confirm that risks are acceptable and allocated resources are adequate Make a "go", "no go" or deferral decision to continue the project	Six Sigma project review (tollgate review)

cycle's stages is first described in detail; it is then shown how the Six Sigma Project LC model can be used to enable an effective integration between Six Sigma and the requirements of the ISO 9001 standard.

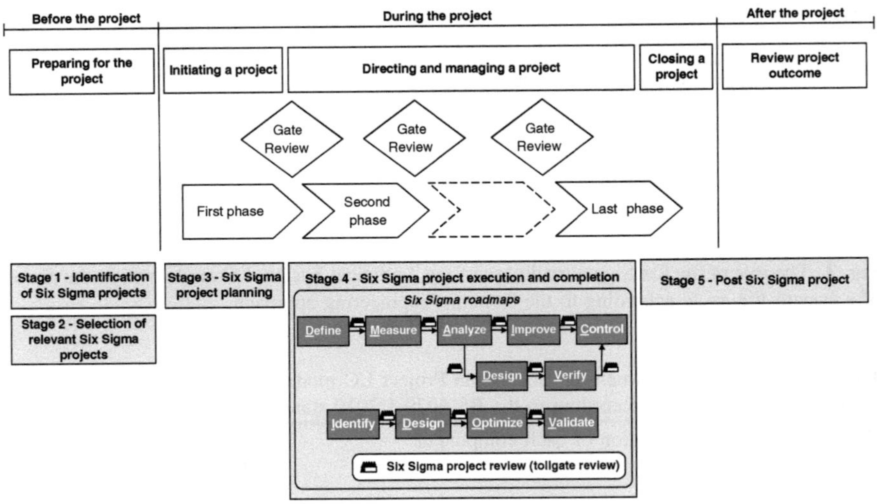

Fig. 4 The role of the stages of the Six Sigma Project LC model within the components of a project life cycle in accordance with the BS 6079-1:2010 standard

3.1.1 The Stages of the Six Sigma Project LC Model

The process under the Six Sigma Project LC model is designed to guide project managers and team leaders to successfully perform, with minimum expenditure of effort and resources, all critical tasks along the path that comprises the five stages described below.

Stage 1—Identification of Six Sigma projects

Potential Six Sigma projects derive from opportunities for improvement and for innovation. The former kind of opportunities usually leads to the identification of possible Six Sigma DMAIC type projects, while the latter normally results in the identification of potential DFSS projects. By its turn, the identification of opportunities of both kinds takes place by analyzing relevant data with origin in several possible sources that can be arranged in a two by two matrix. This matrix organizes different instances of possible sources of data, by placing them into one of four regions (Table 3), according to their relative positioning on the following categories:

- Internal sources of data—Data that is available or need to be proactively gathered, generated internally in the enterprise.
- External sources of data—Data that is available or need to be proactively gathered, from outside of the enterprise.
- Retroactive sources of data—Existing or historical data inside or outside of the enterprise.
- Proactive sources of data—Data usually not readily available either inside or outside of the enterprise, thus requiring proactive actions to be obtained.

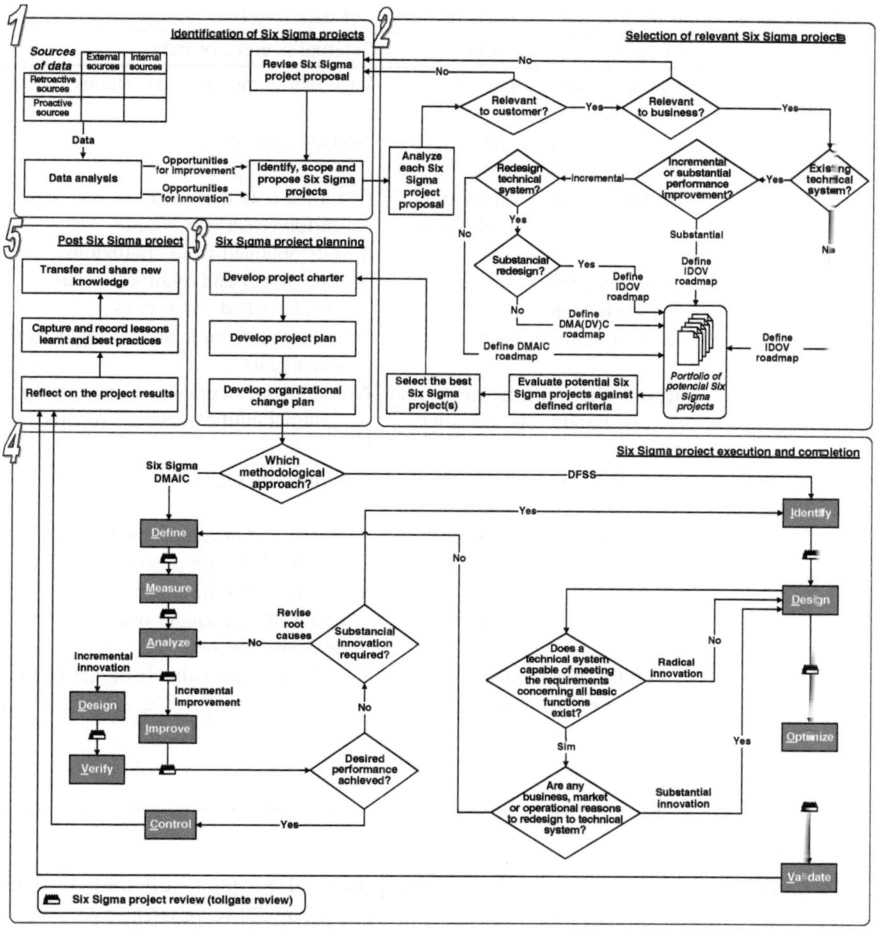

Fig. 5 The Six Sigma Project Life Cycle model

Stage 2—Selection of relevant Six Sigma projects

Potential projects that were proposed in the previous stage of the model, need to be checked in terms their relevance to the business and to the customer. Any project proposal that does not meet both requirements needs to return to the previous stage to be revised.

Each proposed project labeled as relevant needs to be properly framed, in order to determine which methodological approach best fits to its scope, namely the definition of the most appropriate roadmap. This task is fulfilled by taking advantage of reasoning provided by the Technical System Life Cycle model, depicted in Fig. 8, in which the choice of the best methodological roadmap depends on the level of innovation inherent to the potential Six Sigma project.

Table 3 Organization of different instances of sources of data into four regions

	Internal sources of data	External sources of data
Retroactive sources of data	Key performance indicators of the enterprise	Customer complaints and claims
	Internal technical and/or commercial reports	Product warranty systems
	Results of product tests and inspections carried out by the enterprise	Historical data derived from previous customer surveys
	Help desk and customer service data	Results obtained from continuous assessment and qualification of suppliers and contractors
	Internal auditing results	Data derived from legal and regulatory documents
	Results derived from measuring and monitoring activities	Available documentation produced by external entities (studies, reports, etc.)
	Etc.	Etc.
Proactive sources of data	Ideas from internal brainstoring sessions	Benchmarking studies
	Suggestions from personnel	Research and development projects in partnership with universities
	Product life cycle analysis/ studies	Third party auditing results
	Internal auditing results	EFQM Self-assessment results (external factors)
	EFQM Self-assessment results (internal factors)	SWOT analysis results (external factors)
	SWOT analysis results (internal factors)	Etc.
	Etc.	

Regardless of the chosen roadmap, the proposed projects are added to the current Six Sigma project portfolio of the enterprise. This portfolio of project candidates, approved by the enterprise's top management, need to be continuously managed and updated. The evaluation of the potential Six Sigma projects contained in the portfolio, and the selection of the most promising one(s), shall be driven by their positive impact on customer satisfaction and on business results.

Stage 3—Six Sigma project planning

Before starting a selected Six sigma project, it needs first to be properly planned, which includes the development of the:

1. Project charter—It is a single page document, stating the main information about the project, namely its mission, scope, importance, goals, team members, technical leader (i.e. Black Belt), together with the description of the opportunity(ies) that led to the identification of the project.
2. Project plan—It is a detailed description of the project schedule and milestones. The project plan generally makes use of a Gantt chart to depict the duration of the project tasks, and of a RACI (Responsible for, Accountable for, Informed,

Consulted) matrix to define and communicate the roles and responsibilities of project team members.

3. Organizational change plan—This plan aims to ensure that the enterprise is prepared to support the project. Often, it contains actions and strategies to minimize risks concerning issues such as the availability of resources, problems in communication, and resistance to change.

Stage 4—Six Sigma project execution and completion

In the fourth stage of the Six Sigma Project LC model, the Six Sigma project is carried out in accordance with the corresponding roadmap that was determined in the second stage of the model.

As observed in Fig. 5, a methodological roadmap, regardless of its acronym (IDOV, DMAIC or DMA(DV)C), is comprised by phases that are made up by a set of activity stages targeting the accomplishment of specific goals or purposes. A wide and complete set of tools and techniques are available to the Six Sigma project team, to help them to collect, manipulate, analyze and interpret quantitative or qualitative data, enabling them to make decisions based on facts. At the end of each phase of a roadmap, and before moving to the following one, a formal project review, also known as tollgate review, shall be conducted. A tollgate review is a checkpoint where the project Sponsor (a representative of the enterprise s top management), the Black Belt, and various team members meet to determine whether the project activities has been performed as planned or required.

The main activities performed within each phase of the three roadmaps, as well as the most relevant set of tools and techniques employed to support those activities, are indicated in Figs. 6 and 7.

Finally, this stage of the Six Sigma Project LC model provides mechanisms to take advantage of existing synergies and symbioses between the different Six Sigma methodological approaches, thus enabling project teams to move, if necessary, from one roadmap to another one during a Six Sigma project.

Stage 5—Post Six Sigma project

The post-project stage plays an important linkage with the knowledge management process of an enterprise. It starts by fostering the Six Sigma team members and managers to reflect on the results achieved by the project and its impact on the enterprise. Such reflection shall culminate with the capture of the lessons learnt from the project, and recommendations to support further success of future projects and other initiatives within the enterprise.

Post-project reviews provide an opportunity to link the effectiveness in meeting project goals, efficiency in utilizing the resources assigned to the project, and transfer of the special knowledge gained in performing the project to other projects, which is essential to the overall performance improvement of current and future projects, project management processes, and the organization as a whole (Anbari et al. 2008).

Phase	Description	Tools / Techniques
DEFINE	· Identify key customers of the technical system. · Analyze the technical system and prioritize areas for incremental improvement or incremental innovation. · Collect voice of the customer (VOC) data and determine their corresponding needs and wants. · Establish critical to quality characteristics (CTQCs) and their associated operational definitions (specifications).	- Pareto chart - Customer surveys - SIPOC diagram - Value stream mapping - CTQC tree and matrix - Etc.
MEASURE	· Decide on the metrics to use to assess the baseline performance of the technical system regarding the established CTQCs. · Analize and validate the measurement system used to collect data. · Gather relevant data and determine the baseline performance of the technical system. · Define goals regarding the new levels of performance for the technical system.	- Data collection plan - Gage R&R studies - Capability studies - FMEA - Descriptive statistics - Etc.
ANALYZE	· Identify potencial causes affecting the CTQCs of concern. · Analyze the suspected causes and their interactions in order to understand their impact of the poor levels of performance. · Determine the significant causes affecting the CTQCs of concern. · Identify the root causes that lead to the determined significant causes. · Verify if incremental improvement of the technical system is enough to reach the targeted new levels of performance, or if, on the contrary, redesign (incremental innovation) is required.	- Ishikawa diagram - Cause & effect matrix - Hypothesis testing - Analysis of variance - Five "why's" - Etc.
IMPROVE	· Brainstorm possible corrective and improvement actions. · Select those improvement actions necessary to achieve the targeted performance levels. · Design and implement an improvement plan. · Determine the new levels of performance for the CTQCs. · Conclude about the effectiveness of the improvement plan · Validate the best practices derived from improvement plan.	- Brainstorming - 5S - Design of experiments - FMEA - Prioritization matrix - Value stream mapping - Poka-Yoke - Etc.
DESIGN	· Identify functional requirements that are associated to the CTQCs. · Determine the design parameters at the leaf level affecting the functional requirements. · Redefine the choices about the design parameters on the leaf level and optimize conceptual decisions.	- Creativity tools - Axiomatic design - Design of experiments - TRIZ - Simulation techniques - Process reengineering - Etc.
VERIFY	· Perform tests on prototype(s) of the technical system and predict the levels of functional performance. · Optimize the functional performance of the technical system. · Verify and confirm the optimized levels of performance and validate the technical system.	- Simulation techniques - Process management tools - Control charts - FMEA - Design scorecard - Etc.
CONTROL	· Put mechanisms on place to provide ongoing feedback and prevent backsliding. · Ensure continuing measuring and monitoring activities to control the key input variables that significantly affect the CTQCs. · Implement preventive actions to ensure consistence performance of the technical system in terms of its CTQCs	- SPC - Poka-Yoke - Auditing - Process flowcharts - Control plan - Etc.

Fig. 6 Description of the DMAIC e DMA(DV)C roadmaps

3.1.2 Six Sigma Project LC Model to Integrate Six Sigma with ISO 9001

Figure 8 exhibits a framework that scopes the life cycle stages of a Six Sigma project, from its identification to the post-project phase, with the clauses of requirements from the ISO 9001 quality management systems standard. It provides a useful solution for companies to easily integrate their quality management system (QMS) with a Six Sigma program.

In the first stage of the Six Sigma Project LC model, a wide amount of data can be obtained from various sources resulting from the planned QMS processes, including those related to the monitoring and measurement of product, processes, customer satisfaction, and evaluation of suppliers. The analysis of factual data provided by these sources, together with the performance of gap analysis, enable the assessment of the suitability and effectiveness of the QMS and the identification of opportunities for improvement and/or innovation, thus leading to the identification of potential Six Sigma projects.

The evaluation, prioritization and selection of the Six Sigma projects is periodically take place in the context of the management review process. Only

IDENTIFY	· Identify key customers and customer segments of the technical system. · Determine the positioning of key customers in the value chain, and their relationship. · Collect voice of the customer (VOC) data and determine their corresponding needs and wants. · Identify customer requirements and other requirements applicable to the technical system (e.g. legal and regulatory requirements). · Establish critical to quality characteristics (CTQCs) and their associated operational definitions (specifications).	**Main toolbox** - Customer value chain analysis (CVCA) - Personal interviews - Affinty diagram - Kano model - QFD - Etc.
DESIGN	· Define design requirements, namely functional requirements and design constraints. · Generate a range of different design concepts that are able to satisfy the design requirements. · Select the best concept (or concepts) for further design and analysis. · Define design parameters to independently satisfy the functional requirements · Develop the high-level design. · Develop the detailed design.	**Main toolbox** - Pugh method - Axiomatic design - TRIZ - Creativity tools - FAST technique - DFx family of tools - Etc.
OPTIMIZE	· Optimize conceptual robustness of the design · Test the design to predict functional performance and to check if the operational definitions established for the CTQCs are met. · Perform risk analysis on the design. · Optimize the functional performance of the technical system and minimize the chances of failure. · Assess process(es) capability and develop a process management plan, when required.	**Main toolbox** - Taguchi robust design - Simulation techniques - Process management tools - Design scorecard - DFMEA and PFMEA - Etc.
VALIDATE	· Build one or more protypes of the technical system and conduct pilot test(s). · Evaluate and review the pilot results. · Verify the design success. · Validate the design and plan for design implementation	**Main toolbox** - Control charts - Poka-Yoke - Design scorecard - Process management tools - Design of experiments - Etc.

Fig. 7 Description of the IDOV roadmap

potential projects that are likely to be relevant to the business and to the customer shall be considered. In addition to this, before an evaluation takes place, every proposed Six Sigma project should have already been contextualized in the best methodological roadmap.

Each selected Six Sigma project needs to be planned; however such planning need be carried in line with the decisions made, under the QMS planning, to meet the quality objectives. To ensure this, the goals defined for each Six Sigma project shall derive and be aligned with the quality objectives. Six Sigma projects involving efforts towards incremental improvement or innovation need follow the left path under the fourth stage of the Six Sigma Project LC model, indicated in Fig. 5; on the other hand, Six Sigma projects involving substantial or radical innovation are developed in accordance with the IDOV roadmap of the DFSS approach.

As a consequence of a Six Sigma project, corrective and/or preventive actions are usually proposed, implemented, tested and validated, in order to eliminate the causes of a known problem or to prevent it to occur. All successfully completed projects shall contribute to the continual improvement of the ISO 9001 QMS and the lessons learnt during a Six Sigma projects shall contribute to increase the effectiveness of future projects.

The clauses of requirements of the ISO 9001 standard that are indicated on the bottom part of the integration framework, depicted in Fig. 8, provide the necessary support to the activities along the stages of the project's life cycle.

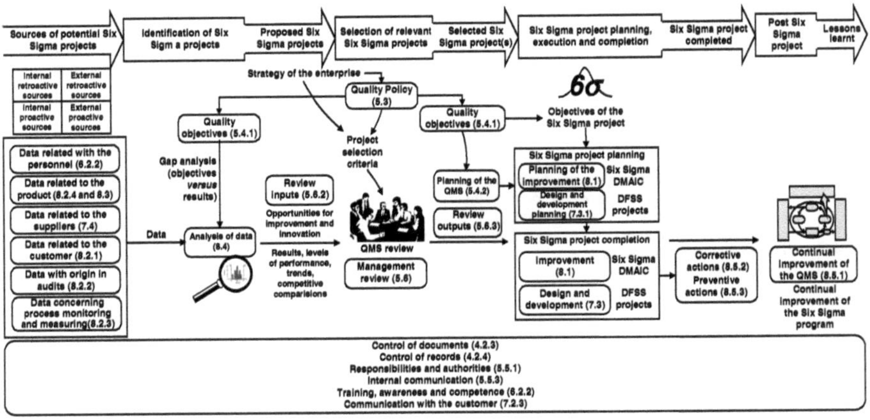

Fig. 8 The integration of Six Sigma and the clauses of requirements of the ISO 9001 standard, using the Six Sigma Project LC model

3.2 Technical System Life Cycle model

A technical system is any man-made entity intended to perform a set of functions in order to achieve predefined and useful outcomes. The life cycle stages of a technical system are usually modeled according to the S–shaped curve, where the system's maturity over time is represented.

The Technical System LC model, exhibited in Fig. 9, recognizes that the methodological approach, to be followed by a Six Sigma project, strongly depends on the conceptual maturity of the technical system targeted by the project, that is to say, on the degree/level of innovation inherent to the Six Sigma project.

Mature technical systems have well-established design concepts, so Six Sigma projects targeting these systems are mainly focused on incremental improvements of their performance, using the DMAIC roadmap. On the opposite, DFSS-type projects involve the development of new or substantially modified technical systems.

Furthermore, the Technical System LC model also recognizes that a Six Sigma project can be deployed to improve, design or redesign any type, or morphology, of technical system. The morphological dimension is strongly related to the type of innovation involved in a Six Sigma project.

The four levels along the maturity axis of the model's matrix correspond to the steps of the ICRA value creation strategy, thus to the life cycle stages of a technical system. By their turn, the four morphological categories considered in the model are in line with the areas of application of the ICRA strategy and with the types of innovation mentioned in the Oslo Manual. The relationship matrix obtained demonstrates that Six Sigma projects can be developed under the context of any type and degree of innovation.

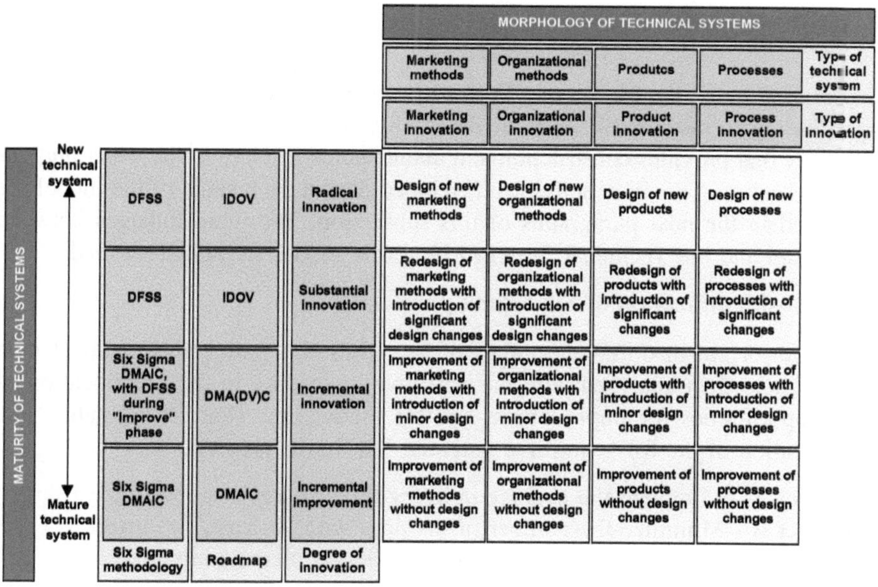

Fig. 9 The technical system life cycle model

This model provides support to the Six Sigma project selection and execution stages of the Six Sigma Project LC model, since it helps Six Sigma managers and team leaders to properly scope any kind of project, particularly for decision-making purposes towards the choice of the methodological roadmap to be adopted.

4 Case Studies

Two distinct case studies illustrating practical applications of the Six Sigma Life Cycle approach are described in the next subsections. Their scope is summarized in Table 4.

Table 4 Scope of the two case studies presented in this section

	Case study 1	Case study 2
Technical system	Purchasing and inventory management process	Specialized transport service for trade fairs and events
Maturity (degree of innovation)	Incremental improvement	Substantial innovation
Morpholopgy (type of innovation)	Process innovation	Product innovation
Methodological approach	Six Sigma DMAIC	Design for Six Sigma
Roadmap	DMAIC	IDOV

4.1 Case Study 1

In this case study, the Six Sigma Project LC model was employed with the aim of contributing to the improvement of the quality management system (QMS) of a company that provides construction and maintenance services in the water and gas utilities sectors. The main activities developed during each stage of the model are described in the next paragraphs of this subsection. The main linkages between this model and the Technical System Life Cycle model are also described.

Stage 1—Identification of Six Sigma projects

Data with origin in different sources of the QMS was gathered, as indicated in Table 5, and then analyzed with the objective of leading to the identification of both opportunities for improvement and for innovation. That effort enabled the identification of the following five potential Six Sigma projects:

- Project A—Minimize the average delivery time of service reports.
- Project B—Minimize the number of invoices with, at least, one error.
- Project C—Minimize the current inventory levels of materials.
- Project D—Optimize efficiency levels of the service realization process.
- Project E—Minimize the occurrences concerning nonconforming service.

Stage 2—Selection of relevant Six Sigma projects

All the five potential Six Sigma projects were considered relevant to the business and to the customer, by the senior leadership of the company. None of the projects were expected to require any redesign efforts; therefore, and attending to the Technical System LC model, the DMAIC methodological roadmap was defined for all these potential Six Sigma projects.

Table 5 Main sources of data, with origin in the QMS, that led to the detection of potential Six Sigma projects

	Internal sources of data	External sources of data
Retroactive sources of data	Key performance indicators used to assess the effectivenesse of the QMS	Customer complaints and claims
	Key performance indicators used to assess processes efficiency	Historical data derived from previous customer surveys
		Reports from the external inspection body
	Customer service data	Results obtained from periodical assessments of suppliers
	Results from actions to eliminate detected nonconformities, including nonconform service	
Proactive sources of data	Internal auditing results to the QMS	Suggestions from customers
	Results from actions to detect and eliminate potential nonconformities	Third party auditing results to the QMS
		Results from customer interviews
	Results from technical supervisions	Results from focus groups sessions with customers and with suppliers
	Results from internal interviews	

Table 6 Evaluation of the potential Six Sigma projects using a prioritization matrix

Potential project	Criteria 1	Criteria 2	Criteria 3	Criteria 4	Overall score
Project A	4	3	3	4	144
Project B	4	4	3	2	96
Project C	3	5	4	4	240
Project D	5	5	2	4	200
Project E	5	3	3	4	180

The evaluation of the potential Six Sigma projects took place during the management review of the QMS. As illustrated in Table 6, the five potential projects were evaluated using a prioritization matrix, against the following four criteria:

- Criteria 1—Project's contribution to the achievement of the quality objectives.
- Criteria 2—Project's contribution to cost reduction.
- Criteria 3—Probability of success of the project.
- Criteria 4—Positive impact on other processes of the company.

To estimate the contribution of the potential projects on the satisfaction of each criteria, rates from 1 (lowest rate) to 5 (highest/rate) were assigned in the prioritization matrix. The product of the five rankings provided a combined score for an overall ranking. Based on the overall ranking results, project C was selected.

Stages 3 and 4—Six Sigma project planning, execution and completion

Before starting the activities under the DMAIC roadmap, the selected project was planned in accordance with the third stage of the Six Sigma Project LC model. The main deliverables of the project planning were the project charter, a Gantt chart, and a RACI matrix.

This Six Sigma project, summarized in Table 7, targeted the improvement of the existing company's purchasing and inventory management process. The improvement goals were consistent with the quality objectives defined for this process in the QMS.

The project team initiated the Define phase of the DMAIC roadmap by defining the elements within the project boundaries, using the In-Scope/Out-of-Scope tool. A system diagram was also constructed to assess the expected impact of this project on the QMS. To better understand the set of activities involved in the existing purchasing and inventory management process, the team mapped the flow of this process using a SIPOC (Suppliers-Inputs-Process-Outputs-Customers) diagram. Then, a Pareto chart was used to determine which inventoried items were most relevant (class A items); based on the Pareto analysis five class A items were identified, namely five different types of pipes. This phase ended with the definition and organization of the CTQCs necessary to measure the success of this process, through the use of a tree diagram. Three main CTQCs derived from this:

- Average levels of inventory for class A items.
- Service levels for class A items.
- Inventory-associated costs (carrying and ordering costs) for class A items.

Table 7 Summary of the main tools/techniques employed in each phase of the DMAIC roadmap, and the purpose of their use

Roadmap phase	Tools and techniques employed	Main objectives
Define	In-Scope/Out-of-Scope	Clarify which elements were within the boundaries of the project
	System diagram	Understand the impact of the Six Sigma project on the QMS
	SIPOC diagram	Map the current purchasing and inventory management process
	Pareto chart	Determine and prioritize the most relevant inventoried items to be studied in the project
	CTQC tree	Define and organize the critical to quality characteristics of the inventory management process
Measure	Cost analysis	Estimate the inventory-associated costs
Analyze	Ishikawa diagram	Identify possible causes for the high levels of inventory, poor service level and high inventory-associated costs
	Interrelationship diagram	Understand the relationships between the possible causes identified in the Ishikawa diagram
	Five "why's"	Determine the root causes of the mentioned problems
Improve	Brainstorming	Generate a large number of ideas in terms of improvement actions that could contribute to decrease the inventory levels, ensuring, at the same time, an increase in the service level
	Time series charts	Examine historical sequence of data showing the demand of each of the relevant inventoried items, in order to understand trends and other patters over time
	Forecasting techniques	Forecast future demand rates for all relevant inventoried items, thus enabling the definition of economic order quantities, order points, and safety stocks to each of them
	5W1H	Define and communicate the planned improvement actions to be implemented
Control	Control plan	Document ongoing control actions and process management responsibilities for key inputs and CTQCs of the inventory management process

In the Measure phase, the values for the three CTQCs were estimated, providing a baseline estimation of the performance levels before improvement. Based on the service levels obtained for all class A items, an overall sigma level could be determined for the purchasing and inventory management process. Furthermore, the current inventory-associated costs were also estimated in the phase.

In order to identify potential causes for the poor performance measured in the previous phase, the project team started the Analyze phase by brainstorming several possibilities, then organizing them in a cause and effect diagram (Ishikawa diagram). To better understand the existing relationships among the potential causes, as well as to identify the most relevant ones, the team made use of an interrelationship diagram. This phase closed with the identification of the root causes for the poor levels of performance by employing, to that end, the five "why's" tool.

Through the analysis of the root causes, the following conclusions were taken:

- The inventory-associated costs were high because neither economic order quantities nor optimal order points were usually established.
- The high values observed in the average levels of inventory were mainly caused by the ignorance about the expected demand rates for the class A inventoried items.
- The low service levels observed, which led to a low sigma level for the purchasing and inventory management process, could be explained by the absence of safety stocks.

Then, in the Improve phase, the following sequence of actions was adopted to address the root causes determined in the previous phase of the DMAIC roadmap:

1. Gather historical data concerning the demand rates for each of the five class A items, then depicting their corresponding monthly values sequentially over time. The resulting time series charts were analyzed, thus enabling the detection of trends and other temporal patterns.
2. Periodically forecast the demand rates for all class A items, by employing the forecasting method that minimized the mean square error of the forecast.
3. Periodically determine, for each class A item, the following parameters, based on the most recent forecasted demand rates: (1) economical order quantity; (2) safety stocks; (3) optimal order points.

The implementation of these actions was planned using a 5W1H matrix (What? Who? When? Why? Where? How?). After implementation, the new levels of performance were determined, and compared to those estimated during the Measure phase of the DMAIC roadmap. As demonstrated in Table 8, a significant improvement was reached.

Finally, in the Control phase, a process control plan was developed and established, containing a set of activities and mitigation actions in order to prevent backsliding in the performance achieved by this process.

Stage 5—Post Six Sigma project

In this stage, top management representatives, together with the team leader, reflected on the project results and how these contributed to the continual improvement of the company's QMS. The best practices were recorded and later implemented in other functional areas of the company.

4.2 Case Study 2

In this case study, the Six Sigma Project LC model was applied in a transportation and logistics company that was pursuing business opportunities, either in the continual improvement domain or in the innovation field. Similarly to the previous case study, the Technical System LC model was useful to properly define the

Table 8 Performance levels for the purchasing and inventory management process, before and after the improvement actions were adopted implemented

Inventoried class A items	Average inventory levels		Inventory-associated costs		Service levels	
	Before improvement (m)	After improvement (m)	Before improvement (€)	After improvement (€)	Before improvement (%)	After improvement (%)
Pipe 1	70	62	527	411	97.6	98.7
Pipe 2	106	71	716	545	97.5	99.9
Pipe 3	67	54	590	434	95.6	99.5
Pipe 4	149	82	830	615	99.5	99.9
Pipe 5	53	33	617	436	93.0	98.5
Overall performance						
Inventory-associated costs	Service level		Sigma level			
	Before improvement	After improvement	Before improvement (%)	After improvement (%)	Before improvement	After improvement
	€ 3.28	€ 2.441	84.18	96.54	2.5	3.3

scope of the Six Sigma project candidates, particularly the one that came to be selected.

Stage 1—Identification of Six Sigma projects

Several opportunities for improvement and innovation were detected after analyzing a set of relevant data from the sources indicated in Table 9. Based on such opportunities, various potential Six Sigma projects were identified and proposed.

Stage 2—Selection of relevant Six Sigma projects

From the totality of the proposed Six Sigma projects, five of them were considered relevant to the business and to the customer. These projects were discussed and assessed during a company's board meeting.

The best Six Sigma project candidate was picked by employing the nominal group technique (Fig. 10), a voting procedure where each participant ranked all the potential projects, assigning the most important the value of five (equivalent to the number of project candidates) and the least important the value of one.

The Technical System LC model was used to scope all the Six Sigma project candidates, including the methodological roadmap applicable to each one. The selected project targeted the design and development of a new (to the company)

Table 9 Main sources of data used to identify potential Six Sigma projects

	Internal sources of data	External sources of data
Retroactive sources of data	Company's dashboard of indicators	Customer complaints and claims
	Customer service data	Historical data derived from previous customer surveys
	Results from past improvement initiatives	
Proactive sources of data	Results from internal interviews	Suggestions from customers
	Benchmarketing data from the Marketing Department	Results from customer interviews
	Results from internal brainswriting sessions	Market and benchmarking studies carried out by a consulting firm

	Potencial Six Sigma projects	Roadmap	Participants in the board meeting						Total score
			Participant 1	Participant 2	Participant 3	Participant 4	Participant 5	Participant 6	
1	New service focused on the B2C market	IDOV	1	2	5	2	1	4	15
2	New transportion service for trade fairs and events	IDOV	4	3	4	5	3	5	24
3	Redesignof the security shipment service	DMA(DV)C	2	1	1	4	2	1	13
4	Improve on-time delivery on critical in time service	DMAIC	5	4	3	3	5	3	23
5	Increase percentage of calls answered on time in call center	DMAIC	3	5	2	1	4	2	17

Fig. 10 Evaluation of the potential Six Sigma projects using the nominal group technique

transportation service for trade fairs and events. Since this falls within the scope of the Design for Six Sigma (DFSS) approach, the IDOV roadmap was defined to carry out the execution of this process. In addition, due to the morphology of the technical system (a transportation service), the same model indicated that this DFSS project is a product innovation type.

Stages 3 and 4—Six Sigma project planning, execution and completion

The planning of the selected DFSS project included the development of the project charter and of a project plan, the former containing a Gantt chart and a RACI matrix. This was similar to the first case study described in the chapter.

Table 10 summarizes the tools and techniques that were employed during this project, together with an explanation of why and how they were used. The project execution began with the identification of pertinent customers and other stakeholders, their role in the service life cycle, and the relationships among them;

Table 10 Summary of the main tools/techniques employed in each phase of the IDOV roadmap, and the purpose of their use

Roadmap phase	Tools and techniques employed	Main objectives
Identify	CVCA	Identify pertinent stakeholders, their relationships with each other, and their role in the service's life cycle
	Personal interviews	Collect stated and unmet needs and other requirements, regarding the service to be designed, from the customers and other stakeholders identified in the CVCA
	Affinity diagram	Organize the large number of customer and other stakeholders' requirements into logical groupings
	Kano model	Determine which requirements contribute to customer dissatisfaction, neutrality, or delight
	QFD	Identify and prioritize critical to quality characteristics (CTQCs), together with their specifications, related to the customer and other stakeholders' requirements
Design	Brainstorming	Generate alternative design concepts for the service
	Axiomatic design	Provide a set of principles to ensure that the service design possesses conceptual robustness, and that the design decisions are coherent across multiple levels of the design hierarchy
	FAST technique	Define and decompose the basic functions of the service, in close articulation with the axiomatic design framework
Optimize	Process management tools	Optimize the efficiency of the service related processes (e.g. routing, scheduling), by simplifying their flow and preventing the occurrence of any type of waste
	Design scorecard	Display the predicted and optimized levels of functional performance, regarding the CTQCs of interest
	Service FMEA	Optimize the design of the service, in order to minimize the risks of failure in the service delivery activities
Validate	Design scorecard	Display the results obtained during the conduction of the service-pilot

a customer value chain analysis (CVCA) (Donaldson et al. 2006) was constructed for this purpose. Then, after a preliminary analysis of existing data (e.g. claims, complaints, customer service data) that provided a basic understanding of the main customer concerns, various interviews with customers from the events' industry were conducted to expand the knowledge about their needs, including the latent ones, and expectations. The set of needs that resulted from this process were then organized using an affinity diagram. Such needs were also sorted and analyzed from the perspective of the Kano model, thus helping the DFSS team to ensure that no critical needs were omitted. The Identify phase of the IDOV roadmap ended with the translation of the customer and other stakeholder needs into specific and measurable or observable CTQCs (e.g. on-time delivery; destination single not received, damaged goods during transportation); this was achieved using the House of Quality framework of the Quality Function Deployment (QFD) approach.

In the Design phase, a set of design requirements, in the form of both functional requirements and design constraints, was obtained; the functional requirements derived from the basic functions determined for the service. As a next step, the DFSS project team was encouraged to generate alternative ideas/concepts for the specialized transportation service for trade fairs and events. All concepts, whose combination of design solutions was potentially capable to meet the design requirements, were initially considered. The selection of the best design concept for the service, which appeared in the form of high-level design elements (or design parameters) attended to the principles of the axiomatic design theory. The selected service design was detailed using the zigzagging decomposition from the axiomatic design approach, together with the FAST technique (functional analysis system technique), to ensure a coherent deployment, across the multiple levels of the design hierarchy, of the initial sets of both design requirements and design solutions.

At the beginning of the Optimize phase, all the necessary elements to perform the service (e.g. truck models, trailer types, storage conditions, tracking systems, types of human resources, delivery processes) were accurately known. This phase involved the realization of some tests to predict the operational/functional performance of the service in terms of the most important CTQCs; the results were displayed in a design scorecard. Based on these results, the team introduced some adjustments in the designed service, seeking the optimization of its performance. The processes related to the service delivery (e.g. scheduling, routing, sorting) was also optimized, but in terms of their efficiency, using a combination of lean and process management tools. Finally, a service failure mode and effect analysis (FMEA) was utilized to identify and address risk factors in the service delivery process.

Finally, a full service-pilot was conducted and evaluated, which led to the identification of some vulnerabilities that required the introduction of some corrective actions. The results of the pilot were verified and recorded in a design scorecard. The service FMEA was also updated. After the adjustments were concluded, the designed service was validated by the company's top management.

Stage 5—Post Six Sigma project
After its completion, a report for this DFSS project was prepared. A set of best practices and lessons learnt, including mistakes made along the IDOV roadmap, was added to this report. This report was later shared within the organization, using the company's Intranet system.

5 Results and Discussion

This chapter introduced and described the Six Sigma Life Cycle approach that comprises the following two models:

- The Six Sigma Project Life Cycle model.
- The Technical System Life cycle model.

The two case studies herein presented contributed to illustrate the applicability of the approach. The key findings derived from this work, and their implications, are discussed in the next points:

- The Six Sigma LC provides an effective framework to efficiently manage any Six Sigma project along the required deployment stages, regardless the maturity and morphology of the technical system scoped by that project.
- The Six Sigma Project LC model accommodates the life cycle perspective of any type of improvement and innovation project developed under a Six Sigma program.
- The Six Sigma Project LC model also enables an enterprise to effectively implement a Six Sigma program, by taking advantage of its integration with the existing quality management system of the enterprise.
- Through the use of the Technical System LC model, it is possible to relate the life cycle stage of a technical system, which depends on its maturity, with the most appropriate methodological roadmap to carry out a Six Sigma project.
- The Technical System LC model also proved to be useful to scope any kind of Six Sigma project, in terms of the level and type of innovation.
- The fourth Six Sigma Project LC model provides a useful process to take advantage of the beneficial synergies that exist between the Six Sigma DMAIC and the DFSS methodological approaches. Unfortunately, it was not possible to further explore, through a case study, the synergies between different methodological roadmaps of Six Sigma.

6 Conclusions

In this chapter, the concept of Six Sigma Life Cycle was precisely studied with the presentation and description of the two models comprising this framework. These models, the Six Sigma Project Life Cycle and the Technical System Life Cycle, were first discussed, and then illustrated in two cases studies showing two distinct applications. In addition, the integration of a Six Sigma program with a quality management systems based on the ISO 9001 standard, by taking advantage of the Six Sigma Project LC model, was also explored.

Six Sigma is a popular and well-proven approach to foster innovation and business improvement. The Six Sigma Life Cycle link together many of the main and emergent topics on the subject, including the ICRA value creation strategy, the Design for Six Sigma methodology, and the integration of Six Sigma with other management systems. The Six Sigma Life Cycle framework, with its two models herein proposed, provide one of the first attempts to frame Six Sigma into a life cycle perspective.

References

Anbari FT, Carayannis EG, Voetsch RJ (2008) Post-project reviews as a key project management competence. Technovation 28:633–643

Antony J, Bañuelas R, Kumar A (2006) World class applications of Six Sigma: real world examples of success. Butterworth-Heinemann, Burlington

Bendell T (2006) A review and comparison of Six Sigma and the lean organizations. TQM Mag 18(3):255–262

BSI (2010) BS 6079-1:2010: Project management—Part 1: Principles and guidelines for the management of projects. British Standards Institution, London

Donaldson KM, Ishii K, Sheppard SD (2006) Customer value chain analysis. Res Eng Des 16:174–183

Harry MJ, Linsenmann DR (2006) The Six Sigma fieldbook: how DuPont successfully implemented the Six Sigma breakthrough management strategy. Currency Doubleday, New York

INCOSE (2004) systems engineering handbook—a "what to" guide for all SE practitioners. International Council on Systems Engineering, Seattle

Kumar M, Antony J, Antony FJ, Madu CN (2007) Winning customer loyalty in an automotive company through Six Sigma: a case study. Qual Reliab Eng Int 23:849–866

McCarty T, Bremer M, Daniels L, Gupta P (2004) The Six Sigma black belt handbook. McGraw-Hill, New York

Montgomery DC (2005) Generation III Six Sigma. Qual Reliab Eng Int 21(6):3–4

Montgomery DC, Woodall WH (2008) An overview of Six Sigma. Int Stat Rev 76(3) 329–346

Pfeifer T, Reissiger W, Canales C (2004) Integrating Six Sigma with quality management systems. TQM Mag 6(4):241–249

Schroeder RG, Linderman K, Liedtke C, Choo AS (2008) Six Sigma: definition and underlying theory. J Oper Manage 26(4):536–554

Watson GH (2004) Six Sigma for business leaders. GOAL/QPC Publishing, Salem

Zu X, Fredenball LD, Douglas TJ (2008) The evolving theory of quality management: the role of Six Sigma. J Oper Manage 26:630–650

On the Influence of Material Selection Decisions on Second Order Cost Factors

Marco Leite, Arlindo Silva and Elsa Henriques

Abstract Life cycle cost of manufactured parts and the selection of the manufacturing equipment to produce the parts are of the utmost importance in today's highly competitive automotive industry. In the context of materials substitution for high volume production, a problem is often encountered on how to accurately predict the cost of manufactured parts. First order cost factors, like the cost of the raw material itself, are normally easily available, but second order factors, like tools and dies cost, amount of scrap, rework and others, are quite difficult to predict to support the substitution decision. Nevertheless, they play a major role in defining the overall life cycle cost. It becomes even harder when the new proposed material is similar to the incumbent material. In these cases, a predictive cost model will have to be sensitive to changes in the material properties to be able to correctly estimate these second order costs. The problem is that material properties are seldom directly related to manufacturing cost parameters in sufficient detail. Very often, relying on empirical models calibrated with historical data represents the only available alternative. This chapter presents discussion around these issues and follows four industrial examples where a methodology is proposed for predicting cost in the presence of alternative materials with the same manufacturing process, using empirical engineering models together with process-based cost models. The relevant manufacturing cost factors are identified and discussed, and conclusions are drawn. A generalization of the methodology is also discussed, enabling further work in different industrial situations.

M. Leite · A. Silva (✉)
ICEMS, Instituto Superior Técnico, Universidade de Lisboa, Lisboa, Portugal
e-mail: arlindo.silva@tecnico.ulisboa.pt

E. Henriques
IDMEC, Instituto Superior Técnico, Universidade de Lisboa, Lisboa, Portugal

E. Henriques et al. (eds.), *Technology and Manufacturing Process Selection*,
Springer Series in Advanced Manufacturing, DOI: 10.1007/978-1-4471-5544-7_4,

1 Introduction

This chapter discusses the influence of material properties on the production cost of a given part on a materials substitution framework using a traditional cost versus weight objective evaluation.

For a given part that can be made of different alternative materials, without changing the main technologies involved in its production, say, for example, a sheet metal stamping part with two material options, mild steel and high strength steel, several questions can arise in the materials selection process. What influence do the materials have on cost beyond their direct cost of acquisition? What are the implications of the decision on second order costs and are these secondary costs significant enough to be considered or can simplification assumptions be proposed without loss of quality in the decision?

Since the production site is kept invariant, with the same infrastructure and the same cost factors (energy, labor, ...), can one assume that raw material acquisition is the only cost driver that changes significantly, or do the different materials change the production economic cost and in what manner?

To support the answer to these questions, four cases are described and analyzed for four high production volume technological processes: progressive die sheet metal stamping, injection molding, forging and squeeze casting. The methodology used to estimate the economic cost of parts production relies on technical cost models, or process based cost modeling (Field et al. 2007). Process based cost models are a predictive way of estimating the production costs of a part, by assessing the major cost drivers and their relation with the product description, process modeling and financial parameters.

Each process based cost model is different, as far as they involve different parts and processes, but all share the same construction: an estimation of the necessary labor, energy, material, equipment, tooling, maintenance, and shop-floor space to produce a part with a specific manufacturing process. For these cost models to work properly, a good product description is needed along with a sound process description (Field et al. 2007). All inputs and outputs are computed and product, process and financial inputs are worked to deliver an economical cost.

When selecting the material of which a part is made one of the issues that needs to be understood is the impact of the materials and its properties on the consequent manufacturing process and on the part performance. The part material, the part process and the part functional performance establish a triad of dimensions of the material selection problem, which are not orthogonal as far as each dimension influences and is influenced by all the others, as shown in Fig. 1. One should note that material selection is typically a part decision issue. However, it cannot be dissociated from the product design requirements. In fact, the cascade of requirements from the product system level to the part level constrains the part performance. The design freedom is then normally restricted to find a part design solution for a defined performance, meaning that material and shape are actually the core of the part design decision making. But properly controlled by the

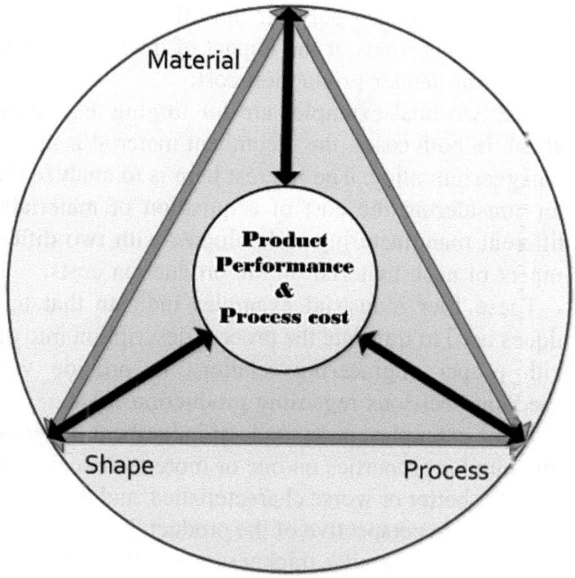

Fig. 1 Materials selection interactions for a part

designer/engineer, even the interactions among the part material, shape and process; can result in quite different process cost and product performance, considered as emergent attributes of the product system.

Designers understand the relation of shape and material with the process but slightly changes in both are normally neglected or considered only based on general rules of design for manufacturing. The impact of material on process is often simplified considering only the different acquisition cost of each material on the final production cost. However, the material of a part is a variable controlled by designers and their impact on shapes and processes, and further on the product technical and economic performance need to be considered at the design stage.

In the following sections these interactions are demonstrated with four industrial examples based on different parts with specific manufacturing processes. For each manufacturing process, different, yet similar material alternatives are selected and costs evaluated.

The first one is for a sheet metal progressive die stamping part, in which the incumbent material is a mild steel grade and the comparison material is one grade of high strength steel. In this example, the cost modeling techniques are developed to visualize the impact of individual decisions open to the designer/engineer on part production cost. The decisions were made under isometric performance conditions, meaning that an equal performance is imposed to the parts, allowing, for example, a reduction in the thickness of the high strength steel part due to the high mechanical properties of the material.

The second example is an injection molding cost model, where the incumbent material is a polyamide and the proponents are a polycarbonate and a polypropylene for an automotive fender. In this case, the decision is driven by the implications of each polymer fluid flow properties on processing parameters such as the cooling times or the impact of thickness and material on gate pressure, and then on the fender production cost.

The two final examples are for forging and squeeze casting of an automotive wheel. In both cases, the incumbent material is an aluminum alloy compared with a magnesium alloy. The interest here is to analyze the materials selection decision not considering the cost of acquisition of materials, but strictly to evaluate the different manufacturing technologies with two different materials and realize the impact of each material on the production costs.

These four industrial examples indicate that traditional cost modeling techniques used to translate the process description into operation cost can be enhanced with proper engineering relations to provide valuable insights on materials selection decisions regarding production costs.

In the examples, parts made of incumbent materials are valued with alternatives with similar properties on one or more technical performance dimensions, leaving room for better or worse characteristics, and these, more than often are related with the life cycle perspective of the product. For example, the use of high strength steel can allow for a similar thickness part with higher performance on strength, which can induce an improved safety of the product, or can allow for a reduction in thickness, which relates to the weight performance attribute and to improved vehicle fuel consumption. In either case, the models allow the designer/engineer to establish a fundamental metric of how much production cost one is "willing to pay" for different alternatives, whose benefits arise later in the use phase in the automotive lifecycle.

It should be noted that the trade-offs presented are made between different stages of the product lifecycle (different timeframes) and different stakeholders: the manufacturing and use phases, the tier or OEM and the end-user. The comparison of materials and processes, designs or other conditions depend ultimately on the objectives of the study, but the purpose is to inform technical decisions concerning materials and technology selection. The fundamental aspect is that the interactions among material, process and design can alter both manufacturing cost and product performance and that these interactions must be made explicit before the parts are actually made.

The relevance of this work resides on the ability of the methodology presented to feed the existing lifecycle tools in the manufacturing and use phase of the assessment with more meaningful and reliable data.

2 Industrial Examples

In this section, the four examples mentioned above are presented. These examples were taken from an industrial context, and the objective is to characterize four high volume production technologies used in the automotive industry where the pursuit for new materials for applications is active.

2.1 The Sheet Metal Stamping Example

In the "stamping world" there has been a trend to substitute mild steels with high strength steels. These new chemistries of steel are designed to improve performance: generally, the objective is to reduce weight, but with an expected cost associated. Therefore, the substitution of mild steel for high strength steel is often not a "win–win" solution for the decision maker. When in the process of evaluating the possible substitution of a material, it is important for the designer to understand not only the potential benefits to the product (Holmberg and Thilderkvist 2002), but also the economic impacts of substitution on manufacturing costs. Figure 2 displays an automotive component with a set of parts built using progressive stamping, that in this case, are welded together to form a cross car beam structure. The use of engineering tools to predict accurate costs in materials substitution cases is of high importance.

The engineering relations proposed by sheet metal forming bibliography were considered to link the material and part shape to the operations and resources requirements. In addition the effect of design features on the process was discussed with industry experts to capture their reasoning for empirical relations and empirical coefficients (examples include the effect of material in part rejects or in the number of parts produced per unit of time).

The substitution of the incumbent material mild steel by high strength steel in a progressive die stamping part changes significantly the costs of the required tooling (dies). Progressive dies are used for mass production of sheet metal components and offer high productivity and part quality with low economic process cost. The most important material selection decision for the progressive die is the one related with active elements of the tool, those in contact with the part: the punch, die and inserts (also referred as active materials, other materials present on the tools, but not in contact with the part are called inactive materials). Kumar and Singh (2007)

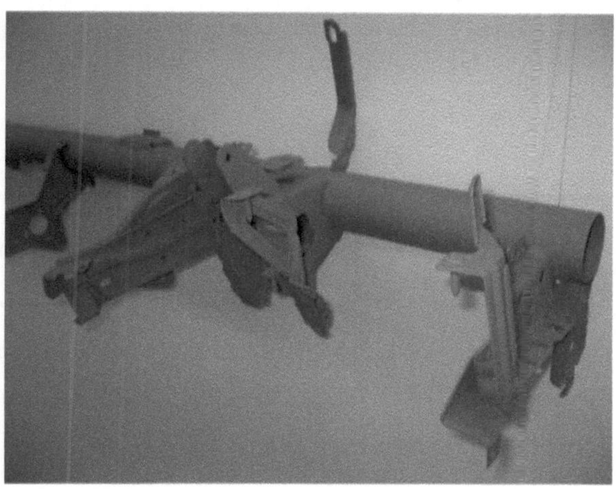

Fig. 2 Progressive stamping parts used in a vehicle cross car beam

proposed an intelligent system for selection of these active materials in progressive die stamping. This selection can be based on recommendations from industry (Uddeholm tooling AB 2011), or on research literature on tool wear (Cora et al. 2009) and tool die design (Prasad and Somasundaram 1992). The selection of stamping tool steel and respective actives surfaces coating depends on part material properties (strength, strain hardening), on the thickness and complexity of the part forming geometries, and on the production volume. Another important aspect of tool dimensioning is the spring-back[1] behavior of high strength steel parts under forming conditions in each station of the progressive die (Banu et al. 2006). In some situations, this behavior can mandate an extra tool station to solve that specific issue, but in others, an extra press force and holding time can correct the issue.

Based on the operations necessary to deliver a specific stamping tool, a decomposition of costs is proposed for the progressive die tool, from preliminary design up to testing, with a special focus on the part material properties that effectively have an impact on its cost (Fig. 3). The costs of designing the tool, the raw materials acquisition costs, the fabrication costs, until assembly and testing were then modeled. At a glance, the part material properties can appear irrelevant in some cost drivers, but they are far from that. For example, the operation of testing a tool for high strength steel part is more burdensome than for a similar mild steel part (part accuracy is more difficult to control due mainly to spring-back phenomena). Some of operations can also be more difficult, because of lack of experience with the new material, but in most cases the tool active elements that compose the tool must be more robust, increasing the tool costs.

For progressive stamping, Leite (2012) considers an automotive bracket part with an envelope area of 560 × 185 mm. The production process of the part currently made of DC03 s steel commercial grade is the baseline to economically evaluate the process for the same part but made of DP600 high strength steel commercial grade. A production volume of 250,000 is considered. In this example, the part is deformed on a five station progressive tool. The thickness of the current part is 1.2 mm (the thickness of the coil) and the part is considered to be of medium complexity for this process. To keep the same mechanical performance, the thickness of the new part can be reduced to 1.1 mm and can be considered of medium to high complexity.

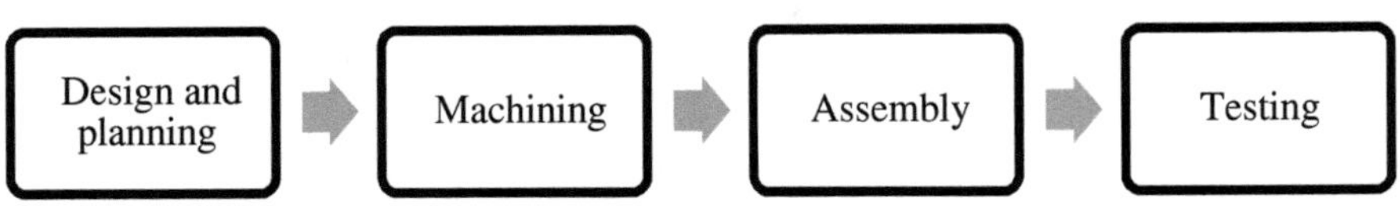

Fig. 3 Decompositions of a progressive dies stamping tool cost model

[1] Spring-back is the recovery of elastic deformation that occurs during stretch-forming of parts.

For complexity factors, empirical data captured from tacit knowledge of industry experts together with explicit data from literature (ASM Handbook Committee and Semiatin 1988) are used to estimate the added process difficulty of a part and its material compared with simple geometry made of mild steel material.

Preliminary tests revealed that the most important design variables affecting the production cost were the part material, the part complexity and the part thickness, which determines the number of tool stations and press load. Some of the relevant material and process data is shown in Table 1. To establish the tool cost, two different cases were created. Case 1, with the incumbent DC03 mild steel, which is a conventional stamping material, and Case 2 where a high strength steel, DP600, is proposed for material substitution. Besides changes in the part thickness, the material substitution introduces changes in the tool design that can impact the production cost. The tool designer has to review the part complexity factor together required number of tool stations together. The review is largely determined by the tacit knowledge, experience and even subjective preferences of the designer. The designer must also calculate the new necessary load force to stamp the part that together with the tool dimensions constraint the required press. Table 1 displays these four alternatives (2.1–2.4), as "what if" scenarios to estimate the impact of the design choices on the production cost. Notice that an increase on tool length results from the increase of number of stations, and that the higher pressing load depends by itself on the ultimate tensile stress and thickness.

In terms of part performance, assuming that no major part redesign is made and similar mechanical behavior is intended for the part (stiffness, strength, dent resistance, natural frequencies), the part thickness could be reduced to 1 mm. However a conservative reduction was pondered to 1.1 mm, resulting in the part weight reduction of 8.3 %. From part description, 48 % of the raw material becomes process scrap and the production generates 1 % of rejected parts.

Alongside with the data presented in Table 1, the process cost model was fed with other data for the cost modeling was inserted to calculate the traditional estimation

Table 1 Some geometry, material and process variables for tool cost estimation

	Case 1	Case 2.1	Case 2.2	Case 2.3	Case 2.4
Part geometry (mm)	560×185×1.2	560×185×1.1			
Coil weight (kg)	1.177	1.079			
Part weight (kg)	0.562	0.515			
Material designation	DC03	DP600			
UTS (MPa)	180	600			
No. of stations	5	5	6	5	6
Complexity (Field et al. 2007; Holmberg and Thilderkvist 2002; Kumar and Singh 2007)	2	2	2	3	3
Tool length (mm)	1400	1400	1600	1400	1600
Tool width (mm)	1000				
Tool height (mm)	375				
Press load (kN)	201	396			

of the necessary labor, energy, material, equipment, tooling, maintenance and shop-floor space to produce a part in a process. Results for the stamping tool cost model obtained by Leite (2012) based on Portuguese 2010 data are presented on Fig. 4.

The analysis of the cost results shows that several cost drivers remain unchanged with the material properties: from Case 1 to Case 2.1, where only part material changes, all cost drivers remain invariant with the exception of the tool active materials, which to shape the high strength steel part need to be upgrade to a more wear resistant material. This change slightly increases the cost of the tool. Nevertheless, if, with the change in material, an additional stamping step is introduced in the process to better compensate spring-back glitches, then a new station is required in the tool and that influences the tool final cost. Also, if the part is asserted with a higher complexity factor to introduce the higher forming difficulty of the high strength steel, then some other cost drivers are affected, like tool

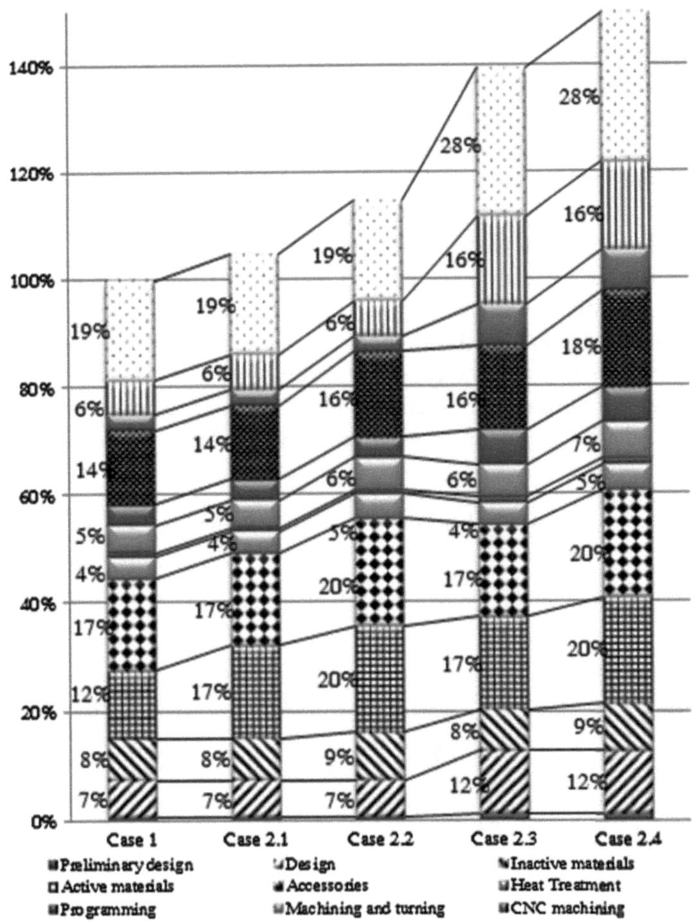

Fig. 4 Tool cost breakdown for different cases

design and testing, and the tool cost increases. A change in the material influences the manufacturing costs. For a benefit of 8 % on part weight, tool costs can rise by 50 % alone, not to mention that high strength steels may increase the stamping process rejects and the stamping cycle time, increasing the allocation of machines and labor to the part manufacturing process.

For the specific case of the progressive die stamping process, the example illustrates the impact of parts materials substitution on the economic cost of the tool. Following a similar procedure for all the other cost drivers, materials, labor, energy, equipment, shop-floor space and overheads, the impact of materials substitution can be estimated for the processing costs of the part (Fig. 5).

As the stamping infrastructure (current stamping line) used for stamping a part made of mild steel is compatible with the larger tool dimensions and stamping force requirements of the high strength steel material, the impacts on the part cost are mainly observed on the material and tooling cost drivers. The effect of the materials substitution on labor and main machine costs is due to the increased line rate factor[2] associated with the high strength steel.

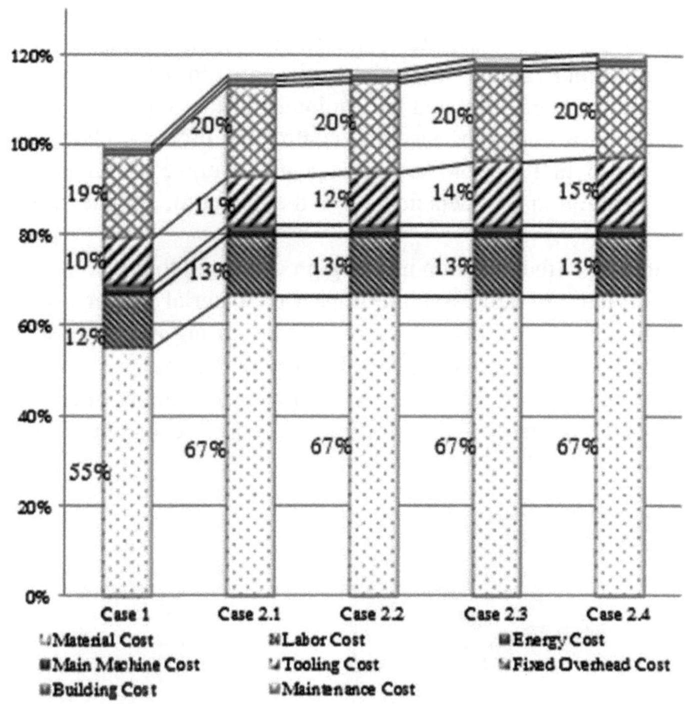

Fig. 5 Cost breakdown for progressive sheet metal stamping cost model

[2] Line rate factor is a factor that multiplies estimated cycle time for each material. A line rate factor above one will increase cycle time.

2.2 Plastics Injection Molding Example

The second example deals with the impact of materials substitution in the manufacturing cost of plastic parts produced by injection molding. The number of polymers, with quite different properties, commercially available for injection molding process is growing, enlarging the domain of the materials substitution problem. So, at the design phase it is important to have insights on what material proprieties most influence production costs of an injected molded. The following analysis keeps constant the geometry of the part and considers injection molding as its primary manufacturing technology. Engineering relations, different from the previous example, were created to modeling the impacts of different materials on costs. In particular, the engineering relations identified by Leite (2012) for equipment selection, cycle time and tool costs were used to link material properties to the production cost The part selected for the study is a car fender studied in the EDAM pilot project (Cunha et al. 2007), presented in Fig. 6.

The injection molding process can be a highly technical and depends on an intricate net of process parameters. However, it can be described in a simple way: a thermoplastic raw material is prepared in a dryer, melted and dosed in the injection-molding machine and then injected at a given temperature and pressure into the molding tool (mold), where it cools and solidifies into the shape of the final part. Besides the material and the equipment, other inputs are the energy and the direct labor to set-up, supervise and control the process. The outputs are the part itself, shaped in the mold cavity at a given range of temperatures (from injection to ejection ones) dependent on the material, engineering scrap and rejects.

The parameters of the injection molding process need to be estimated to prepare technical cost model to addresses the impact of material properties on cost. The required clamping force to keep the mold closed during injection, which, for the same part geometry, is highly dependent on part material is a main factor that determines the selection of the injection machine. The number of molding cavities (number of parts produced per injection shot) also constrains the equipment

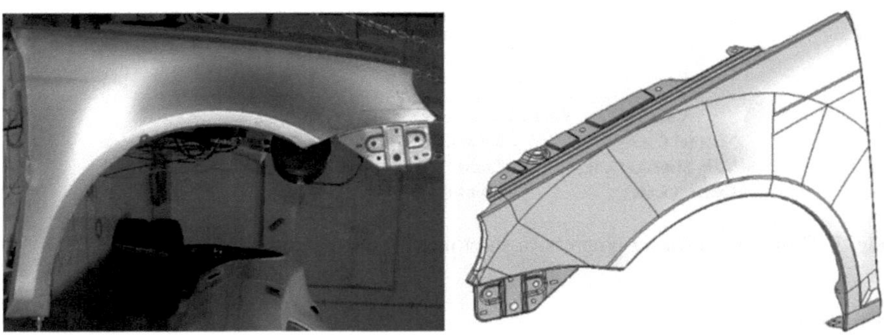

Fig. 6 Fender component (real and CAD model) (Cunha et al. 2007)

selection, since it deeply affects the clamping force as well as the mold size. For a fixed part, the number of cavities determines attributes of the equipment: its platen dimensions and shot capacity. The clamping force relies on two different parameters: the necessary pressure to hold the material inside the mold and the necessary injection pressure to fill the mold with melted polymer.

The necessary injection pressure at the mold gate is a function of the part thickness, flow length of the material to form all part features and material properties, like its viscosity, shear properties and thermal properties:

$$\text{Gate pressure} = f\left[\left(\begin{matrix}\text{flow}\\ \text{lenght}\end{matrix}\right), \left(\begin{matrix}\text{material}\\ \text{properties}\end{matrix}\right), \left(\begin{matrix}\text{part}\\ \text{thickness}\end{matrix}\right)\right] \quad (1)$$

Being clearly material dependent, data for solving Eq. 1 is usually provided by plastics manufacturers.

The injection cycle time, the time required to shape a part (or a set of parts in a multi-cavity mold), affects the consumption/use of resources like labor, equipment and energy and, in fact, it largely determines important cost drivers of injection molding based production. The cycle time relies, not only on part geometry, but also on materials properties and even mold characteristics. Along with part cooling time that represents from 60 to 80 % of cycle time for small sized parts, other time parcels contribute to the cycle time, such as the time to inject the polymer into the mold cavity, the time to open and close the mold and, the time to eject the part from the mold. For one fixed part geometry thermal diffusivity and melting temperature of the polymer, together with the part ejection and the mold temperatures regulate the cooling time of the injection cycle. Equation 2 gives the estimated cooling time for an injection molding flat plate with a thickness smaller than 3 mm.

$$t_c = \frac{h^2}{\alpha\pi^2} \ln\left[\frac{4}{\pi}\left(\frac{T_m - T_w}{T_e - T_w}\right)\right] \quad (2)$$

where t_c is the time necessary for the centerline of the mold to reach the ejection temperature T_e; h is plate thickness; α the thermal diffusivity of part material; T_m is the polymer melt temperature and T_w is mold temperature.

Although part material properties affect the injection machine regulation and the processing parameters (Yarlagadda and Teck Khong 2001; Liang and Ness 1996), the influence of the part material can be considered neglected in all the other time parcels of the part injection cycle.

To consider the contribution of the tooling cost driver for the part production cost, the initial investment on the mold must be ponder together with its life expectancy. Mold life varies widely from mold to mold and even for the same mold it is a function of the part materials selection: the same polymer, but reinforced with glass fibers increases the abrasiveness of the material, affecting the life expectancy of the mold. Therefore, even if the mold expected life can be difficult to foreseen in a material substitution scenario, it is an important input to consider, due to its impact on the final part unit cost. In case of lack of information to rely on a sensitivity analysis can supply better insights on the impact of mold life. An estimate of the initial investment in the mold was retrieved from a mold cost estimator (Mold cost estimator).

With these three parameters from Eq. 1 decomposed, it is possible to link the influence of the part material with requirements of the production process and further on with the part production cost. Some of the part and process parameters used for the technical cost model are presented on Table 2.

A production volume of 50,000 fender pairs per year is assumed for the analysis. The thickness is held constant for each material and the polyamide (PA + PPO) is considered the incumbent material. The polycarbonate (PC + PBT) material option shows an obvious disadvantage, as it has higher raw material acquisition cost and higher machine requirements. However, this disadvantage needs to be traded-off with the benefit of the smaller injection cycle time that results in a lower time use of equipment and labor for the same production volume.

The technical cost model results for the incumbent material and alternatives are presented on Fig. 7. The production cost of a pair of fenders made of policarbonate is 23 % higher than the one achieved for the Poliamide fenders, for the assumed production volume. However, a cost reduction of 17 % is expected for the polypropylene reinforced glass fiber alternative, being the best alternative in a production economics perspective. For all alternatives, material represents 30–50 % of total cost. Equipment and tools represent another significant share of the costs, while energy, labor with maintenance and facilities account for less than 10 % of the total costs.

A material substitution decision generally instigates simultaneously a change in the geometry of the part. However, specially if the part is an element of a quite complex product already in production, the geometry change can be constrained to small adjustments that do not interfere with the whole design of the product. A change in the part thickness is quite current in these cases. Based on the developed technical cost model of the process a sensitivity analysis can be done showing the impact of the fender thickness on its final cost. As the thickness decreases, the part small volume demands for less materials acquisition. Though, less material does not mean necessarily lower costs, since part thickness both influences injection

Table 2 Injection molding major production parameters (Cunha et al. 2007)

Material trade name	Noryl GTX	Makroblend	Themylene 20 %GF
Material	PA + PPO	PC + PBT	PP + GF
Material cost (€/kg)	2.84	3.40	2.12
Part thickness (mm)	2.5	2.5	2.5
Part weight (kg)	0.931	1.033	0.889
Injection pressure (MPa)	81	131	76.1
Total cycle time (s)	76	69	74
Minimum clamping force (kN)	28000	44000	26000

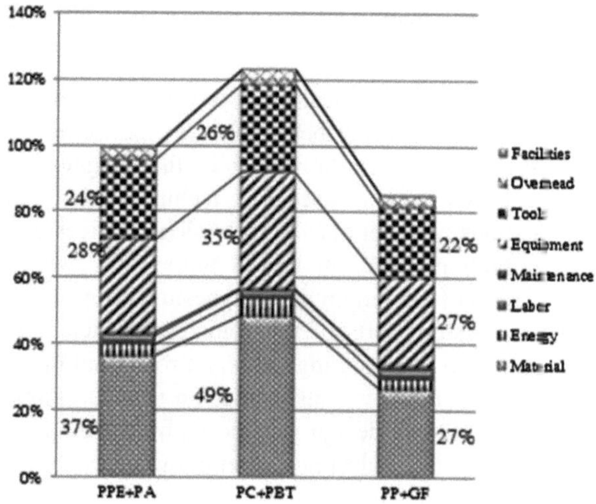

Fig. 7 Injection molding cost drivers variation (100 % is for incumbent material) for a 50 k pairs of fenders per year

pressure and cooling time. With part thickness reduction, an increase in the necessary injection pressure occurs and thus also the size of the machine, particularly as regards to its maximum injection pressure. In the opposite direction, thicker parts, requiring more time for cooling, increase the cycle time and thus the resource allocation for the same production volumes. This behavior is exhibited in Fig. 8.

In this technical cost model, developed for injection molding, the thermal properties, along with the rheological properties of the material flow were linked with the process requirements allowing the estimation of the impact of materials substitution on equipment selection and cycle time. Moreover, as far as the model is sensitive to part design critical features, the effect of design changes triggered by material options can also be evaluated to support informed decisions.

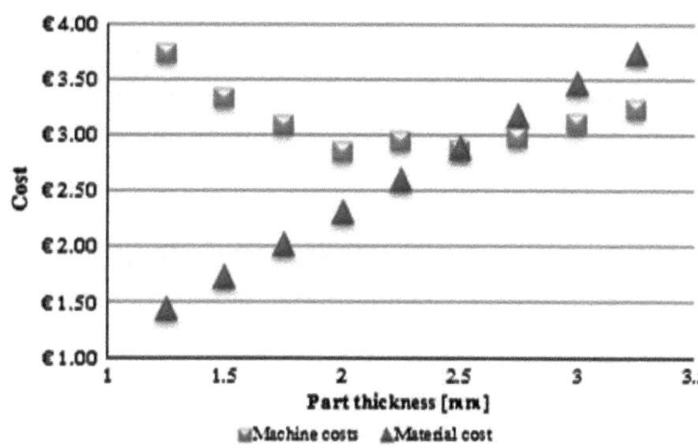

Fig. 8 Cost sensitivity analysis to part thickness for 50 k pairs of fenders per year

2.3 Magnesium and Aluminum Forging Example

Forging is a technological process where materials acquire excellent mechanical properties, allowing for better technical performance and lightweight components with low-density metallic materials, like magnesium and aluminum alloys.

Similarly to previous sections, technical cost models are used to understand the implications of materials and technology choices on economic cost of a product in its production stage. A forged automotive wheel was selected to study the implication of aluminum and magnesium alloys in the processing cost. Wheels are important safety parts of the vehicle and their attributes affect the vehicle performance due to un-sprung mass and rotational inertia. In addition, being a visible part of the vehicle, wheels contribute to its aesthetics.

From a product design standpoint, one of the issues with magnesium wheels is their lower rotational fatigue strength when compared with aluminum alloys for a similar wheel design. This is the reason why only 10–15 % weight reduction is obtainable, from a potential of 35 % weight reduction, when comparing a magnesium forged wheel with an aluminum forged one (Becker and Fischer 2004). One possible solution is to increase the thickness of the spokes' cross section to allow the verification of rotary fatigue requirements (Becker and Fischer 2004; Shang et al. 2008a). It is assumed that there must be an increased thickness of the spoke cross section of 20 %, relative to the aluminum solutions due to the rotary fatigue (Shang et al. 2008b). Also related with attributes of the magnesium wheel is the galvanic corrosion behavior of magnesium in contact with other metals. Magnesium wheels demand for surface coatings to extend their life in service.[3]

One issue related with magnesium is the safety of the production facilities. Magnesium alloys ignite and burn fiercely when heated to a point near the melting temperature. Certain precautions need to be taken, when handling and processing them, namely in machining processes. Specific cutting tools geometries and machining parameters are necessary. For example, when compared to aluminum higher feed rates and cutting speeds are recommended to generate larger chips with less chance of ignition. General cleaning practices are mandatory.

Under the material substitution framework, again, the objective is to evaluate and understand the effect of a material change in the economic cost of part manufacturing. The wheel forging operations are studied for an aluminum alloy and a magnesium alloy. The boundaries of the study do not include all the process of producing a wheel (the production of automotive wheels is complex and different processes are involved), but only the forging and its predecessor operations. For the forged wheel, it is assumed that billets are presented ready for the initial cutting operation, followed by a forging station, which includes the pre-heating of the billets and dies to the desired temperature (the forging operation is required to be done at "almost" isothermal conditions), a three step forging operation and the final trimming of the flash generated. Therefore, for the analysis, the production

[3] This cost penalty was not studied.

process needs to be broken down into its six different steps: (1) sectioning of a billet, (2) pre-heating, (3–5) forging steps and (6) trimming (Fig. 9).

After sectioning, the billet goes to an induction heating setup to elevate its temperature to forging conditions. Induction heating offers an attractive combination of speed, consistency and control needed for the forging process. The model uses materials and process parameters to estimate the power needed to heat at the cycle time of the forging operation.

The parameters that influence the choice of the forging press size are the tonnage necessary and the dies size. The estimated tonnage for each part was estimated based on analytical models available in literature (Kalpakjian and Schmid 2008). The necessary load depends on the material flow stress, the part projected area, the friction coefficient between part and die surfaces and on a complexity factor (Semiatin 2005). The choice of a hydraulic press is necessary in order to enable a good control over the strain rate and to ensure a controlled pressure over the cycle time.

The cycle time is based on an estimate of the strain rate and on the necessary time for all the movements: opening of the die, positioning of the pre-heated billet, applying pressure when closing, holding pressure and opening the die and retrieving the forged part to begin a new cycle. Coupling the processing conditions such as deformation rate and temperature to the desired microstructure of the material, processing maps are useful tools for the establishment of hot working processes for a wide range of metals and their alloys (Slooff et al. 2010). So, to estimate the hot working conditions of each alloy, processing maps were used to predict process parameters that feed the cost models. The aluminum and magnesium alloy data was retrieved from literature (Slooff et al. 2010; Prasad and Sasidhara 1997; Wang et al. 2007). The load necessary to forge the part, which is related with the flow stress, was estimated based on the slab method for a cylindrical work piece (Kalpakjian and Schmid 2008).

Table 3 provides a comparison of some relevant design features of the wheel alternatives. The incumbent material is Al6061, a precipitation hardening aluminum alloy, containing magnesium and silicon as major alloying elements. The alternative is AZ80, a high-strength wrought magnesium alloy.

Table 4 shows relevant process data necessary to estimate the needed forging force. The force increases with the sequence of forging steps and the values are higher for aluminum due to its higher flow stress.

Results from the technical cost model developed for the forging of automotive wheels show that the different materials and their inherent properties affect the process cycle time and other secondary cost drivers. Figure 10 shows the cost

Fig. 9 The forged wheel process flow

Table 3 Design parameters

	Forged wheel	
Material family	Aluminum	Magnesium
Specific alloy	Al6061	AZ80
Weight of the wheel [kg]	9.55	8.36

Table 4 Estimating forging force needed for each of the three forging steps

	Units	Magnesium alloy			Aluminum alloy		
		1 step (blocker)	2 step (edging)	3 step (finish)	1 step (blocker)	2 step (edging)	3 step (finish)
Initial diameter	mm	203	412	434	203	412	434
Final diameter	mm	412	434	457	412	434	457
Weight	kg	10.6			12		
Coef. friction		0.1					
Flow stress	MPa	80			100		
Av. pressure	MPa	104	108	114	132	138	145
Press force	kN	13 926	16 026	18 636	17 696	20 405	23 787
Complexity		2	3	3	2	3	3
Estimated force	kN	27 853	48 079	55 911	35 393	61 215	71 361

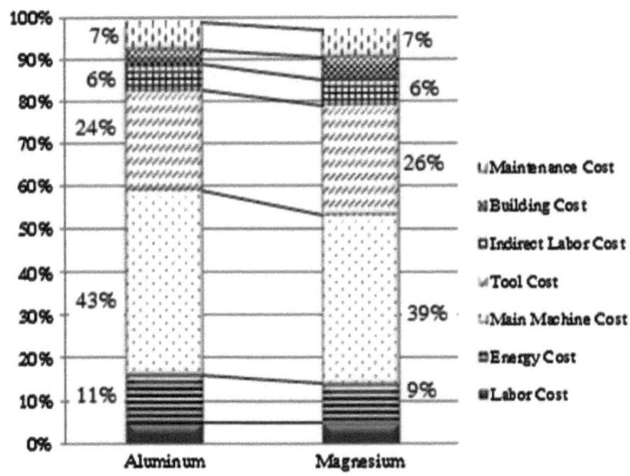

Fig. 10 Cost breakdown for a production volume of 400 k forged wheels per year (100 % is for incumbent material). Material cost not presented

summary and cost drivers for each process, without the material acquisition costs. The magnesium alloy wheel has a lower production cost than the aluminum wheel In particular, and the equipment costs are lower, due to the lower cycle times needed in the machining operations and the lower temperatures and easier forming in the forging operations, and it compensates for the higher tooling cost. The shop-floor building costs are always higher for the magnesium alternatives due to a safety premium applied. However, in current market conditions, the magnesium alternative is economically penalized due to the costs of raw material acquisition.

For the component studied, material substitution decisions cannot be made based only on the economic cost criterion, since automotive wheels have a strong impact on performance, light-weighting and visual aspect of the vehicle. Some engineering designs can be made to reduce the manufacturing costs associated with the reduction of wheel weight within the same process. In fact, the engineering study of the magnesium wheels showed that the lower density of the magnesium could not be taken to its full advantage, due to its lower fatigue response.

2.4 Magnesium and Aluminum Squeeze Casting Example

The automotive wheel example was further studied with the squeeze casting technology. Squeeze casting is a generic term to specify a casting technique where solidification is promoted under high pressure within a reusable mold. It is a metal-forming process, which combines permanent mold casting with die forging into a single operation where metal solidifies under applied pressure (Ghomashchi and Vikhrov 2000). Generally, the squeeze casting is able to produce components in a variety of shapes and sizes, which, unlike other casting techniques, are fine grained with excellent surface finish and almost no porosity. This allows reaching improved mechanical properties over those of conventional castings, although a cost penalty is typically associated. In this example, the developed cost model examines the potential cost competitiveness of the process with two alternative low density metals: aluminum and magnesium alloys.

To manufacture a squeeze casted wheel, a process flow composed of three steps major was considered: (1) melting of the material (2) squeeze casting and (3) trimming (Fig. 11). Subsequent steps of thermal treatment and surface coating are not studied. The squeeze casting step involves itself a set of quick internal stages (Ghomashchi and Vikhrov 2000):

- A graphite solution is spread over the die cavity to ensure insulation.
- A fixed amount of molten metal is poured into a preheated die cavity.
- The press closes the die cavity and starts to pressurize and shape the liquid metal. This is carried out very quickly, rendering solidification of the molten metal under pressure.
- The pressure is held on the metal until complete solidification. This not only increases the rate of heat flow, but also eliminates macro/micro shrinkage porosity.
- Finally the hydraulic is withdrawn and the component is ejected. A new cycle can then start.

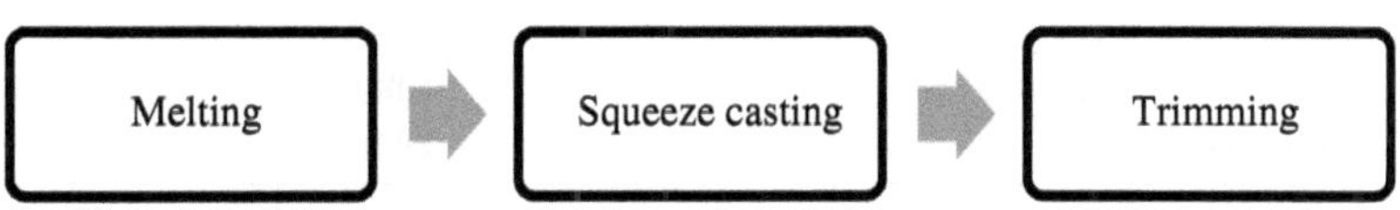

Fig. 11 Process flow for squeeze casting

In the technical cost model, a steady state flow of heat from the part to the die is considered to estimate the solidification time:

$$t_s = \frac{\rho V[H + c(T_P - T_1)]}{Ah[T_1 - T_o]} \tag{3}$$

where ρ is the density of molten metal; H is the latent heat of fusion; V is the volume of casting; A is the area of die surface; c is the specific heat of molten metal; h is the heat transfer coefficient; T_P is the molten metal superheat temperature; T_1 is the metal mean melting temperature; and $T_0$is the initial mold temperature.

From Eq. 3 and data from Table 5, solidification time was calculated. The magnesium, although having a higher volume (due to increased thickness of the wheel spoke) and lower heat transfer coefficient, benefits from lower liquid density and lower latent heat of fusion. From this evidence, a similar cycle time between both alloys can be assumed, although in this case the magnesium proves to solidify slightly faster.

Figure 12 presents the cost summary and cost drivers for each process without the material acquisition costs. Results without raw material costs and correspondent recycling credits show that, for the squeeze casting operation, the higher cycle time of the aluminum alternative increases its equipment allocation, but the difference is not enough to separate the alternatives. As mentioned, shop-floor building costs are always higher for the magnesium alternatives due to a safety premium applied.

Table 5 Data used for the solidification model for squeeze casting

	Aluminum	Magnesium
Pouring temperature (°C)	650	675
Melting temperature (°C)	570	595
Die temperature (°C)	210	210
Part thickness in the direction of the heat flow (mm)	20	24
Thermal conductivity of the molten metal (W/mK)	94	78
Die thickness in the direction of the heat flow (mm)	5	5
Thermal conductivity of the die (W/mK)	28.6	28.6
Heat transfer coefficient (W/m^2K)	2580	2072
Density of the molten metal (kg/m^3)	2385	1810
Latent heat of fusion (kJ/kg)	396	370
Volume of the casting (m^3)	0.0044	0.0048
Area of the die surface (m^2)	0.183	0.183
Specific heat of the liquid (J/kgK)	1080	1350
Solidification time (s)	29.8	28.4

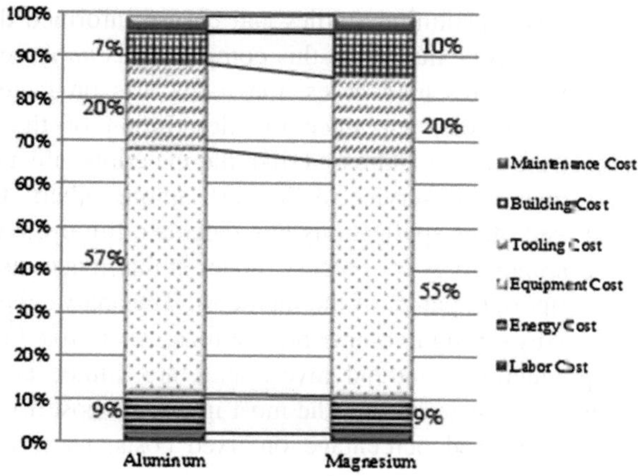

Fig. 12 Cost summary without materials acquisition

3 Conclusions and General Discussion

This chapter contributed to realize that material selection and material properties are of importance not only on part performance, but also on the manufacturing cost of a part on a common industrial process. This will, in turn, influence the lifecycle cost of both part and manufacturing process. For that purpose, a selection of examples from industry was studied. In all the examples presented, the general assumption is that when considering materials substitution, apart from thickness, the overall size and shape of the part remain unaltered: the size of the wheel remains constant in the model, while the thickness of rim and spokes increased. The Direct materials substitution is frequently approached in product improvement contexts focusing lightweight strategies that include leaner designs and the exploitation of new technologies. However, the impact of shape, by changing the length or width of the part, on both process and material was not up for discussion in this chapter, although the models can address it.

The proposed technical cost models are tools able to distinguish differences between materials and to provide insights on materials substitution regarding process cost, but require an extensive understanding and modeling of the process. The models prove, for all cases that the material impact cannot be neglected in the manufacturing process and that materials substitutions can and must be evaluated. Significant changes in production costs are mandated from material properties like ultimate tensile stress or material strain hardening coefficient in the sheet metal stamping case, for example.

Another related aspect is that due to different standardization and raw material availability around the world, it is possible that some translation issues occur from the initial design phase to the design "request for quotation" phase or detail design phase. A part can be originated in France, have a quotation in Portugal for manufacturing and assembly in Brazil and/or Spain. To the benefit of all stakeholders, materials substitution studies can ensure informed decisions regarding materials selection problems where this complexity is an issue.

For the presented cases, the focus was on materials impact on weight and production cost, but since the description of the process is extensive, other objectives can be addressed, like the materials effect on life cycle issues or on the performance on service of the part, among others. Results from the models can also feed other applications that calculate life cycle environmental impacts.

Common to every process model is that materials acquisition cost is normally one important parcel of the final cost. The analyzed cases were focused on typical processes for high volume productions, where fixed costs, resulting from previous or production front-end investments are diluted and the variable costs emerge, turning the material into the most important cost driver. For lower volumes, there is an increased percentage on fixed costs, namely on the ones resulting from tooling requirements, but even for these, there is also an impact on equipment allocation, since materials affect the cycle time of the process, and therefore the impact of material on process costs.

In this work, the methodology presented allows the evaluation of the impact of interactions between part shape, material and technological process on process cost and product performance of a given part in a direct materials substitution framework. One benefit of this methodology is that, even though the designer cannot use technical cost models as a black box, it can be customized to be used in different industrial situations to evaluate the impact of materials changes in a product.

Acknowledgments The authors want to thank the MIT Portugal Program for funding of this research.

References

ASM Handbook Committee, Semiatin SL (1988) Forming and forging, 9th edn. ASM International, Metals Park, Ohio

Banu M, Takamura M, Hama T, Naidim O, Teodosiu C, Makinouchi A (2006) Simulation of springback and wrinkling in stamping of a dual phase steel rail-shaped part. J Mater Process Technol 173(2):178–184

Becker J, Fischer G (2004) Manufacturing and potential of extruded and forged magnesium products. In: Prof. Dr Kainer KU (ed) Magnesium—alloys and technology, Wiley, Weinheim Germany, p 90–105

Cora ÖN, Namiki K, Koç M (2009) Wear performance assessment of alternative stamping die materials utilizing a novel test system. Wear 267(5–8):1123–1129

Cunha A, Roth R, Cirincione R, Camanho PP, Neves PC, Esteves S, Correia N, Silva A, Henriques E, Reis L, Sousa L, Pecùas P, Baptista R, Dias GR, Viana JC, Simões R (2007) Pilot-project EDAM—MIT, Portugal

Field F, Kirchain R, Roth R (2007) Process cost modeling: strategic engineering and economic evaluation of materials technologies. JOM 59(10):21–32

Ghomashchi MR, Vikhrov A (2000) Squeeze casting: an overview. J Mater Process Technol 101(1–3):1–9

Holmberg S, Thilderkvist P (2002) Influence of material properties and stamping conditions on the stiffness and static dent resistance of automotive panels. Mater Des 23(8):681–691

Kalpakjian S, Schmid SR (2008) Manufacturing processes for engineering materials, 5th edn. Pearson Education, Upper Saddle River

Kumar S, Singh R (2007) A short note on an intelligent system for selection of materials for progressive die components. J Mater Process Technol 182(1–3):456–461

Leite M (2012) Techno economic evaluation in materials selection for multiple parts under OEM-Tier relations. Instituto Superior Técnico, Lisboa

Liang JZ, Ness JN (1996) The calculation of cooling time in injection moulding. J Mater Process Technol 57(1–2):62–64

Prasad YVRK, Sasidhara S (1997) Hot working guide: a compendium of processing maps. ASM International, Materials Park, OH

Prasad YKDV, Somasundaram S (1992) CADDS: an automated die design system for sheet-metal blanking. Comput Contr Eng J 3(4):185–191

Semiatin SL (2005) ASM handbook, vol 14A—metalworking: bulk forming. ASM International, Novelty

Shang R, Altenhof W, Hu H, Li N (2008a) Rotary fatigue analysis of forged magnesium road wheels. In: International S (ed) SAE international world congress, reprinted from: magnesium technologies, 2008 (SP-2205), Detroit, Michigan, 14–17 April 2008

Shang R, Altenhof W, Hu H, Li N (2008b) Rotary fatigue analysis of forged magnesium road wheels. SAE Technical paper series 2008-01-0211

Slooff FA, Dzwonczyk JS, Zhou J, Duszczyk J, Katgerman L (2010) Hot workability analysis of extruded AZ magnesium alloys with processing maps. Mater Sci Eng A 527(3):735–744

Uddeholm tooling AB. (2011) Tooling solutions for advanced high strength steels, SSAB

Wang CY, Wang XJ, Chang H, Wu K, Zheng MY (2007) Processing maps for hot working of ZK60 magnesium alloy. Mater Sci Eng A 464(1–2):52–58

Mold cost estimator. UFE Incorporated

Yarlagadda PKDV, Teck Khong CA (2001) Development of a hybrid neural network system for prediction of process parameters in injection moulding. J Mater Process Technol 118(1–3):109–115

Aircraft Industrialization Process: A Systematic and Holistic Approach to Ensuring Integrated Management of the Engineering Process

José Manuel Lourenço da Saúde and José Miguel Silva

Abstract Aircraft production is a complex and lengthy process which involves a judicious selection of materials in combination with the option of various production processes. The objective of this chapter is twofold: the first part encompasses the evolution of aircraft materials and related techniques, underlying their impact in the complete life cycle of an aircraft; the second part describes the industrialization process to allow manufacturing of airplanes. It essentially deals with the specific process where engineering is involved namely in terms of the various steps that need to be taken to generate the information essential for the process. It includes also the connection to specific ERP modules such as MRP II and CRP which are essential to the management of aircraft production.

Acronyms

APU	Auxiliary power unit
BOM	Bill of materials
BTP	Build to print
BTS	Build to spec
CFRP	Carbon fiber reinforced polymer
CRP	Capacity resources planning
EBW	Electron beam welding
ERP	Enterprise resources planning
FAI	First article inspection
FAL	Final assembly line
FRP	Fiber reinforced polymer
FSW	Friction stir welding

J. M. L. da Saúde (✉) · J. M. Silva
Department of Aerospace Sciences, University of Beira Interior,
6201-001 Covilhã, Portugal
e-mail: jmsaude@ubi.pt

J. M. Silva
e-mail: jmas@ubi.pt

E. Henriques et al. (eds.), *Technology and Manufacturing Process Selection*,
Springer Series in Advanced Manufacturing, DOI: 10.1007/978-1-4471-5544-7_5,

GTAW	Gas tungsten arc welding
HAZ	Heat affected zone
LBW	Laser beam welding
MRP II	Manufacturing resources planning
OEM	Original equipment manufacturer
OoA	Out of autoclave
Qty	Quantity
PBS	Product breakdown structure
RFI	Resin film infusion
RTM	Resin transfer moulding
TC	Type certificate
VARTM	Vacuum assisted resin transfer moulding
WBS	Work breakdown structure

1 Aeronautical Materials and Related Technologies

1.1 Introduction

Engineers are offered a broad range of materials for the design and manufacturing of components and systems for a myriad of applications. Jointly, structural metallic alloys and non-metallic engineering materials (such as polymer based composites) ascend to a number of over 100,000 different available options which have to be adequately weighted during the design phase of a new aircraft. This together with the countless existing manufacturing processes, coupled with the complex relationships between the different selection parameters, pose engineers a huge challenge regarding the right selection of a material or related manufacturing process for a certain application, namely in the domain of aircraft design and production.

As a consequence, the selection of materials is an endeavoring task which is often supported by a systematic approach based on quantitative and qualitative selection procedures, many of them developed as aiding computational tools which are of paramount importance during the preliminary design phase (Ashby et al. 2004).

In the particular case of aeronautical applications, the role of the recent advances regarding materials and manufacturing technologies has largely been responsible for major performance improvements in many aerospace systems. The distinctive premise which makes a candidate material suitable to be used in an airframe component relies upon the need for maximizing the strength-to-weight ratio as the reliability and damage tolerance capabilities are kept as high as possible to meeting the mandatory airworthiness requirements.

The need to accomplish this design principle is in the base of a significant gap in terms of the strategies defined for the development of new types of aircraft with improved performance and efficiency when compared to other forms of vehicles which are used in other transportation modes. Parallel to this, all the design and manufacturing efforts carried out within the aeronautical industry have been increasingly driven by cost effectiveness and environmental concerns which have shifted the industrialization processes to other types of paradigms. In any case, flight safety has always surpassed the other design drivers, which forces engineers to follow a judicious and balanced approach during the design and manufacturing phases in order to obtain an economically viable product.

Table 1 presents a summary about the main design drivers of aircraft structures since the early years of commercial aviation up to the present.

Having this in mind, the main objective of the present section is to provide the reader a brief summary about the idiosyncrasies of aeronautical structures and materials, aiming at contextualizing their impact in the aircraft industrialization practices currently followed by OEMs. However, it should be stressed out that it is not the authors' intention to undertake an exhaustive analysis about the different physical and mechanical properties of the distinct classes of aeronautical materials and related manufacturing processes. That is a matter which is conveniently addressed in depth in literature which can be consulted if further information is required.

1.2 Historical Overview

The dawn of the air transport era in the beginning of the twentieth century was a corollary of the existing technology at that time. In fact, early aircraft were built from the lightest materials available, which were dominated by wood based materials for structural purposes, like bamboo, balsa wood or spruce (Cutler 1999). Being an emerging type of vehicle, the structural configuration of the pioneering aircraft followed the existing knowledge regarding civil engineering, in particular the concepts of bridge building, consisting of a combination of beams and trusses.

Table 1 Evolution of structural design drivers for civil aircraft

Period	Main design objective
Early 1990's	Static limit strength and weight
1930's–1960's	Static limit strength, fatigue and corrosion resistance
1970's–1980's	Specific-strength/stiffness, tailored mechanical properties (composites), damage tolerance capabilities, increased weight savings
1990's–present time	Specific-strength/stiffness, tailored mechanical properties (including 3D requirements), optimization of costs (fabrication and maintenance), maximization of flight safety, environmental concerns (recycling capacities, introduction of green materials and wastage reduction), multifunctional capabilities

However, the need for a lighter structure encouraged the development of new types of configurations, such as the wire-braced structures. In this case, rectangular structural frames were prevented from collapsing by wires stretched from corner to corner. Biplane aircraft, which dominated aircraft design for many years, also benefited from the lightness of wire bracing, which combined with interplane vertical struts provided an effective solution to sustain bending and twisting loads.

On the other hand, the demand for weight reduction forced designers to cover the different structural parts with an organic based fabric surface. This structural arrangement can be found in many wings of ancient airplanes, where a set of ribs maintain the airfoil shape and transfer the local loads on the fabric covering to the wing spars. The success of the wire-braced type of construction was engaged with the simplicity of the related production techniques, most of them common to wood construction used in routine applications, which explains why this structural feature was adopted in aircraft construction for many years.

As the aircraft kept increasing their performance, soon engineers realized that a more robust material had to be used to withstand greater aerodynamic and maneuvering loads. As a result, the next step in aircraft construction was the replacement of wood by metallic alloys. Despite this change, wire-bracing structures were still used until the middle of the 1930's. However, the fabric covering was then replaced by very thin metal skins designed to carry a part of the load, which made possible to reduce the weight of the underlying structure.

Aluminum alloys have been the dominant material for airframes since this period due to their concomitant superior strength-to-weight ratio and reasonable cost (Megson 2007; Liu and Kulak 2000). Yet, on a first stage, the change from wood based materials to aluminum alloys had a limited impact in the former structural configurations as the wire-braced and truss constructions were still kept for some years.

The need for faster and lighter aircraft together with the need for passenger carrying capabilities impelled designers to consider higher stress fields in the metal skins, which had to be reinforced by means of stiffening elements (stiffeners or stringers) to prevent buckling and undesired deformations that could compromise either the structural integrity or the aerodynamic performance of the airplane.

As a consequence, airframe configurations evolved from the archaic biplanes with internal struts and bracing wires into cantilever monoplanes, which had a superior performance as a result of a lower amount of drag-producing protuberances. This was the beginning of the semi monocoque construction which represented an unquestionably major stride forward in aircraft design since the earliest days of flying.

In fact, the current design of almost all commercial airplanes is based in a semi monocoque configuration as a result of the need for maximizing the available space inside the fuselage, where the most part of the payload is going to be placed in. In simple words, the typical structural arrangement of a semi monocoque fuselage consists of a long tube with a considerable number of regularly spaced longitudinal reinforcements (stiffeners, stringers or longerons) combined with transversal frames that provide a correct load transfer to the skin and the maintenance of the cross-sectional shape of the fuselage.

The change to a semi monocoque structural arrangement using a predominant metallic construction, parallel with the raise of the overall aircraft dimensions, had a direct consequence in the applied manufacturing processes. For purposes of construction (as well as the result of production done in partnership), aircraft are normally divided into a number of sub-assemblies which have to be joined to form the final component. One key issue was to identify the most efficient technique capable of joining all parts of a large civil aircraft while keeping the reliability and safety standards as high as possible, providing that manufacturing costs should be dragged down to a minimum.

Riveted joints appeared as an immediate solution due to the inherent advantages of this type of technique: it is a low cost process; it can join entirely dissimilar materials without prior complex preparation stages; it does not influence the material microstructure; and it allows for a subsequent disassembling operation of the different parts with a minimum impact in their integrity.

Additionally, the damage tolerant properties of riveted joints have been recognized over the years since the presence of a row of rivets can act as a crack arrester under fatigue loading. Despite the fact this technique does not require highly skilled operators, the application of a large number of rivets is a labor-intense process which involves several tasks, such as hole drilling, countersinking, debarring, riveting, shaving and sealing. Some recent progresses haven been made by manufacturers to make riveting an expedite process through the use of automated equipments (Rooks 2001; Holden et al. 2007), although some accuracy issues and access difficulties to inner parts still limit its application to some specific components.

Adhesive bonding is another alternative technique to join different structural components. In fact, as opposed to riveting, adhesives have been used since the early days of aeronautical construction as wood aircraft depended mainly on glued joints for their strength. After transition to metallic construction, adhesives went out of favor for other forms of joining processes which allowed higher volume productions. However, in a recent context, adhesives seem to be gaining room either in metallic and non-metallic joints, especially in the interface regions between stringers/skins and core materials/face sheets in sandwich type components (Higgins 2000). Also, the advent of composites acted as a catalyst for the increasing use of adhesive bonding, where high performance resins have been applied to bond various laminates and to reinforcing elements to the skins. One major advantage of adhesive bonding follows from the integral (or continuous) type of joint which prevents the occurrence of stress concentrations and fretting which are likely issues of rivet joining.

Welding is a joining process widely used in different industrial sectors. This technique was a major breakthrough in the transition from wood based to steel tubular structures used in aircraft construction. However, the extension of this joining process to aluminum parts has to be weighed against other alternative techniques, as there are some important drawbacks regarding welded joints of some types of aluminum alloys.

In fact, the first form of welding applied in airframe components was based in a high heat addition process using oxyacetylene torches, which can cause some significant problems, such as the formation of a Heat Affected Zone (HAZ) in the neighborhood of the welding line with reduced mechanical properties. The occurrence of large distortions induced by temperature effects, especially in long and thin components, is another possible problem that can compromise dimension control in large parts of the aircraft. More recently, new types of welding processes with improved characteristics have been adopted by aeronautical manufacturers in certain types of applications. More accurate temperature control of advanced techniques have made possible the use of welded joints in prior low-weldability precipitate hardened aluminium alloys commonly used in aircraft components. Gas Tungsten Arc Welding (GTAW), Electron Beam Welding (EBW) and Laser Beam Welding (LBW) are assuming a raising position in the construction of aeronautical parts due to the possibility of using highly concentrated energy beams which induce small extensions of HAZs and distortions (Mendez and Eagar 2001).

Composites were the most probable leap forward in aeronautical construction. In particular, fiber reinforced polymer based materials (FRPs) started capturing the attention of designers and manufacturers in the 1960s as an alternative material to metallic alloys due to their prospective weight benefits. In fact, during the last decades composites have gained an eminent position in either civil or military aircraft, especially high performance FRPs made of carbon (CFRPs) or aramid (KFRPs) fibers (Soutis 2005). The explanation for this success relies upon the considerable advantages regarding high performance FRPs, namely: higher strength (or stiffness)-to-weight ratios than conventional materials; the ability of tailor made properties as a result of the proper selection of different lay-up stacking sequences; excellent resistance to fatigue and to corrosion damage; possibility of designing parts with complex geometrical features (like double curvature).

However, the winner of the metal *vs* composites challenge is still far from being known, as the latter have some important drawbacks that have been slowing down the pace of the change between the two types of materials and related technologies. If we combine the complexity of aircraft systems with the heterogeneity of composite materials, we will easily come to a large number of variables which are not readily quantified and measured, such as the material behavior during the life cycle, the impact of the environmental conditions, the real maintenance costs, among other uncertainties. More specifically, recent concerns about potential problems resulting from the generalization of composite materials in aircraft components are related to high manufacturing and processing costs, low damage tolerance properties, difficulty in the inspection and in repair, dimensional tolerances and a reduced level of knowledge about the material behavior in service during its life cycle.

Mindful of these problems, aircraft manufacturers soon realized that the massive introduction of composites urged the parallel development of new technologies which tried to minimize risks to an acceptable level. For example, the development of hybrid materials, like fiber metal laminates, were adopted as a viable solution to the damage tolerant issues of conventional FRPs, bringing fatigue resistance and residual strength after low energy impact to unparalleled levels (Alderliesten and Homan 2006).

Another problem of thermoset composites is their difficulty of recycling, which in the context of the current environmental friendly and green era poses serious challenges to manufacturers in terms of the development of alternative polymer based composite materials with improved recycling capabilities.

Recent efforts have been made in this line of research, with the introduction of bio-derived fibers (like hemp, kenaf and flax) combined with thermoplastic polymers, bio-derived thermoset resins and other bio-derived materials (Zhang et al. 2005; Rajesh et al. 2008). Figure 1 illustrates the application of a cork based material as a core in a sandwich component which was developed envisaging its application in the interior of a fuselage. It has been demonstrated that besides providing greater performance in terms of acoustic/thermal insulating capabilities and damage tolerant properties under impact loading (Castro et al. 2010) cork based composites also allow obtaining complex geometries which are not possible with other types of rigid core materials.

1.3 Prospective Evolutions

The increasing demand for transport combined with a harsh global competition between airlines and manufacturers forces them to cut down on the aircraft life cycle costs. This goal can be accomplished following different paths, ranging from the development of new materials and processes, revolutionary design configurations of the aircraft and optimization of the operational procedures. In this context, focus is put in the first of these options, trying to identify prospective evolutions regarding materials and fabrication processes which will contribute to the development of more efficient aircraft.

It is a fact that the development of advanced composite materials during the last couple of decades has triggered structural performance. However, stringent goals

Fig. 1 CFRP sandwich panels for structural applications using a cork-based core material with enhanced mechanical properties

have been defined for the aeronautical sector in the short-medium term, which imply a life cycle cost reduction, shortened development times and environmental compatibility. The need for more efficient aircraft (or, in other words, with lower fuel consumption and pollutant emissions) will force a significant weight reduction of airframes and systems. The recent development of nano-reinforced composites (like those using carbon nano-tubes) opened a large window with a sight of materials with outstanding mechanical properties, especially in terms of specific strength and fatigue resistance. However, although this is a very promising path of evolution, its current technology readiness level is still far from practical viable applications in an industrial context, as many issues concerning nano-technologies have yet to be solved in the near future. Aware of these problems, manufacturers are presently trying to answer to the challenges facing composite construction through more mature and sound solutions.

The current phase is being directed towards affordable processing methods such as non-autoclave processing. Autoclave curing is routinely used for the fabrication of composite aircraft components since this method allows obtaining optimized mechanical properties for superior structural performance. However, it also has some important drawbacks in terms of production costs, as it involves expensive equipment (requiring a considerable amount of energy), highly skilled hand-labor and a time-consuming process, including the preparation of the prepreg material and curing cycle. Notwithstanding the advantages of prepreg in terms of specific strength, it requires a very strict and controlled process, as this material has to be stored at low temperatures in refrigerated chambers and its manipulation takes place in a very neat and clean atmosphere under controlled humidity and temperature conditions.

Thus, manufacturers have been considering out-of-autoclave (OoA) alternative methods to obtaining composite parts with comparable mechanical properties but at lower costs. Recent progresses regarding low pressure moulding processes, such as RTM (Resin Transfer Moulding), VARTM (Vacuum Assisted Resin Transfer Moulding) or RFI (Resin Film Infusion), have been shown to be viable and more expedite techniques to obtain composite components with adequate quality standards in the context of the aeronautical industry (Soutis 2005; Noor et al. 2000). The key benefits of OoA processes are mostly related with promising increased manufacturing rates, greater production flexibility and competitive costs. Additionally, the possibility of using prepregs with low temperature and pressure curing cycle capability allows to obtaining large FRP components with complex geometries which are difficult to process with conventional autoclave processes. Further cost reduction when manufacturing with composites can be achieved by reducing the assembly cost of conventional metallic construction techniques, in particular those regarding riveted joints, which involve a time consuming process (drilling of thousands of holes followed by fastener insertion and sealing).

Despite of the increasing use of composite parts, aluminum alloys still represent a competitive solution as a material for airframe structures. Recent progresses have been achieved by the main aluminum alloys producers envisaging the improvement of the mechanical properties as a response to the superior advantages of composite materials in terms of specific-strength (Starke and Staley 1996; Williams and Starke 2003).

Among the new types of materials, aluminum–lithium (Al–Li) alloys are particularly attractive because of their weight-saving potential and improved stiffness, which can range between 10 and 18 % depending if strength-critical or stiffness-critical structures are under consideration (Immarigeon et al. 1995). Significant improvements in damage tolerance and corrosion resistance properties can also be attained by using Al–Li alloys, with obvious implications in the life cycle cost of aircraft through potential reductions in the amount and complexity of the inspections required during maintenance procedures. However, lithium is an expensive material and its prospective use in electrical batteries will continue to inflate its price, which make Al–Li alloys costly. Moreover, the key benefits of these materials are strongly influenced by the processing conditions, and therefore product quality is more difficult to control than for conventional alloys.

The possibility of using advanced metallic alloys offers some considerable advantages in terms of joining parts through welding processes. In fact, welding is widely applied in structural construction due to the high joint efficiency without weight penalty. In the particular case of aeronautical structures, welded joints are attractive solutions to other conventional joining methods (like fastening) since they do not imply any drills or overlapped parts, having important advantages regarding the minimization of stress concentration regions and weight reduction.

Advanced welding techniques have been adopted by aeronautical manufacturers for specific applications, as the Electron Beam Welding (EBW), Laser Beam Welding (LBW), Gas Tungsten Arc Welding (GTAW), Plasma Arc Welding (PAW) or Variable Polarity Plasma Arc Welding (VPPA) (Mendez and Eager 2001). All these processes involve a significant heat input which eventually leads to a reduction of the mechanical properties of the base material.

However, in the 1990s a new solid-state joining process called Friction Stir Welding (FSW), which is based on the insertion of a non-consumable rotating tool into the abutting edges of sheets or panels, has been introduced with remarkable advantages when compared to conventional welding techniques, namely: low welding distortion and good dimensional stability, improved joint strength (static and fatigue), low residual stress, fully mechanized process with lean characteristics and possibility of application in alloys and materials which are not welded by conventional processes (Mishra and Mahoney 2007).

FSW has also some major benefits in the perspective of costs, since it involves low energy consumption, lower processing times, significant reduction of material and consumable wastage and low defect rates (which means that there are potential cost savings regarding quality control of the welded joints).

The attractiveness of FSW explains why this technique has been experiencing a huge growth during the last years with applications in multiple domains such as ship, automotive, train, aeronautical and aerospace structures. In the particular case of the aeronautical industry, FSW has been used to join an increasing number of components in recent aircraft, as in the interface between stiffeners and frames with fuselage panels.

A considerable reduction in the production times has been reported by manufacturers who have been adopting FSW as an optional technique to riveting, allowing joining different parts six times faster than automated riveting solutions and 60 times faster than manual riveting, all this with higher quality levels. This means that this emerging technique offers striking advantages in the context of metallic construction, either in the machining or assembly phases, providing substantial cost savings in comparison to riveting due to the inherent higher efficiency of the FSW process. Therefore, the uncompromised adoption of FSW as a common technique in aeronautical construction is a most likely scenario in the near future, being in line with the urging for new solutions regarding process innovation.

Another promising avenue for the improvement of aircraft performance with a potential reduction in the total life cycle costs relies upon the application of smart materials and structures. In simple words, smart structures sense external stimuli, process the sensed information and respond accordingly with active control in real or near-real time.

There are several types of materials that change their physical properties when subjected to electric, magnetic or thermal loads. Typical smart materials include shape memory alloys/polymers (SMA/SMP), piezoelectric materials, magneto- and elector-strictive materials and electro-magneto-rheological fluids (Shwartz 2002). Whichever its form, active elements can be either embedded or attached to the host structure. Figure 2 shows an example of an optical fiber sensor embedded in a CFRP laminate, which renders possible the creation of a distributed sensorial system within the structure to detect and evaluate loads and failure.

The possibility of providing a structure for multifunctional capabilities offers great advantages in terms of the life cycle cost of an aircraft. In addition to the enhancement of the performance of air vehicles, which will result into lower fuel consumption, there are important benefits in the costs regarding maintenance procedures or corrective measures for the mitigation of structural damage

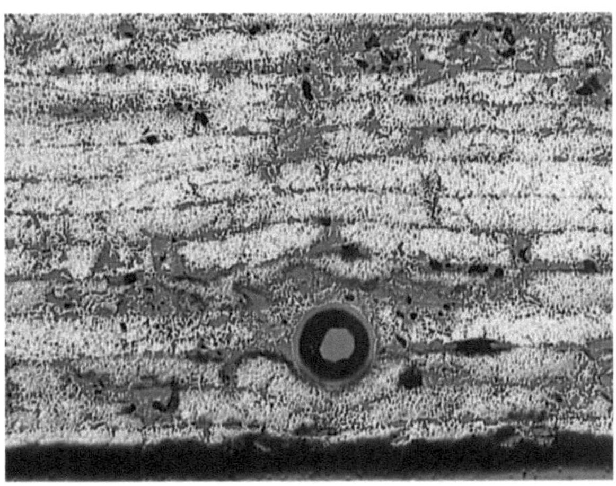

Fig. 2 CFRP laminate with and embedded optical sensor (cross-sectional view)

mechanisms. In fact, typical applications of adaptive composite systems are devoted mostly to vibration suppression and structural health monitoring, having a strong impact in the mitigation of fatigue damage.

Nevertheless, for the adaptive composite systems to see widespread commercial applications, several technical challenges must be overcome before the technology can be incorporated into future operational vehicles. Currently, there are still some issues regarding the overall efficiency of the adaptive systems, the reliability of some smart materials and, above all, several fabrication, maintenance and repair uncertainties which make impractical its use at a reasonable cost (Noor et al. 2000).

Having stated the above, it is clear that structural efficiency is a consequence of a multi-disciplinary approach where each area has a different weight in the final optimized solution regarding the process of selection of materials and related technologies. A broad understanding of the current available technologies for developing a new aircraft is a crucial step to ensure its economical feasibility, which in turn strongly depends on the total life cycle costs. Therefore, a large number of variables have to be weighed either during the design or the series production phases of an aircraft, this latter called the industrialization process. The purpose of the next section is to provide the reader with a standard and holistic management approach to the process of industrializing of an aircraft from an engineering standpoint.

2 The Aircraft Design Process Versus Industrialization

After the discussion in terms of materials and technologies, before addressing the industrialization it is pertinent to provide some insights in terms of the design process of an aircraft.

To this end, Figs. 3 and 4 are representations of the design process for civil aircraft—the life cycle of military aircraft is similar, except for the terminology adopted, see AAP-20 (2010).

For the sake of coherence, as a remark, although ISO 9001:2008 and ISO/IEC/IEEE 16326:2009 require a recycle phase at the end, the life cycle defined in the above figures only includes the operation and disposal and not the recycle stage.

This option is ***more realistic*** than assuming that an aircraft enters a fully recycling phase which is theoretically possible but cannot be ensured due to various and complex technologies, processes and materials involved in the original manufacturing process.

As seen in Fig. 3 the central part of it includes a "demonstration" stage during the design phase.

This apparently simple task, among various activities, encompasses in fact ***the production of one or more aircraft***. These units, acting as ***prototypes or demonstrators***, are due to fly according to a specific plan and airworthiness requirements, during a certain timeframe, in order to prove that they are airworthy leading

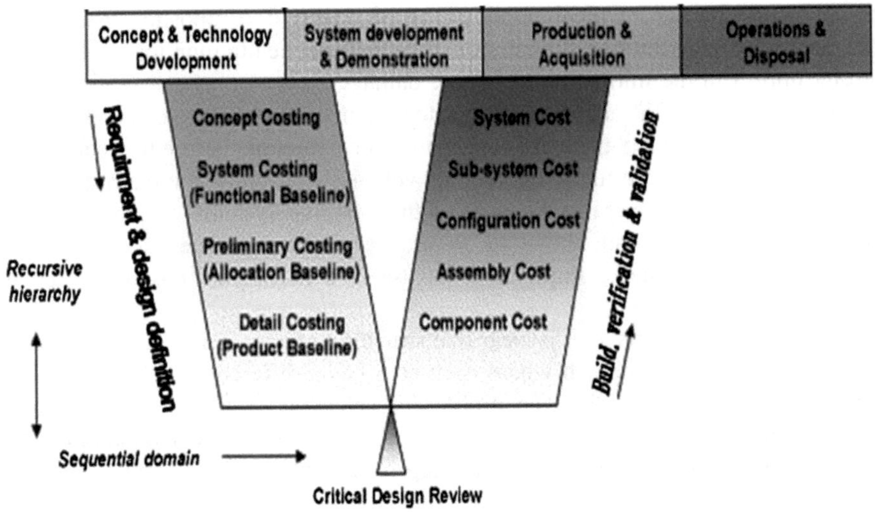

Fig. 3 Aircraft life cycle after Price et al. (2006)

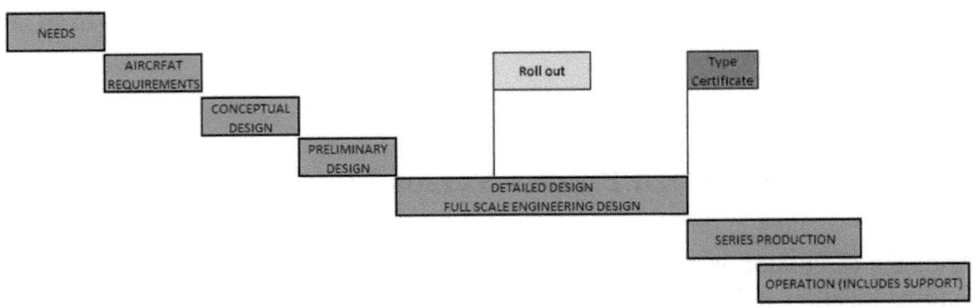

Fig. 4 Simplified aircraft life cycle after (authors)

to obtain the Type Certificate issued by an aeronautical authority under predefined regulations, such as those issued by EASA or by FAA.

This fundamental document is the ***formal guarantee*** that the project of an aircraft can move into series production, thus enabling the aircraft to be sold.

To reach that stage, the aircraft manufacturer (in fact the designer of the aircraft) needs to create certain conditions, namely, financial, engineering, logistics and production to enable the manufacturing of items and assembly of the aircraft.

Under this perspective, the design phase of the aircraft life cycle has then two complex activities, that is, the ***development of parts*** and their ***production***.

The ***development*** of parts corresponds to what is called in AAP-20 (2010) the Full Scale Engineering Development, or in other words, it is the detailed design of parts, components, systems including airframe, engines and the integration in the aircraft.

This is done starting from the aircraft requirements that were generated at the end of the conceptual phase, and using engineering tools (CAD, CAE, etc.) and appropriate aeronautical and engineering standards. As such, in practical terms, the design, either of a part or of a major aircraft assembly, inevitably requires in ***parallel*** the manufacturing of components which at the end of the day are essential to ensure the success of the design stage.

Further to this, in general terms the manufacturing that takes place during the design process is in essence an industrialization process because it enables, among various aspects:

- the manufacturing of ***single parts and assemblies***;
- to ***produce the prototypes*** in view of testing and certification;
- to produce ***specific specimens*** to allow the necessary aeronautical testings (static, fatigue, etc.) as part of the demonstration of compliance with aeronautical regulations;
- to ***set up*** the manufacturing organization to enable series production.

The next section deals then with the process of industrialization of aircraft but from an ***Engineering*** standpoint by describing it and emphasizing the critical aspects.

The reason for this option comes from the fact that by approaching the industrialization from such angle it is then possible to have a ***holistic and full integrated perspective*** about the whole process, instead of being centered on operations, logistics, etc., which is a restrictive approach.

3 Aircraft Industrialization: Concept and Overview

By definition, aircraft industrialization is the group of ***non-recurring activities*** encompassing the process of defining, planning, execution and control of the set of actions, duly organized under predefined airworthiness regulations, carried out in an integrated manner by engineering, quality, production and logistics, aimed at creating conditions to produce single parts and to assemble aircraft.

As seen, in a first moment, the industrialization is then aimed at building up ***one or more prototypes*** (essential to obtain the aircraft Type Certificate (TC)) and later to allow series production.

Pursuant to this approach of the aircraft life cycle, the industrialization is embodied in the detailed project thus it is not possible to split engineering design from it.

Having seen in brief terms the general purpose of the industrialization of an aircraft, the next paragraphs describe ***the critical elements of the industrialization*** that the management need to take in consideration, and the ***various elements that compose the industrialization*** from an engineering perspective.

3.1 Critical Aspects of the Aircraft Industrialization: Overview

The essential goals of the industrialization are two:

- to ***produce one or more prototypes*** of an aircraft in order to obtain the relevant TC;
- to create the ***industrial conditions*** that will permit to perform mass production of the aircraft awarded with the Type Certificate.

Under this perspective, the industrialization can only be considered concluded after the aircraft **Type Certificate** has been awarded and **the capacity to perform series production** capacity is in place.

Pursuant to the above, the management of the industrialization becomes a ***critical factor*** namely in terms of allowing the organization to meet the various goals related to the process, that is:

- Time to Type Certificate;
- Costs of Production;
- Time to customer.

The **Time to Type Certificate**…

… is probably ***the most critical factor*** of the industrialization process because it corresponds to meeting the most important deadline, that is the moment when the aircraft manufacturer can start ***mass production***. In this respect the aircraft design process encompasses a significant portion of ***trial and error*** loops ***to adjust the design specifications*** of parts to the real-production conditions and ***to tune the production process*** to ensure that parts meet the design specification.

The adjustments are performed through various ***First Article Inspections*** (FAI), until production delivers the part matching the design data.

A ***good*** industrialization process is the one that at the end will have had reduced number of iterations ***between design and FAI***.

In this process, the aforesaid loop, the metal transformations such as sheet metal are critical because they require a significant amount of testing.

Unlike the sheet metal transformation, the ***metal cut technology*** in general terms does not require any specific simulation, just because in most situations it is a one-step process to FAI, just requiring the necessary programming of the CNC machines.

Likewise, parallel to the above transformation process, ***heat treatment, electroplating, curing, bonding, welding***, etc. also encompass a ***cycle of tests*** to ensure that the process delivers the parts with the relevant design characteristics, which is carried out under the family of the Special Process (see Sect. 2c).

In terms of the **Costs of Production**…

... the industrialization process needs to ensure that the best practices and technologies are selected enabling the ***recurring costs*** related with the series production of each aircraft will be the lowest possible.

In concrete, at the time of ***delivery to the customer*** of the first aircraft unit, known as Manufacturing Serial Number (MSN) 01, the ***price of the aircraft*** has already been defined, eventually with no margin for additional changes.

In practical terms, the price of an aircraft tends to be established ***firm and fixed*** at the moment of signing the contract between customer and aircraft manufacturer. In this context, a very simple expression defining ***the price of an aircraft*** can be given by the following expression:

$$\textit{Price of Unit of Aircraft} = \left(\frac{\mathrm{NRC}}{\mathrm{N}} + \mathrm{RC}\right) + \mathrm{M}$$

where:
NRC represents the total value of the Non Recurring Costs;
RC represents the total value of the Recurring Costs;
N is the number of aircraft defined to amortize the NRC;
M is a commercial margin.

This expression ***forces*** the aircraft manufacturer, at the time of signing the sale to be as much as possible accurate in predicting ***the future*** costs of the units to be produced and delivered to customers a number of years ahead.

This is done taking into consideration a number of ***in-house and external costing variables***.

The ***internal costing variables*** depend upon the definition of the ***non-recurring activities*** applicable to specific activities, such as:

- the ***engineering effort*** related with the detailed design of each of the aircraft elements;
- the engineering effort ***to establish the sequence of manufacturing processes,*** including preparation of paperwork (routing); definition of the Bill of Material and special processes;
- the ***aircraft full scale ground and flight testing*** aimed at receiving the Type Certificate.

The ***definition of costs*** for each of the above non-recurring activities becomes critical in the sense it will dictate the amount of financial needs to allow to perform the various activities. In this context, a number of aspects need to be considered to reduce and/or control as much as possible the non-recurring costs, such as:

- the ***make or buy*** policies;
- the ***number of prototypes or pre-series*** aircraft to be produced;
- the ***type of manufacturing processes*** applicable to produce the single parts.

In terms of the ***external variables*** that influence the production costs those are related with the acquisition prices of, namely, the following items:

- ***raw material and products***;
- the aircraft systems to be acquired to specific OEM's (*engine, landing gear, APU, inboard systems, interiors, entertainment, mission systems, etc.*)([1]);

Having mentioned the above, the aircraft manufacturer needs then to have specific tools to enable an accurate exercise of costing.

Finally, the **Time-to-Customer**...

... in a certain way this can be seen as one of the facets of productivity. Despite the margin to drop production costs, which are normally contractually associated to a certain ***learning curve process***, (which normally only accounts for the first 100 aircraft unit produced) the option can only be, as much as possible, to select ***automated production*** process.

In this context, the ***interdependence*** between ***design and production*** through the industrialization process becomes obvious and critical in terms of meeting the above-mentioned goals.

The next section describes the sequence of activities that need to be performed to implement the industrialization of an aircraft, using an engineering approach.

3.2 The Industrialization Sequence: An Engineering Approach

Regardless of the information platform used to assist the OEM performing aircraft design and production, there is an engineering process supporting the industrialization which is defined to include the following sequence of activities:

- *Product Breakdown Structure;*
- *Manufacturing Breakdown Structure;*
- *Special Processes;*
- *Methods (routing).*

The next sub-sections describe briefly the structure of each of the 4 steps of the industrialization.

(a) Product Breakdown Structure

The scope of the PBS is to structure in a tree-shape the whole aircraft in terms of parts and assemblies using ***a predefined format*** of information based on certain aeronautical standards, such as, AECMA 1000D, ATA ISPEC 2200 or MIL-STD-1808.

The utilization of such type of standards also allows to establish the ***work breakdown structure*** (which also allows to define the workshare between the

[1] IATA ISPEC 2200 and AECMA 2200 are standards that provide a detailed list of aircraft systems as well as MIL-STD-1808.

design and production partners (in case the aircraft is produced under a partnership, which is today the common situation for the major aircraft OEM—Boeing, EADS, Embraer, etc.) (Filippone 2008).

The PBS also plays a major role in the fundamental activity known as ***control of the configuration*** of the aircraft to be manufactured (as described in point d. of this section).

The PBS includes each aircraft component and assemblies. This means that, the PBS allows to demonstrate the configuration of and aircraft in terms of parts (nomenclature, quantities, etc.) that compose each area/sections (wings, fuselage, etc.).

In concrete, starting on item 21 through 99 (see e.g., MIL-STD-1808, 1996), it is possible to establish a standard rule to assign references to parts using the tree-structure of the above standards.

Presently, given the graphical and computational tools available in the market, every aircraft can be developed using CAE with 3D drawing capacity (e.g., CATIA), which means that the creation of the PBS is progressive and part of the design process, from conceptual to the full scale engineering development.

Moreover, from an ***operations*** (production) standpoint the PBS is an essential tool to establish the ***Bill Of Materials*** (BOM).

(b) Manufacturing Breakdown Structure

This step of the Engineering Process deals with the definition of the *production layout* required mainly for the assembly process.

Unlike the aircraft assembly process that has a predefined sequence, the ***manufacturing of single parts*** does not require any specific production layout as that is done using the various production cells, such as, the stretching, pressing, grinding, gantry n-axis machines, cold rolling as well as the heat and electrolytic treatments, etc.

Moreover, as the production cells cannot be moved around the shop-floor, the manufacturing of single parts tends to be defined as a certain pre-defined ***production process***, such as the sequence regarding sheet metal transformation illustrated in Fig. 5.

The selection of technologies associated with the manufacturing of single parts is then the key element of that phase of the process of manufacturing, which represents the sequence of steps to transform raw material into a part.

For the sake of completeness, Fig. 6 identifies today's typical technologies that can be observed in aeronautics to produce single parts in metal.

Distinct from the manufacturing of single parts, the aircraft assembly requires the definition of the ***sequence/layout of assembly*** from small sections into the final and complete aircraft. Apart from the existence of specific areas, (rest, warehouses, etc.) the decision on the plant layout depends upon a number of factors, being of relevance:

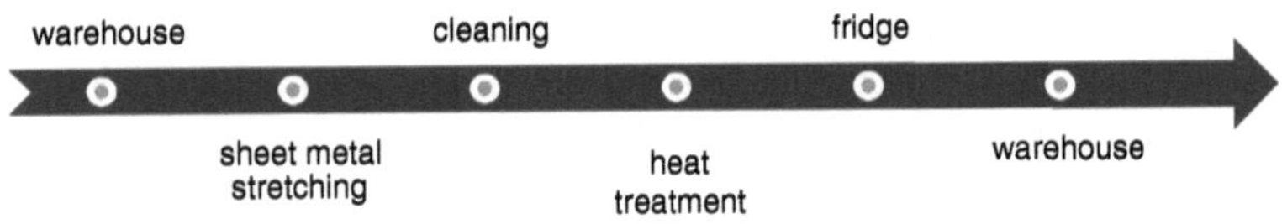

Fig. 5 Typical sequence of sheet metal transformation (aircraft shells)

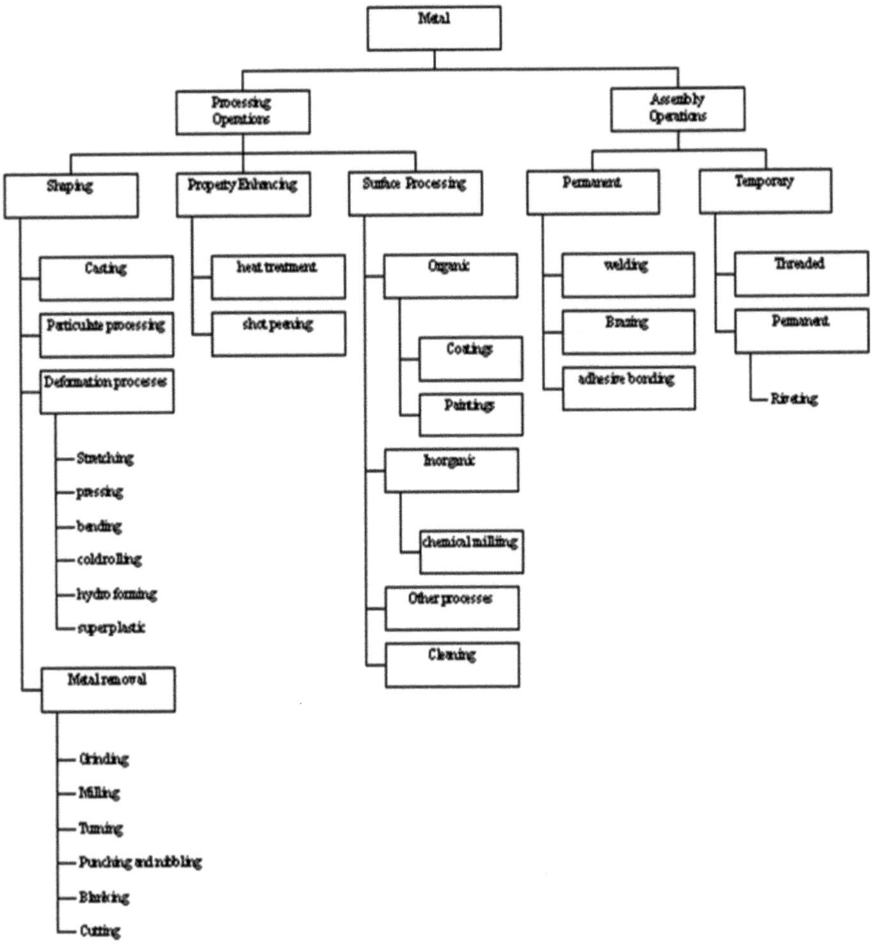

Fig. 6 Manufacturing process (metal)

- the ***space/volume*** for the items to be assembled;
- the number of assemblies to be done ***simultaneously***;
- the ***assembly path***;
- the ***level of automation*** in the assembly process because it requires specific equipment.

The next section deals with a fundamental aspect of the industrialization that is the *Special Processes*.

(c) Special Processes

In aeronautics the ***Special Processes*** (Dreikorn 1995) (SP) represent those industrial activities that require validation before they are released for utilization. In other words they involve approval before they are used during the series production. As per today's state of the art of the aircraft production technologies, the most common SP are electro platings, welding, heat treatment, and chemical milling.

A fundamental aspect of the SP is the selection of the ***applicable standard (commercial or defense) and sequence*** that needs to be followed. Table 2 gives a global perspective of the aspects that need to be defined for some SP's.

Having described in general terms the philosophy behind the SP's, it is then of major importance the management of the SP program applicable to all aircraft parts. This means that the SP is complex and long given the significant variety and quantity of aircraft parts.

In fact, each item to be submitted to SP requires a dedicated definition of the SP structure and sequence to ensure that at the end of it the part meets the design specifications and thus can be approved (through a FAI). To this end, as referred in the Table 2, considering the variables involved in the various SP, its implementation requires a deep knowledge in a wide range of engineering areas, such as, chemistry, metallurgy, materials, welding, composites, etc.

In general terms, the SP represents a critical path of the industrialization.

(d) Routing Process

As defined in the aeronautical regulations, and for the sake of traceability and thus flight safety, the manufacturing sequence needs to be fully documented. This covers the design and production activities applicable to all ***aircraft part*** as well as to all ***jigs and special tools***. By doing this, the aircraft manufacturer can then perform 2 essential management functions, that is:

Table 2 Examples of special process versus aspects to be established

Special process	Aspects to be established
Electroplating (electroless and chemical)	Chemicals Duration of cycle Current intensity
Welding	Type of welding Current intensity Working pressure Duration of cycle
Heat treatment	Heating cycle Duration Cooling cycle
Composites	Tooling Cure cycling

- ***Product Configuration Management;***
- ***Manufacturing Configuration Management.***

The goal of ***Product Configuration Management*** is to plan, perform and control each item (part, component, system, etc.). This type of activity is implemented by ensuring that the ***production working documents*** include the elements of identification of every part and assembly.

Similarly, the ***Manufacturing Configuration Management*** is aimed to plan, perform and control for each aircraft item the sequence of production. This is done by documenting in detail the sequence, the materials, the approvals during the transformation of raw material into a new part.

The ***routing process,*** which uses ***working cards***, represents then the essential information that describes the whole sequence of production, which also enables to generate the BOM applicable to an aircraft, to plan the utilization of each working cell through an ERP (see next section).

The structure of the routing process includes in addition to the necessary identification of items (e.g., part number, serial number, drawings, standards) and the sequence of transformation, at the least information regarding the "quantity of and specification of raw material and products"; the "manufacturing set up" and "machine times"; the identification of "resources" (human, equipment, production cells, etc.).

The next section approaches the way the four Engineering Steps can be used in the whole aircraft industrialization thus giving an holistic perspective about the process.

(e) Data Integration—the ERP approach

A standard methodology to integrate information when discussing aircraft manufacturing is to use the concept of Enterprise Resources Planning (ERP) which in general terms include various modules, e.g., Engineering, Quality, Finance, Supply Chain, Human Resources.

The objective of these ERP modules is to make available for the company a set of functionalities enabling it to run its business in an integrated manner.

Regardless of the importance of each individual ERP module, the Supply Chain is probably the most fundamental component, because its object is to ensure the organization is supplied with the goods and services enabling the manufacturing of single parts as well as the assembly of the aircraft.

In this context, a fundamental feature of an ERP is the ability to plan the production work to be performed. The implementation of this capacity depends on planning factors which on its turn are related with the requirements in terms of number of aircraft to be delivered to customers (normally reflecting customer back log orders) as well as on the in-house production process itself (defined by the company).

Given the description of the sequence of the engineering process related with the aircraft industrialization, that is, *Product Breakdown Structure, Manufacturing*

Breakdown Structure, Special Processes and Methods (*routing*), it results from there that the information that permits the ERP module to work comes from the data contained in the "Working cards (routing)". In other words, the "Working cards" centralize all data which enable the utilization of two important and specific ERP modules, that is, the MRP II and the CRP.

Regarding the MRP II this module is fundamental because it allows to determining, at a certain moment in time, the quantity of materials and products to be acquired for the manufacturing of single parts and for the assembly of an aircraft, thus enabling the production to satisfy the customer delivery needs. This is done by using the information available from the "Working cards", namely in what respects to the type and quantity of materials and production times.

Figure 7 provides a global and integrated graphical perspective of how MRP II receives information from the production working cards.

Likewise, the CRP enables to plan the utilization of production means for the manufacturing of single parts and sub-assemblies. To this end, the CRP algorithm uses the information defined in the "Working cards (routing)" namely, in what matters the utilization (type and time) of production equipments and cells as well as human resources (essentially used to perform assembly of aircraft).

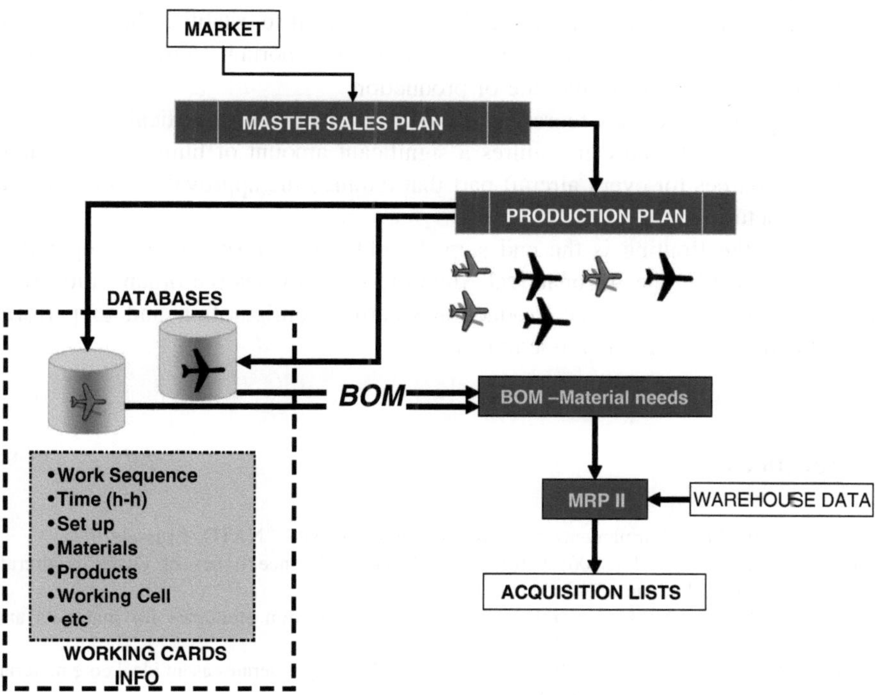

Fig. 7 MRPII data cycle

To conclude this brief description about the integration of the information, it is worthwhile to refer the importance of the information generated also for other company functions such as Finance in terms of the annual budgets.

3.3 Aircraft Industrialization: Synthesis

The aircraft industrialization starts as soon as the design organization releases the first aircraft data for the production to manufacture single parts and ***ends*** at the exact moment a TC is issued by an aeronautical authority for a certain aircraft. From that point onwards the OEM is ready to enter series production.

From a ***management perspective***, the industrialization of aircraft can be organized in 4 essential and universal engineering steps, i.e., *Product Breakdown Structure, Manufacturing Breakdown Structure; Special Processes and Routing*.

These four steps help in building up two decisive aspects of the whole process, that is, the ***informational*** structure and the **hardware** assisting the production of single parts and the assemblies.

The *Product Breakdown Structure* is a fundamental step because, among various aspects, it allows to perform the ***Control of the Configuration of the aircraft*** and establish the aircraft system and parts tree and thus the Bill of Materials to allow the procurement.

The goal of *Manufacturing Breakdown Structure* it to establish the production layout namely at the level of aircraft assembly, an important element closely tied to efficiency as well as to the rate of production.

The Special Processes represent up to a certain extent a critical path of the industrialization, because it requires a significant amount of human and technological resources for every aircraft part that requires the approval before released to manufacturing.

Finally, the Routing is the end step. In addition to responding to regulatory aspects defined by the Aeronautical Authorities it provides the organization with the capacity to perform the Production Control Configuration and to perform manufacturing of parts and assemblies.

References

AAP-20 (2010) Phased armaments programming system (PAPS). NATO, Brussels

Alderliesten R, Homan J (2006) Fatigue and damage tolerance issues of Glare in aircraft structures. Int J Fatigue 28(10):1116–1123

Ashby MF, Bréchet YJM, Cebon D, Salvo L (2004) Selection strategies for materials and processes. Mater Des 25:51–67

Castro O, Silva JM, Devezas T, Silva A, Gil L (2010) Cork agglomerates as an ideal core material in lightweight structures. Mater Des 31:425–432

Christner B (2003) Enabling technology for an aircraft alternative. Mech Eng Mag 125(3):D14

Cutler J (1999) Understanding aircraft structures, 3rd edn. Blackwell Publishing, Oxford
Dreikorn M (1995) Aviation industry quality systems: ISO 9000 and the federal aviation regulations. Amer Society for Quality, Milwaukee
Filippone A (2008) Comprehensive analysis of transport aircraft flight performance. Prog Aerosp Sci 44(3):192–236
Higgins A (2000) Adhesive bonding of aircraft structures. Int J Adhes Adhesives 20(5):367–376
Holden R, Haworth P, Kendrick I, Smith A (2007) Automated riveting cell for A320 wing panels with improved throughput and reliability (SA2). AeroTech Congress and Exhibition (L.A., CA, USA), SAE International, Los Angeles
Immarigeon JP, Holt RT, Koul AK, Zhao L, Wallace W, Beddoes JC (1995) Lightweight materials for aircraft applications. Mater Charact 35(1):41–67
ISO/IEC/IEEE 16326:2009 (2009) Systems and software engineering -life cycle processes - Project management. ISO-International Organization for Standardization Internacional, Geneva
Liu J, Kulak M (2000) A new paradigm in the design of aluminium alloys for aerospace applications. Mater Sci Forum 331–337:127–142
Price M, Raghunathan S, Curran R (2006) An integrated systems engineering approach to aircraft design. Prog Aerosp Sci 42(4):331–376
Megson THG (2007) Aircraft structures for engineering students, 4th edn. Butterworth-Heinemann, Oxford
Mendez P, Eagar T (2001) Welding processes for aeronautics. Adv Mater Proc 159(5):39–43
MIL-STD-1808 (1996) Department of defense interface standard: system subsystem sub-subsystem, Washington
Mishra R, Mahoney M (2007) Friction stir welding and processing. ASM International, Ohio
Noor A, Venneri S, Paul D, Hopkins M (2000) Structures technology for future aerospace systems. Comput Struct 74(5):507–519
Rajesh D et al. (2008) Bio-based structural composite materials for aerospace applications. In: 2nd SAIAS symposium (14–16 September) stellenbosch, South Africa
Rooks B (2001) Automatic wing box assembly developments. Industr Robot Int J 28(4):297–302
Shwartz M (2002) Encyclopedia of smart materials. John Wiley and Sons Inc., New York
Soutis C (2005) Carbon fiber reinforced plastics in aircraft construction. Mater Sci Eng, A 412:171–176
Starke EA Jr, Staley JT (1996) Application of modern aluminium alloys to aircraft. Prog Aerosp Sci 32:131–172
Williams JC, Starke EA Jr (2003) Progress in structural materials for aerospace systems. Acta Mater 51:5775–5799
Zhang MQ, Rong MZ, Lu X (2005) Fully biodegradable natural fiber composites from renewable resources: all-plant fiber composites. Compos Sci Technol 65:2514–2525

Material Flow Cost Accounting: A Tool for Designing Economically and Ecologically Sustainable Production Processes

Ronny Sygulla, Uwe Götze and Annett Bierer

Abstract In the context of designing modern, competitive production processes, companies face the challenge of integrating the growing ecological demands of customers and other stakeholders as well as resource scarcity on one hand and the dominant need for economic success on the other hand. An approach to meet ecological *and* economical goals is the improvement of the material and energy productivity. This is strongly supported by the method of material flow cost accounting (MFCA). It aims at the identification of processes' material and energy related inefficiencies and the (monetary) quantification of their effects on the overall process chain. This chapter firstly introduces the basic methodology of MFCA. Afterwards, refinements and enhancements concerning the modeling of loops and stocks, the integration of energy and the design of a prospective MFCA are proposed. Concluding, aspects of MFCA's practical implementation are discussed.

1 Introduction

The design of production processes has to meet several objectives, primarily economical. Here, a special focus is on the demand for materials which typically cause the largest share of industry's production costs by far.[1] Moreover, these materials as well as the used energies largely affect the environmental performance of industrial companies. Due to a rising public debate on the need for more ecological sustainability, concrete governmental regulations and customers' awareness of ecological matters, companies face the challenge of integrating

[1] The German industry spent 45 % on average for input materials in 2009 (Statistisches Bundesamt 2011).

R. Sygulla · U. Götze (✉) · A. Bierer
Chair of Management Accounting and Control, Chemnitz University of Technology, Chemnitz, Germany
e-mail: uwe.goetze@wirtschaft.tu-chemnitz.de

E. Henriques et al. (eds.), *Technology and Manufacturing Process Selection*, Springer Series in Advanced Manufacturing, DOI: 10.1007/978-1-4471-5544-7_6,

economical and ecological objectives. Regarding material and energy, both objectives can be achieved by increasing the resource productivity. This requires the analysis of the resource consuming processes and their complex interrelations. Here, the sum of processes of a particular production is named as *process chain* in the following.[2]

Since in practice the improvements primarily have to be economically advantageous, the evaluation of an existing process chain and of developed alternatives must be based on a monetary valuation. The corresponding appraisals typically refer to economic data that are recorded, analyzed, and reported by a company's cost and management accounting system. Traditional management accounting theory already provides a large spectrum of methods for cost analysis, dealing with several types of costs or focusing on specific objects, e.g., the products or the cost centers. But, since these traditional methods largely fail to identify material and energy inefficiencies (see also Sect. 2.1), several so called flow oriented concepts were developed. The most 'mature and comprehensive' (Loew 2003) is the material flow cost accounting (MFCA) whose general principles were recently published as ISO standard 14051 (ISO 2011).

In the following, firstly, the basic approach of MFCA is introduced, compared to traditional cost accounting, and critically reviewed (Sect. 2). On basis of the identified weaknesses, refinements regarding the modeling and appraisal of loops and stocks as well as the integration of energy are proposed (Sect. 3). Afterwards, the method is enhanced to a future-oriented 'Plan-MFCA'—integrating a technical analysis and allowing the ex ante comparison of existing and alternative process chains (Sect. 4.1). Moreover, the integration of the developed Plan-MFCA into methods for appraising long-term (up to life cycle wide) economical and ecological effects is discussed (Sect. 4.2). The examination of implementation related aspects of MFCA reveals that, in particular, the data acquisition, the analyses' level of detail, and the adoption of MFCA thinking at all management levels are key elements of a successful implementation strategy (Sect. 5). Finally, the results of the chapter are summarized (Sect. 6).

2 Basic Approach of MFCA

2.1 Idea

MFCA is a specialized accounting method which aims at the identification and monetary valuation of inefficiencies in material use. Generally, it can be applied to a wide range of systems—single companies, value chains or even geographic

[2] The definition of the specific processes may vary with the relevance of the respective 'area'. So, a single production process can be limited to one very resource intensive production step or include a whole assembly line affecting the total resource demand only marginally.

regions. For reasons of simplification and due to the fact that MFCA was primarily implemented at industrial facilities so far, in the following, the analysis of industrial process chains as well as the included single processes will be described.

As mentioned, MFCA was developed on basis of the criticism of traditional cost accounting's treatment of material costs. Typically, the traditional methods primarily aim at the more or less exact assignment of costs to the final products (the primary cost objects). Furthermore, the methods allow specifying cost data of the cost centers in order to evaluate their performance. Here, in particular, the indirect costs which cannot be clearly attributed to the single products but (largely) to the cost centers, are in the focus of analysis. However, the material costs whose importance has been already mentioned in the previous section are direct costs for the most part. Thus, they are directly assigned to products and so, 'bypass' the cost center accounting and the pressure for cost reduction that is imposed at the level of cost centers. As a result, the material efficiency of production processes and, even more important, the economic effects of *in*efficiencies largely remain unknown (BUM and UBA 2003).

MFCA takes up the traditional methods' weakness regarding the evaluation of material use. It considers the production of goods as a system of movements of materials—the material flows—which are assessed quantitatively and monetarily. Additionally, the flows are distinguished in desired material flows (movements of production's input raw materials, operating supplies, intermediates, products, etc.) and in undesired material flows which represent the movements of processes' unintended material outputs such as clippings, rejects or used lubricants. As mentioned, the specific focus is on the material inefficiencies of production processes and therewith, on the undesired material flows. Due to MFCA's differentiated view on processes' material output, its cost analysis reports *all* costs incurred in the context of 'producing' material losses. Furthermore, its quite detailed quantity and cost models highlight the single inefficiencies. So, engineers can focus on the technical and/or organizational improvement of the processes with the highest potentials for cost reduction. The subsequent decrease of the total amount of undesired outcomes entails a reduced demand for input materials and, therewith, results in positive economic *and* ecologic effects (Strobel and Redmann 2002; Hyršlová et al. 2008; ISO 2011).

The original concept of MFCA bases on the work of the German 'Institut für Management und Umwelt' (institute for management and environment/ecology) in the late 1990s which initialized few pilot projects in the German industry (see Strobel and Wagner 1999; Wagner et al. 2010). However, MFCA's breakthrough had been in Japan. Due to the great success of first implementations in the year 2000, the Japanese Ministry for Economy, Trade and Industry strongly promoted MFCA and more than 300 Japanese companies adopted the method by now (Nakajima 2010; Schmidt 2012). In parallel, MFCA's methodology was refined and the work on the corresponding ISO standard 14051 (ISO 2011) was initialized (Kokubu et al. 2009). The final version of the standard was published at the end of 2011 and summarizes the actual state of the art of MFCA.

2.2 Methodology

In the following, the basic methodology of MFCA will be described and demonstrated using a simple example.[3] The general procedure of MFCA can be divided into three steps of flow modeling (Sygulla et al. 2011):

- modeling the flow structure,
- quantifying flows in physical units, and
- monetarily appraising the flow system.

The initial task of *flow structure modeling* is the description of the system to be analyzed (the process chain). This entails the boundaries' and quantity center definition. The quantity centers are elements of the system for which material in- and outputs can be physically quantified.[4] Among others, production processes and storage areas may correspond with a quantity center. The final task of flow structure modeling is adding the regular material flows between the quantity centers (see Fig. 1).[5] Regarding the modeling of the material flows, it has to be noted that the analysis of *all* used material is commonly not recommendable. In order to ensure a meaningful cost-benefit ratio of the analysis, only the materials which represent a relevant share of costs should be included.

The second step of the MFCA procedure is the quantification of the identified regular material flows and the changes of quantity center stocks within a defined time period (e.g., a month or a year). To ensure the consistency of the analysis the material in- and outputs of every quantity center have to be balanced, considering possible changes in stock. Consequently, the quantities of flows and stocks should be expressed in a single physical unit of mass[6] or at least in convertible units. By adding the quantities to the flows and quantity centers of the flow structure model, it is enhanced to a *flow quantity model*.

The analysis of quantities may already highlight inefficiencies in material use, but their negative economic effects are still unknown. So, MFCA's final step is the monetary appraisal of the material flows and the stocks resulting in a *flow cost model*. Here, the flows and stocks are perceived as cost collectors to which the incurred costs are assigned. The corresponding tasks of cost accounting are the identification, the quantification, and the assignment of costs. For the identification

[3] The explanations in Sect. 2.2 largely refer to the recent ISO standard 14051 (ISO 2011). So, only direct citations of ISO (2011) as well as the use of other sources will be explicitly specified.

[4] ISO's (2011) definition of a quantity center additionally includes the demand for a monetary quantification of the in- and outputs. However, as the following explanations will demonstrate, the monetary valuation of in- and output flows is *based* on their physical quantification which, therewith, is the sole requirement for defining a quantity center.

[5] Here, only the directions and the characteristics (desired or undesired) of the flows are determined. The quantification of flows follows in the second step.

[6] For the intended analysis, the mass of materials can be presumed as constant. So, the inputs, outputs, and changes in stocks of the quantity centers can be listed in form of a material balance.

Fig. 1 Flow structure model

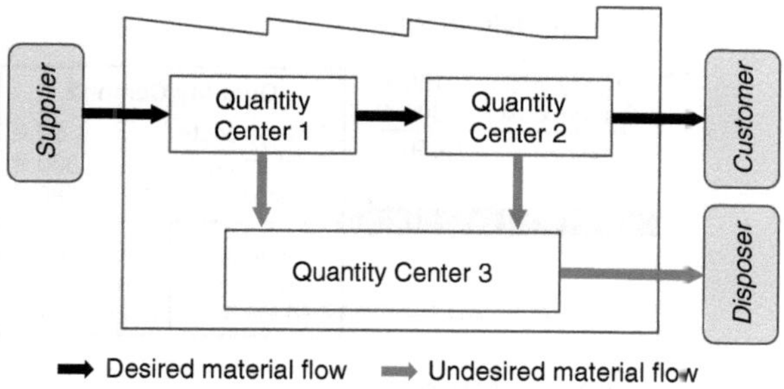

and quantification of the single cost items, the existing cost data of the regular (traditional) cost accounting may be used. But, it has to be regarded that MFCA-specific definitions and terms of costs may differ from that of traditional cost accounting. ISO (2011) defines the following major cost categories:

- *material costs*—costs for the materials that are used and/or consumed in a quantity center
- *energy costs*—costs for the energy used to enable operations
- *waste management costs*—costs for handling material losses generated in a quantity center
- *system costs*—all costs incurred in the course of in-house handling of the material flows except for material, energy, and waste management costs (e.g., depreciation or costs of labor, maintenance, or transport)

The material costs are based on fix prices per unit and—in the context of MFCA—are all direct costs in relation to the flows. So, they can be easily traced (directly assigned) to the material flows and, as a specific characteristic of this cost category, to the stocks of the quantity centers. In contrast, the other cost categories include solely indirect costs which cannot be directly attributed to the material flows but to the quantity centers or to superior cost centers. In the latter case, these costs are perceived as indirect quantity center costs and the first task of cost assignment is the allocation of the cost center costs to the quantity centers—here, the known methods of traditional cost (center) accounting can be used. In both cases, the quantity center costs have to be allocated (indirectly assigned) to the material flows—the final cost collectors. The preferred allocation base is the mass ratio of the material output flows. Sole exceptions are the waste management costs which are obviously only caused by the undesired material output flows and therefore are fully assigned to them.

A numeric example of the described cost assignment is given in Fig. 2, for reasons of simplification it is limited to one material flow. The example also illustrates the so called 'cost carryover between quantity centers' (ISO 2011). Since the physical material flow is considered as a cost collector, it 'carries' the cumulated costs which were assigned to it in the previously passed quantity center 1. So, within the task of allocating indirect costs, the energy and system costs of

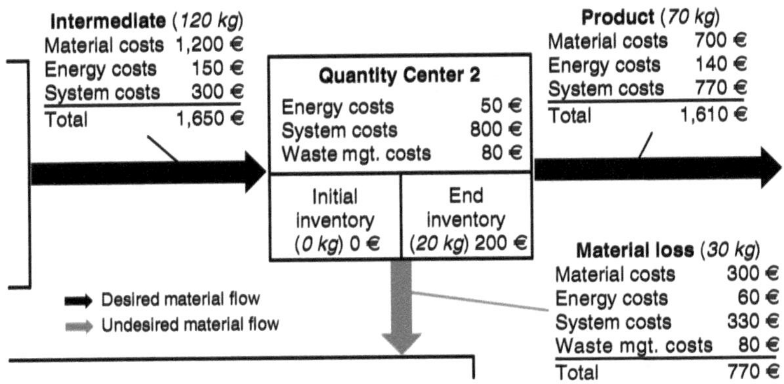

Fig. 2 Calculating cost flows of a single quantity center

the input flow are summed up with the corresponding costs of the actual quantity center 2 and are assigned to the output flows according to their mass ratio.

A second specific characteristic of the given example is the including of a stock in quantity center 2 which serves as a buffer for desired output materials. As described above, the initial and end inventories are valued by the corresponding material prices only. The system and energy costs of this quantity center and the previous ones are not allocated to them but (totally) to the output flows. In cases of stock changes, this rule of cost assignment distorts the results of the analysis. In the example, the energy and system costs incurred for the creation of the stored intermediates are assigned to the output flows (product and material loss). So, costs of *desired* materials (of the stored intermediates) are partly allocated to *undesired* material flows. For a discussion of this shortcoming and a solution see Sect. 3.1.

The results of the cost analysis can be visualized by the flow cost model. It is generated by adding the costs to the flow quantity model. As Fig. 2 implies, such cost models can rapidly grow and become very complex. So, usually, only the total cost values are displayed. As a result, the use of flow cost models is commonly restricted to providing an overview and to pointing at the most inefficient quantity centers. A systematic analysis of the material inefficiencies calls for a more detailed examination of the according cost structures. An appropriate visualization is provided by the so called *flow cost matrix.* Table 1 shows its general structure and extends the above used numeric example to all three quantity centers of Fig. 2. The evaluation of the total cost of desired and undesired outputs of the process chain (grey section on the right bottom) reveals that nearly the half of the costs incurred for the creation of losses. The matrix also reports that the largest share of loss costs is caused in quantity center 2. Consequently, the corresponding production process has the largest potential for cost reduction by improving its resource use.

Table 1 Flow cost matrix

(all values in €)		Quantity center 1					Quantity center 2					Quantity center 3				
		Material costs	Energy costs	System costs	Waste mgt. costs	Total in QC	Material costs	Energy costs	System costs	Waste mgt. costs	Total in QC	Material costs	Energy costs	System costs	Waste mgt. costs	Total in QC
Input	Input from previous QCs	-	-	-	-	-	1,200	150	300	-	1,650	500	85	380	130	1,095
	New input in QC	1,400	175	350	50	1,975	-	50	800	80	930	-	15	120	70	205
	Total in each QC	1,400	175	350	50	1,975	1,200	200	1,100	80	2,580	500	100	500	200	1,300
Output	Desired output of QC	1,200	150	300	-	1,650	700	140	770	-	1,610	-	-	-	-	-
	Change in QC stock (desired)	-	-	-	-	-	200	-	-	-	200	-	-	-	-	-
	Undesired output of QC	200	25	50	50	325	300	60	330	80	770	500	100	500	200	1,300
Total	Total cost of desired output	1,200	150	300	-	1,650	700	140	770	-	1,610	700	140	770	-	1,610
	Total change in stocks (desired)	-	-	-	-	-	200	-	-	-	200	200	-	-	-	200
	Total cost of undesired output	200	25	50	50	325	500	85	380	130	1,095	500	100	500	200	1,300
	Total cost	***1,400***	***175***	***350***	***50***	***1,975***	***1,400***	***225***	***1,150***	***130***	***2,905***	***1,400***	***240***	***1,270***	***200***	***3,110***

2.3 MFCA and Traditional Cost Accounting

Before reviewing the approach of MFCA critically in the Sect. 2.4, the relationship between MFCA and traditional cost accounting is examined here. As mentioned in Sect. 2.1, MFCA is a (quite) specialized accounting method applied in the context of the evaluation and the (re)design for resource efficiency of production processes. Nevertheless, its objectives overlap with those of traditional cost accounting (Schweitzer and Küpper 2011):

- mapping and documentation of business processes,
- supply of information for planning, monitoring, and controlling of business processes,
- behavioral control of managers and staff,
- appraisal of products, intermediates, and self-made assets.

But, traditional cost accounting's function of management support is more general, its (cost) information is used for a wide range of decisions. In contrast, MFCA focuses on process design. In particular, in the context of the last of the abovementioned objectives the difference becomes visible. While traditional cost accounting's appraisal is commonly used for product costing, evaluation of short-term company success, and inventory valuation, MFCA primarily appraises the processes' desired and undesired outcomes (products, intermediates, and losses) in order to derive conclusions about the processes' resource efficiency.

From the perspective of the company's cost accounting *system* MFCA can be perceived as a specialized part of this system. But, MFCA differs from traditional cost accounting in its methodology to some extent. Examples are the specific cost categorization basing on individual definitions of cost types (e.g., the material costs), the monetary appraisal of undesired outcomes, and the general use of mass relations for cost assignment. So, in order to minimize the effort and enhance the benefits of the whole accounting system, a useful concept for integrating MFCA has to be chosen.

Figure 3 presents a possible concept for integration on the data level. Here, the cost-type and the cost center accounting of traditional cost accounting provide input cost data for MFCA. This implies firstly the 'labeling' of the single cost items regarding the needs of both accounting methodologies (Fig. 3 visualizes the costs regarded by MFCA as dashed lines). Secondly, the cost and quantity center structure must be harmonized, so that the cost-type accounting can collect and provide data at the required level of detail and the cost center accounting can be used for calculating the indirect quantity center costs (see Sect. 2.2).[7] Afterwards,

[7] For reasons of simplification the quantity center structure can also be derived from that of the cost centers. But, it has to be ensured that every quantity center is attributed to only *one* superior cost center, if the quantity center structure is more differentiated than that of the cost centers. In case of the cost center structure is more detailed in some areas, subordinate cost centers must be attributed to the superior quantity centers correspondingly.

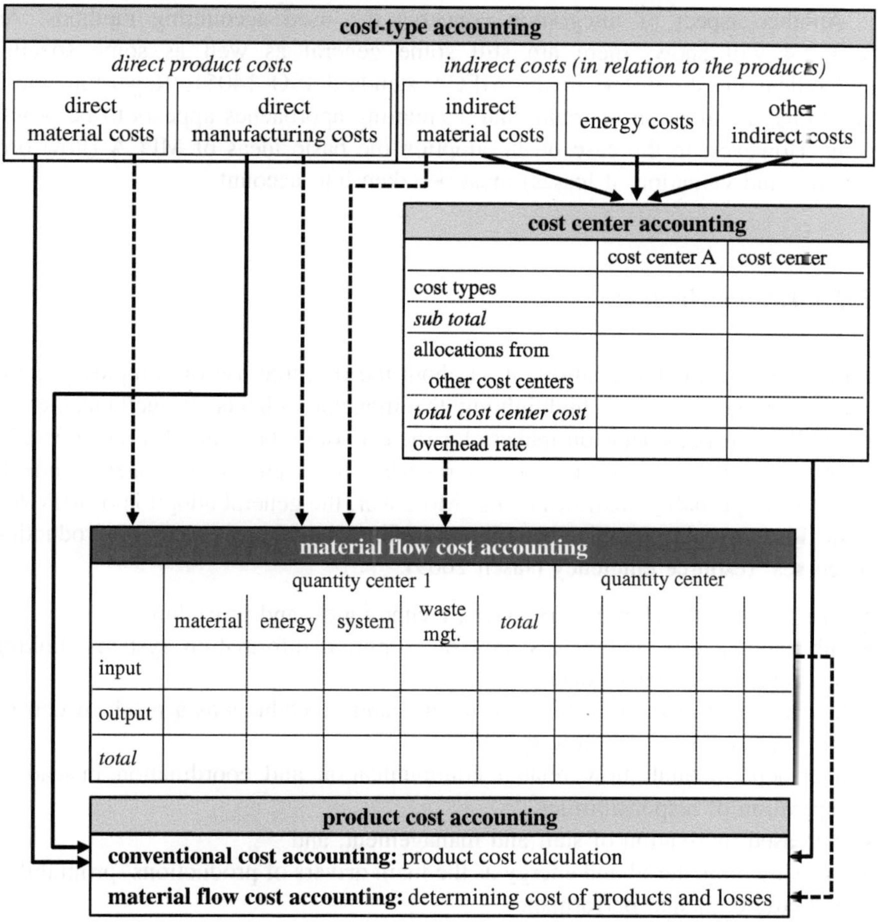

Fig. 3 MFCA as part of the cost accounting system (on basis of Bierer and Götze 2012)

the actual flow costs are calculated in MFCA (as a separated part of the whole cost accounting system) using the described methodology.

Finally, the results of MFCA can be contrasted by those of other (traditional) cost studies, as it is indicated in Fig. 3 in the box of product cost accounting. Due to the harmonized data base, the comparability of the results is ensured and the studies can be understood as analyses of the same object (the production) from different perspectives. So, enhanced or even new information for decision making will be available by integrating MFCA with the existing (traditional) accounting system. Furthermore, it has to be noted that the existing data base of traditional cost accounting is expected to be enhanced by MFCA's usually more detailed assessment of material and (if regarded) energy movements. So, the quality of traditional cost studies will be improved simultaneously.

Another aspect of integration concerns the used accounting methods. As Sect. 2.4 will show, there are still some general as well as some specific methodical questions left in the MFCA standard ISO 14051. Regarding these questions, an adoption of traditional accounting approaches appears to be promising. However, in the case of an adoption the basic ideas of MFCA (flow orientation and valuation of losses) must be taken into account.

2.4 Critical Review

MFCA provides detailed information about the resource use of complex process chains. In particular, by highlighting the monetary effects of inefficiencies, it generally sensitizes decision makers for the economic benefits of a resource efficient production and, furthermore, it reveals wastes and their monetary effects which were probably unknown so far. Moreover, the general adoption of MFCA's basic idea reveals potential benefits exceeding the improvement of production processes' resource efficiency (Jasch 2009):

- incentives to develop new products, technologies, and procedures,
- enhanced quality and consistency of corporate information systems, linking physical and monetary data,
- improvement of organizational structures and procedures as a result of consistent referencing to the flow system,
- inter-departmental, flow related communication and coordination instead of separation of responsibilities,
- increased motivation of staff and management, and
- focusing on material and energy as the main drivers of productions' profitability and environmental performance.

The enhancement of the quality of existing accounting systems by integrating MFCA as a partial system was described in the Sect. 2.3 already. But, MFCA analysis' benefits have to be contrasted with the necessary efforts for its implementation. The very detailed flow models require an appropriate system of control points for measuring the physical quantities of the single flows. Additionally, in comparison with the traditional methods of cost accounting the necessary recording is now extended to the undesired outcomes of the production processes. But, regarding the ratio of efforts and benefits, practical experiences show that especially the initial analysis of the material flows typically highlights much inefficiency in the process chain and partly even shortcomings of the existing material management system. The subsequent elimination of the inefficiencies (and the shortcomings) resulted in significant cost savings (Strobel and Wagner 1999; Hyršlová et al. 2008; Nakajima 2009; Kokubu and Kitada 2010; ISO 2011).

Irrespective of the achieved success of MFCA's practical implementation, its methodology has to be questioned from a theory perspective in order to evaluate the significance of its results. *Firstly*, the use of mass ratios as allocation bases for indirect costs (the quantity center costs) implies a corresponding correlation of the masses and the cause of costs which may be doubtful, e.g., for the depreciation of a machine. However, when considering the desired and undesired outputs of production processes as joint products, traditional methods 'fall back' on using physical allocation bases such as the relation of masses as well. Moreover, the mass relation does not necessarily refer to *all* output materials; it can be adapted to the individual conditions of the quantity centers. For instance, if an auxiliary material has a disproportional high mass, the mass relation used for cost allocation may only regard the basic material (ISO 2011).

Secondly, there are stock-related questions about flow modeling—the inappropriate treatment of inventories' production costs in case of stock changes was already mentioned in the Sect. 2.2. Closely connected to this is the allocation of storage costs. Although they are obviously caused by the stored materials, these costs are assigned to the output flows, the 'not-stored' materials. Again, the results are distorted and the storage costs are assigned to the desired *and* undesired output flows regardless of the stored materials' characteristic (see Sect. 3.1).

A *third* unsettled detail is the modeling of material loops typically occurring in the context of internal recycling. Here, the main problem is an interdependency of material in- and output flows complicating their cost appraisal (for details of criticism and a proposal for solving the problem see Sect. 3.1).

Fourthly, there is a lack of knowledge concerning flow modeling of energy. Due to a growing economic and ecologic importance of this resource, it should be integrated with the existing material analysis of MFCA. Recent MFCA literature already mentions the general possibility of modeling energy flows (e.g., ISO 2011) or even provides examples including at least energy input flows (e.g., Viere et al. 2011). However, the identification of energy related inefficiencies requires the full adoption of the MFCA methodology, including the differentiation of desired and undesired energy flows. Thus, it seems reasonable to enhance MFCA to an integrated material and energy flow cost accounting considering the partly differing characteristics of material and energy flows. For example, the allocation of costs has to be discussed because energy flows cannot be quantified in units of mass (see Sect. 3.2).

Finally, the general information value of MFCA's results for process design has to be questioned. So far, the method evaluates the process chains' as-is state. It points at existing inefficiencies and reports their monetary effects. Subsequently, technical and/or organizational improvements can be developed. But, a meaningful decision making requires the analysis of these improvements and their (monetary) effects on the whole process chain as well. In order to ensure the comparability of the as-is state's and alternative process chain's appraisal, the evaluation must be based on the same method. Consequently, MFCA has to be enhanced for the evaluation of future states of process chains (see Sect. 4.1). Moreover, the improvement of processes often requires new equipment and/or affects the type of use of the existing. So, the improvements must be considered as an investment and long term decision, respectively. In this regard, MFCA's limitation to a single time period is a general shortcoming, and MFCA analyses should be complemented by methods of investment appraisal, including life cycle costing (LCC).

In the context of life cycle wide analyses another benefit of MFCA gets visible: Literature already states that MFCA and life cycle assessment (LCA), a method for appraising the ecologic effects of material and energy use, are based on quite similar quantity models (Kokubu et al. 2009; ISO 2011). This argumentation can be extended to investment appraisal and LCC. So, a harmonization of these models may decrease the total effort of appraising monetary and environmental effects of process chain improvements and therewith, support the design of an economically and ecologically sustainable production (see Sect. 4.2).

3 Refinement of Flow Cost Modeling

3.1 Pulling Out Costs of Loops and Inventories

Complex process chains usually include recycling processes, e.g., the remelting of internally generated scrap or the reconditioning of cutting fluids. These recycling processes lead to *material loops* which often appear economically (and ecologically) favorable because the amounts of input materials as well as of materials to dispose are reduced. But, every material loop is an indication for the generation of material losses and therewith, a potential starting point for technical improvements to increase the material efficiency. Furthermore, the recycling processes raise additional costs which have to be considered in the cost analysis.

In the context of modeling material loops, the handling of cost related interdependencies of the cycled material flow is a specific challenge. Figure 4 shows a simplified example. Here, a single material is used in a production process. The losses are fully recycled and returned to the production process.[8] This raises the question how to appraise the 'cycle' flows—the costs of the material loss flow influence the costs of the raw material substitute flow and vice versa. To solve the problem, literature suggests three options:

1. calculating the total cost of every material flow,
2. considering only the additional loop costs, and
3. reporting loop costs as an extra cost category.

The *first option* follows MFCA's intention to assign costs to *every* material flow. But, this entails that the costs of the raw material substitute flow and of the material loss flow have to be calculated simultaneously. This can be realized by building up and solving a corresponding linear equation system (see ISO 2010 for

[8] In this regard, it has to be noted again that the model records the material quantities flown within a defined time period. Since in Fig. 4 the product flow is the only flow leaving the loop, all occurred system costs are finally assigned to it.

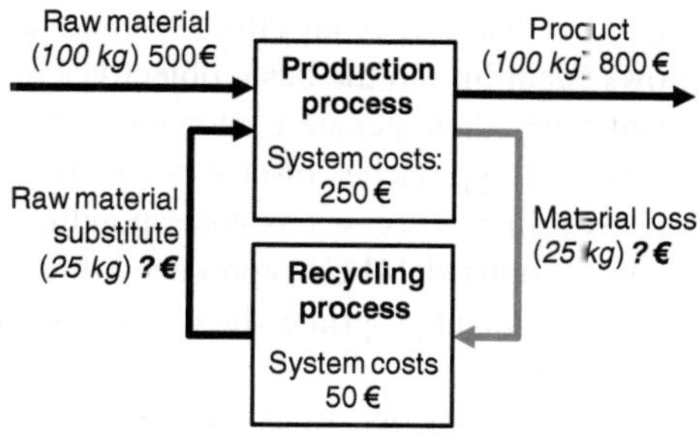

Fig. 4 Cost interdependencies in material loops (on basis of Viere et al. 2011)

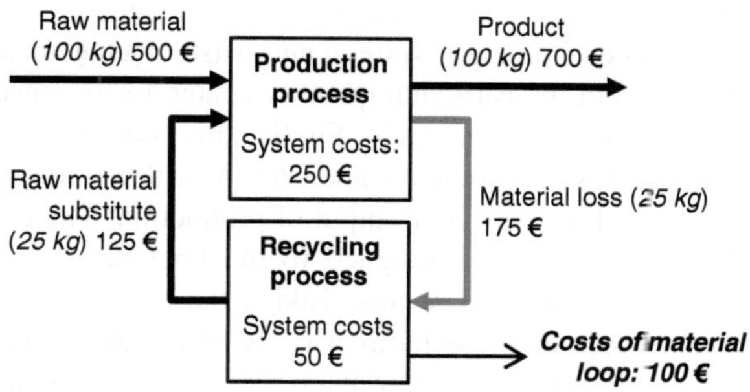

Fig. 5 Reporting loop costs as an extra cost category (on basis of Viere et al. 2011)

a numeric example). A much more simple way is ignoring the material costs of the cycled flows (*second option*), since they end up in the outgoing product in any case. So, only the additional costs of the recycling process (here 50 €) are carried over as a cost input of the production process (ISO 2010; Viere et al. 2011).

Both of the above presented options finally assign the loop costs to the material output flow (product) and therewith, neglect the fact that these costs are originally incurred by the handling of an undesired material (flow). Consequently, the material inefficiencies of the production process are 'invisible' in the MFCA analysis of the overall process chain.[9] A solution is presented by Viere et al. (2011). They suggest to 'pull out' the costs of the material loop and to report them as an extra cost category (*third option*). Figure 5 refers to the example above and illustrates the results of using the 'pull out option'. The costs assigned to the product are decreased by 100 € which are caused by the material loop and now reported separately as costs of handling undesired material flows.[10]

This third option may conflict with MFCA's basic methodology since the costs pulled out of the material loop are a cost flow without a physical material flow. On

[9] In case the material loss is not fully recycled and there is an undesired material output flow leaving the loop, the criticism still concerns the share of loop costs assigned to the raw material substitute flow.

[10] The value includes 50 € of the production process' system costs (assigned to the material loss flow) and the system costs of the recycling process, 50 € as well. The flow of the raw material substitute is valued by its material costs only.

the other hand, it is an effective (and simple) method to evaluate material loop flows according to the basic objectives of MFCA—highlighting inefficiencies and identifying their negative monetary effects. The procedure is similar to that of traditional approaches of cost accounting which would report the system costs of the recycling processes as corresponding cost center costs. The additional benefit of the proposed MFCA approach (third option) is the cumulation of loop costs including the appropriate share of costs arisen within the production process (flow orientation).

Quite similar questions are raised in the context of appraising *inventories* and dealing with *stock changes* and *storing costs*. According to basic MFCA, the materials stored in a quantity center are valued by their material costs only. The indirect costs (energy, system, and waste management costs) of the material input flows and of the particular quantity center are assigned to the material output flows only (ISO 2011, Sect. 2.2). Firstly, this procedure entails that intermediates and material losses produced and stored within the analyzed time period are not considered as a relevant output of production. The costs incurred in the context of their production are assigned to the final desired and undesired outputs which distorts the analysis' results. Taking the numeric example of Sect. 2.2, within the quantity center 2 the indirect costs of the stored intermediate—a *desired* material—are partly assigned to the *undesired* material output flow, the material losses. Secondly, it has to be questioned whether the assignment of storage costs (included in the quantity center's system costs) to materials leaving the quantity center adequately reflects the cause of these costs. For example, the occupancy costs of storage areas (including depreciation, costs of cooling and lightening, etc.) are obviously more closely related to the stored materials than to the outgoing. So, the assignment of these costs to output flows may distort the results of MFCA as well.

The methodical solution suggested here is the 'pull out' of costs again. Therefore, the stock inventories—of desired and undesired materials—are considered as cost carriers, equal to the material flows, and all costs of their production and storage (in the examined time period) are assigned to them. This solution closely follows the monetary appraisal of inventories in traditional cost accounting. Besides, the inventories are a 'physical basis' for cost assignment. So, the pull out of inventory costs may be considered as less conflicting with the basic idea of MFCA than the pull out of material loop costs.

Table 2 seizes the above mentioned example of Sect. 2.2 and illustrates the new distribution of the total cost of the process chain considering the separation of inventory costs (possible storing costs are not regarded yet). Contrasting the results of Table 2 with those of Table 1 makes the aforementioned distortion of results caused by using the 'traditional approach' visible. The energy and system costs of the inventory, amounting to 216.66 €, were (wrongly) assigned to the desired and undesired material output flows of quantity center 2 (see Table 1). By defining the inventories as cost carriers as described above, these costs are reported separately now (see Table 2).

Table 2 Pull out of inventory costs (total cost of the process chain, extract of flow cost matrix)

(All values in €)		...	Quantity center 3				
		...	Material costs	Energy costs	System costs	Waste mgt. costs	Total in QC
	...	...	...	...	...	...	...
Total	Total cost of desired output	...	700.00	116.67	641.67	–	1 458.34
	Total change in stocks (desired)	...	200.00	33.33	183.33	–	416.66
	Total cost of undesired output	...	500.00	90.00	445.00	200.00	1,235.00
	Total cost	...	1,400.00	240.00	1,270.00	200.00	3,110.00

As an additional benefit of pulling out inventory costs, the monetary effects of production and/or logistic concepts and strategies can be compared. So, for instance, the general suggestion of the Lean approaches to design systems with minimized stocks can be monetarily evaluated for the concrete process chain and compared with other available design alternatives (for an appropriate enhancement of MFCA for modeling future states of process chains see Sect. 4.1).

3.2 *Integrated Material and Energy Analysis*

In the context of designing an ecologically sustainable production the evaluation of the internal use of the resource energy is vitally important. In particular, the CO_2 emissions—one of the most commonly used environmental performance indicators—are determined by the production's energy demand for the most part.[11] Regarding economic objectives, statistics report industry's energy costs as a marginal share in the total production cost (e.g., Statistisches Bundesamt 2011). But, these statistics only take into account the costs of external sourcing and neglect those of the complex internal infrastructure for energy conversion and supply. Thus, a more detailed analysis of energy use may reveal unknown potential for cost savings as well. Consequently, it seems to be useful to enhance the methodology of MFCA by an analysis of the internal energy flows in order to detect material *and* energy inefficiencies and to support improvements with respect to both types of resources (Prammer 2009; Nakajima 2011). MFCA literature usually mentions the general importance of considering energy (e.g., Nakajima 2011; ISO 2011). But, there are only few contributions dealing—more or less detailed—with possible designs of such integrated analysis of material *and* energy

[11] For example, Volkswagen recently communicated its strategy for decreasing production's CO_2 emissions including the concrete goal of reducing the energy demand by 25 % until 2018 (Viehmann 2012).

use so far (Nakajima 2008; Prammer 2009; Sygulla et al. 2011; Götze et al. 2012). In the following, the characteristics of such integrated models are discussed and approaches for overcoming the inherent difficulties are proposed.[12]

For the integrated material and energy analysis, the general procedure of MFCA (modeling the flow structure, quantifying flows in physical units and monetarily appraising the flow system) is kept. However, in the context of modeling, the specific characteristics of energy raise several questions concerning:

- the relationship of energy carriers and energy,
- the desired energy 'output' of production processes,
- the dimension unit of energy, and
- the cost assignment to material and energy output flows.

Within the first step of flow modeling, the relevant regular energy flows are identified. In this regard, it has to be noted that the used forms of energy are always linked to a material energy carrier. The adequate modeling of parallel energy and material flows may be reasonable for some carriers such as gas or coal and if the material losses of energy conversion (ash, waste gases, etc.) are relevant for the analysis. But, for the most forms of energy (e.g., electricity or compressed air) the use of the sole energy flow is sufficient (Sygulla et al. 2011).

For the most part, the energy flows are modeled in the same way as the material flows. An exception is the desired energy output of production processes. A closer look at their general energy use reveals that a specific share of the input energy is needed to perform the desired operation—in the following, this share is named as process energy. The rest of the input energy leaves the process as energy losses (waste heat, vibrations, etc.). However, from the physics perspective the process energy commonly leaves as waste heat as well. So, it cannot be measured separately. The amount of process energy must be calculated or estimated for the particular process. Furthermore, in the context of modeling, the definition of a separated process energy flow does not seem to enlarge significance but complexity. Consequently, for reasons of simplification and basing on the presumption that the material outputs are the primary outputs of a production process, the process energy is 'assigned' to the material output flows (Sygulla et al. 2011; Götze et al. 2012). Figure 6 exemplarily illustrates the flows of an energy conversion and a production process.

The second step of flow modeling is the quantification of flows. Here, a quite obvious fact is that energy cannot be quantified in units of mass but of energy, preferably in joule or watt hours. By using appropriate measurement equipment, the flown quantities of most input energies can be recorded.[13] But, as mentioned

---

[12] In the strict sense, the approach is enhanced to a *material and energy flow cost accounting*. For reasons of simplification, the name material flow cost accounting (MFCA) is still used in the following.

[13] For quantifying the energy content of 'material based' energy carriers, such as gas or coal, the specific heating value can be used.

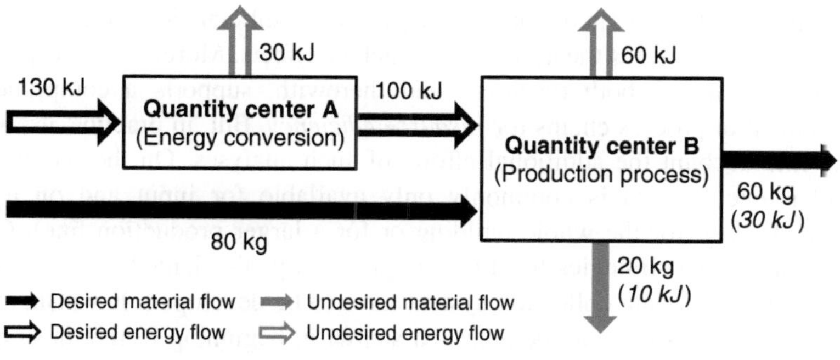

Fig. 6 Material and energy flow quantity model

above in the context of process energy, the amounts of energy output flows will often have to be calculated or estimated. At this, the use of energy balances ensures the completeness of quantification (Sygulla et al. 2011; Götze et al. 2012).

Regarding the final step of flow modeling (appraising the flow system), the integration of energy flows, on the one hand, entails that energy costs turn from indirect to direct costs and can now be traced (directly assigned) to the corresponding flows. On the other hand, the involved definition of a second quantification unit necessitates the revision of MFCA's allocation rules. In the basic methodology the quantity center costs were assigned to the material output flows according to their mass ratio (see Sect. 2.2). Now, some quantity centers have material *and* energy output flows which are quantified in nontransferable physical units (see quantity center B in Fig. 6), the only shared unit is the monetary. Correspondingly, a first (simple) solution is the use of the flows' direct costs ratios (material to energy costs) as allocation base. However, it is debatable whether the energy output of a production process causes the quantity center costs in the same way as the material outputs do. So, the second solution is based on the initial distinction of quantity centers according to their primarily intended type of output (material or energy). Subsequently, the quantity center costs are assigned to these output flows only using the ratio of the physical amounts. Taking the example of Fig. 6, quantity center B is a 'material quantity center'. Its system and waste management costs are fully assigned to the material output flows. The energy loss flow and the share of process energy (included in the outgoing material flows) is valued on basis of the costs assigned to the input energy flow only, none of quantity center B's costs are allocated to them. In this regard, it has to be noted that the reuse of output energies of production processes has to be considered as an energy loop modeled analogous to material loops (see Sect. 3.1). However, in case of a quantity center cannot be clearly attributed, both types of output are considered as intended and the abovementioned first option can be used for cost assignment. A third solution is the more detailed analysis of cost occurrence and a corresponding more precise attribution of costs to material or energy flows. An appropriate model supporting the analysis of cost drivers is presented in Sect. 4.1 (the ITO-model).

The proposed integrated analysis highlights inefficiencies in material *and* energy use and identifies the according monetary effects. Moreover, it respects the interdependencies of both resources and therewith, supports a comprehensive improvement of process chains for *resource efficiency*. But, in practice, users have to take into account the additional efforts of such analyses. On the one hand, in particular, energy data is commonly only available for input and on a quite aggregated level (for the whole building or for a larger production line). Consequently, individual strategies for a meaningful and goal-oriented enhancement of the internal energy data collection system have to be developed. For instance, the existing data acquisition can be enlarged stepwise, beginning in the most significant areas of energy consumption and should be assisted by selective mobile measurements. On the other hand, companies with energy-intensive processes may use modified traditional approaches already (e.g., the energy cost accounting proposed by Bierer and Götze (2012)). Here, the integration of MFCA with the existing accounting system (see Sect. 2.3) promises to enhance and enlarge the basis of information for decision making.

4 Prospective Approaches of MFCA

4.1 Short-Term Appraisal of Process Alternatives

The MFCA analysis highlights resource inefficiencies by reporting the points of material and energy loss origin and the costs incurred by producing these losses. As a result, the corresponding technical and/or organizational shortcomings of the existing process chain can be analyzed in detail and appropriate alternative processes can be developed. Subsequently, the monetary effects of the available process alternatives on the process chain need to be quantified for decision making. Since these alternatives are not realized yet, they have to be considered as (possible) *future states* of the process chain. In this regard, it has to be emphasized (again) that the cost saving effects of an alternative process cannot be identified on the basis of MFCA's as-is data. More explicit, firstly, the reported costs of losses are not or at least only under specific presumptions fully proportional to their physical amounts. So, assuming a linear relation of loss quantity and costs may cause significant overestimations of the cost saving potentials. Secondly, the improved processes may influence the performance of other, not modified processes of the process chain as well (e.g., if the requirements on the input and/or the quality of the output work piece are modified). Consequently, the quantification of the monetary effects of process alternatives requires a new analysis of the whole modified process chain. But, in contrast to the basic conditions in Sect. 2, this analysis refers to a future state of the process chain. So, the flow quantities and costs have to be computed or estimated on basis of appropriate presumptions or models of the future state. MFCA has to be enhanced to a *Plan-MFCA* using future related data.

The intended forecasting of the total cost of process chain alternatives requires a meaningful categorization of cost items and the analysis of parameters influencing the costs. Regarding the cost categorization, it has to be noticed, that in particular MFCA's definition of the system costs is quite wide-ranging and the several included cost items are very heterogeneous; for instance ISO (2011) itemizes depreciation as well as costs of labor, of maintenance, and of transport. In order to ensure completeness and to avoid redundancies, Sygulla et al. (2011) suggest the definition of sub categories according to the input factors (labor, equipment, capital, and external services). Additionally, this supports the appropriate treatment of the complex network of relevant technical and economical cost drivers.

For the analysis of cost affecting parameters, Götze et al. (2010) suggest an Input-Throughput-Output model (ITO-model) for the description of the quantity center from a technical perspective.[14] It aims at the identification of technical drivers of the transformation of material and energy inputs into outputs in order to deduce correlations of flow quantities and costs (see also Schubert et al. 2011). The ITO-models are necessarily very process specific.

Figure 7 shows a generalized scheme on basis of the examination of a turning process by Götze et al. (2010). As visualized, MFCA's input and output analysis is enhanced by the inclusion of technical data and a throughput perspective is added. The throughput describes the specific process restrictions, basing on the intended operation, and an analysis of the combination of the machine, further equipment, and used tools. Moreover, the major drivers of the transformation of materials and energies can be mathematically described. For the mentioned concrete examination of a particular turning process, the depth of cut, the feed, and the cutting speed were identified as the main technical drivers of energy and time consumption and of total production cost (Götze et al. 2010).

The integration of the ITO-models allows the forecast of the quantity and cost related 'reactions' of single quantity centers to modified inputs, outputs or process parameters as well as to change in prices. As a result, the total cost of the future states of existing processes and of alternatives can be estimated at a quite high level of detail, considering the complex relationships of the single processes (Sygulla et al. 2011). Additionally, the technical-economical analysis can provide helpful information for the task of designing process alternatives, e.g., about the effects of specific process level changes on the overall system or on the final products' technical characteristics. Moreover, the integration of technical and economical aspects in a joint model supports the interdisciplinary communication between engineers, controllers, and managers. Here again, the expected benefits from combining MFCA and traditional cost analyses—in particular in the context of decision support—have to be mentioned (see Sect. 2.3).

---

[14] Alternatively, ITO-models of the underlying technical processes and, if needed, of sub processes can be used for the analysis. For reasons of simplification, in the following, quantity centers and processes are considered as equal.

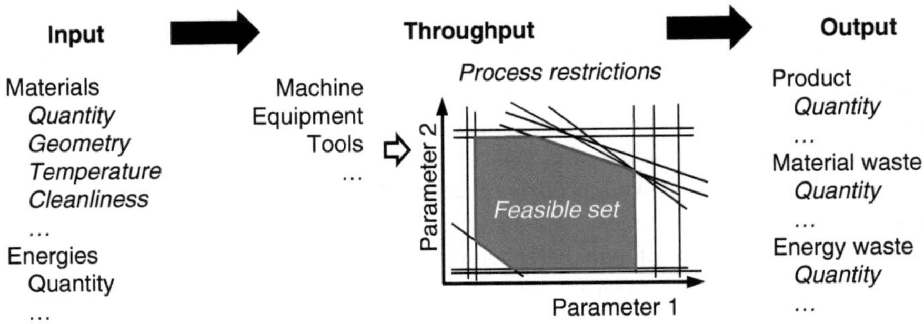

Fig. 7 ITO-model (generalized scheme)

4.2 Long-Term Appraisal: The Life Cycle Perspective

The MFCA based monetary comparison of process alternatives described in Sect. 4.1 refers to the analysis of a single time period. But, the modification of processes (the change of process parameters or even of the underlying technology) may have long-term effects which are not adequately considered yet. Additionally, these effects typically do not only concern the processes but also the characteristics of the final product. So, a meaningful decision making in the context of improving production processes' resource efficiency has to take into account the long-term effects on the processes and on the products. In this regard, the following aspects will be explored in more detail:

- process changes as investment decisions,
- monetary appraisal of life cycle effects (life cycle costing),
- ecologic appraisal of life cycle effects (life cycle assessment).

MFCA supports the design of a sustainable production by highlighting inefficient resource use on the process level and by appraising available alternatives. Different process (chain) alternatives often require new technologies and new manufacturing equipment, respectively. This implies that such alternatives have to be considered as objects of *investment decisions* (Nakajima 2011). In this context, the comparison of the alternatives' total cost of a single time period—as it is intended by the (Plan-)MFCA—may already be interpreted as a (simple) static method of investment appraisal.[15] But, its use requires specific conditions: the revenues of the compared alternatives must not differ and the time period used for the analysis has to be representative for the whole use phase of the appraised equipment. Since the underlying changes of the process chain alternatives may significantly affect the process conditions (e.g., the processing time and therewith,

[15] For details about the methods of investment appraisal mentioned in this chapter see e.g., Götze et al. (2008).

the production capacity) and/or the final products' characteristics (quality), the first requirement is already unrealistic in such cases. Regarding the representative time period, for instance a hypothetical average period can be used. But,—as it is criticized for every static method of investment appraisal—a single time period cannot reflect adequately the timing of costs and fluctuation of prices. So, the significance of the results is quite limited. As a result, it seems to be useful to assess the investments required for the realization of the process alternatives by using dynamic methods. They explicitly regard several time periods and the timing of costs or more precisely the 'time value of money' (Götze et al. 2008)—most familiar may be the net present value method. Additionally, the approaches of investment appraisal provide methodical support for the identification of the optimum date for changing from the existing to a chosen alternative process chain (for appropriate methods see, e.g. Götze et al. 2008). The required data is commonly derived from the (traditional) cost accounting. Here, the introduced Plan-MFCA (see Sect. 4.1) may take over this function and support the multi period assessment by forecasting quantities and costs of the single time periods. Its application promises very detailed and precise results.

The 'traditional' investment appraisal typically focuses on the phase of use of the corresponding object of analysis (machines, tools, etc.). A more sophisticated approach is the so called *life cycle costing* (LCC).[16] Based on methods of 'traditional' investment appraisal—commonly the net present value—it can be used for the evaluation of an investment object with a special focus on its life cycle phases. The underlying life cycle models can be designed more or less detailed. For reasons of simplification, in the following, the classification of creation, use, and end-of-life phase is applied.

Against the background of the analysis of process chains—intended in here—, relevant investment objects have to be discussed. The process chain and the processes are quite abstract elements of a structure model of the real production. To apply the life cycle approach to them, the processes are represented by the underlying manufacturing equipment and the process chain is considered as the sum of processes. So, an LCC analysis of an existing or alternative process chain is primarily an examination of the life cycle phase related costs of the single processes' manufacturing equipment. As mentioned above, the change of processes can also result in changes of the final product's characteristics which may affect the product's life cycle costs. In such cases, the decision about a concrete process (chain) alternative should regard (relevant) effects on the product's life cycle as well.

Since LCC is based on the methods of investment appraisal, the Plan-MFCA can provide methodical support, in particular, with respect to the assessment of the manufacturing equipment's use phase. The detailed forecasting of the process's total cost regarding modifications of parameters, equipment, etc., and of basic conditions (degree of utilization, prices of materials, etc.) enlarge the existing basis

[16] For a detailed description of LCC's methodology see e.g., Horngren et al. (2012).

of information for decision making derived from traditional cost accounting. However, LCC's examination 'from cradle to grave' requires much future related data which is partly highly uncertain.[17] So, the (possibly) impaired significance of the results has to be taken into account as well.

Beside economic considerations, the ecology of production and products comes to the fore. Negative environmental effects become more and more relevant for customers' buying decisions and are subject to governmental restrictions (environmental laws). So, a meaningful process design has to integrate long term monetary *and* environmental effects. An appropriate method for the analysis and appraisal of ecologic effects of industrial production and of products is *life cycle assessment* (LCA).[18] Based on the above mentioned life cycle models, the LCA procedure generally includes the steps of an inventory analysis, an impact assessment and an interpretation of the results. The primary task of the inventory analysis is the life cycle phase specific physical quantification of all relevant material and energy in- and output flows of the examined process or product. The subsequent impact assessment aims at the aggregated reporting of the corresponding environmental effects by using several impact indicators. The probably most popular and most frequently communicated of these indicators is the CO_2-equivalent which expresses the global warming potential.[19] The final interpretation examines the results against the background of the underlying goals, the chosen scope as well as other restrictions (quality of data, etc.) (ISO 2006).

Literature commonly emphasizes that MFCA and LCA analyses are based on a quite similar physical evaluation of material and energy flows. So, an existing MFCA flow quantity model could be an appropriate starting point for an additional LCA (see e.g., Möller 2010; ISO 2011; Viere et al. 2011) and may significantly reduce the effort for LCA's data acquisition. In the context of an integrated approach, there could be developed a shared quantity model regarding flows that are relevant for the economical as well as the ecological analysis. Furthermore, the above mentioned analysis of changed production processes' effects on the products' characteristics may provide helpful information for the ecologic appraisal of the products' use and end-of-life phase.

In summary, it can be remarked, that MFCA itself does not provide any life cycle data. Its focus is on the identification of inefficiencies basing on the analysis of a single time period. However, the suggested enhancements for quantity and cost forecasting as well as the integration of the technical analysis based on the ITO-model can support approaches and methods of monetary and ecological, life cycle wide appraisals of process (chain) alternatives. Moreover, the methodological nearness of MFCA to cost accounting and to LCA strongly suggests the

[17] For instance, future techniques of recycling (end-of-life phase) may be unknown today.

[18] An overview of the methodology is given by ISO (2006).

[19] Recent approaches like the carbon footprint and the carbon accounting (see e.g., Bowen and Wittneben 2011) refer to this indicator only and can be considered as simplified or carbon specific LCA approaches.

consideration of MFCA as a starting point of an integrated approach or at least a tool set for appraising the monetary and environmentally effects of process chain alternatives.

5 Aspects of Implementation

Against the background of international supply shortages and rising prices of particular materials and energy carriers as well as of governmental and companies individual targets for greenhouse gas reductions, the improvement of production's resource efficiency becomes a major management issue. As shown in this chapter, MFCA is an appropriate method supporting a sustainability management regarding economic as well as ecologic objectives (Nakajima 2006; Hyršlová et al. 2008; Nakajima 2011; Schmidt 2012). Therefore, Nakajima (2011) even demands the adoption of 'MFCA thinking' by the ultimately top management.

The realization of MFCA in practice requires an appropriate implementation strategy aiming at a meaningful ratio of efforts and benefits of the analysis. Here, particularly, questions of data acquisition, of analysis' level of detail, and of the possible integration of MFCA with other management areas have to be answered.

The quality and detail of flow data is obviously a main parameter of the significance of MFCA analysis' results. But, on the other hand, an extensive acquisition of material and energy flow data may raise unjustifiable high costs. So, firstly, the existing quantity and cost data of the company should be analyzed whether it can be used for MFCA (flow modeling). In this regard, it has to be noted that the few contributions dealing with practical experiences of material flow data acquisition strongly differ in their results. While Hyršlová et al. (2008) state that most of the required data is almost available from existing corporate databases, Schmid (2001) explicitly names the quality of the existing data as a significant weakness. Assuming that at least not every material and, in particular, energy flow quantity and cost data are available from existing information systems, the analysis should be enlarged stepwise (as mentioned in Sect. 3.2). On basis of the existing data and possible additional mobile measurements, first rough flow models may reveal the most inefficient areas of production. Subsequently, the analysis level of detail and the quality of flow data can be enhanced stepwise in order to identify the causes of material and energy inefficiencies on the level of single technical processes.

Furthermore, the importance of management involvement for a successful and permanent implementation of MFCA is emphasized in literature (e.g., ISO 2011). But, in practice, a 'management by MFCA is far from being generalized yet' (Nakajima 2011). Most companies only use it as a temporary special cost study.[20]

[20] It has to be noted, that at least some Japanese companies successfully utilize MFCA as a technique of daily operational management and benefit from this practice (Nakajima 2010).

But in doing so, they ignore the benefits MFCA can provide when applied in other management areas and in connection with neighbored approaches. Firstly, the integration of MFCA and an existing cost accounting system promises to enhance the information value of both (see Sect. 2.3). Secondly, due to the similar procedure, MFCA can be basis for LCA and vice versa (see Sect. 4.2). Thirdly, MFCA implementation practice often revealed significant inconsistencies of companies' existing materials management system (see Sect. 2.4). So, it seems to be useful to adopt the flow oriented more detailed and consistent[21] recording of material (and energy) movements for such systems. Fourthly, the new focus on losses—which also includes rejected parts and scrap components—may enhance the existing information for quality management. Moreover, process design and quality management can be combined in the context of continuous improvement processes (Nakajima 2008, 2010).

In conclusion, it has to be noted, that the implementation of MFCA commonly demands for a large database and therefore, the analysis may cause significant cost. But, by using the data and results of MFCA in other management areas and for other tasks, respectively, the overall benefits may dominate.

6 Summary

The need for a resource efficient design of production processes is undisputed. In this regard, economic as well as ecologic reasons argue for a more detailed examination of material and energy losses. The introduced method of MFCA pursues this strategy. It considers process chains and the single processes as systems of material and energy flows. Basing on a detailed analysis, it aims at highlighting inefficiencies and reporting the monetary effects. Due to the intention of reducing or even eliminating these inefficiencies, the subsequent redesign of production processes improves the economic *and* the ecologic performance of the whole process chain.

The presented refinements and enhancements of the MFCA methodology increase its usability in the context of process design. The integration of MFCA and the presented ITO-model supports the development of process alternatives and allows the ex-ante evaluation of their quantity and monetary effects. Additionally, the MFCA models can be basis for a long term (up to life cycle wide) economical or ecological appraisal of alternative production processes, including impacts on the products.

Regarding MFCA's practical use, an appropriate and customized implementation strategy is vitally important. Here, the proposed stepwise implementation can assist the overcoming of possible adoption barriers. In particular, the design of

[21] The high consistency arises in particular from the uses of in- and output balances for every quantity center (see Sect. 2.2).

data acquisition and the analysis' level of detail should respect the requirements of an integration of MFCA with the existing cost accounting system as well as with other management areas to tap the full benefits of a company-wide use of MFCA data and results.

Acknowledgments The authors like to thank the European Union (European Regional Development Fund) and Germany's Free State of Saxony for funding the Cluster of Excellence 'Energy-Efficient Product and Process Innovation in Production Engineering' (eniPROD®) as well as the Deutsche Forschungsgemeinschaft (DFG) for funding this work within the Collaboration Research Centre 'SFB 692'.

References

Bierer A, Götze U (2012) Energy cost accounting: conventional and flow-oriented approach accounting. J Compet 4(2):128–144. doi:10.7441/joc.2012.02.09

Bowen F, Wittneben B (2011) Carbon accounting: negotiating accuracy, consistency and certainty across organisational fields. Acc Audit Acc J 24(8):1022–1036. doi:10.1108/09513571111184742

Bundesumweltministerium (BUM), Umweltbundesamt (UBA) (eds) (2003) Leitfaden Betriebliches Umweltkostenmanagement, Berlin

Götze U, Helmberg C, Rünger G, Schubert A, Goller S, Krellner B, Lau A, Sygulla R (2010) Integrating energy flows in modeling manufacturing processes and process chains of powertrain components. In: Neugebauer R (ed) Proceedings of the 1st international colloquium of the cluster of excellence eniPROD. Wissenschaftliche Scripten, Auerbach, pp 409–437

Götze U, Northcott D, Schuster P (2008) Investment appraisal, methods and models. Springer, Berlin. doi: 10.1007/978-3-540-39969-8

Götze U, Schubert A, Bierer A, Goller S, Sygulla R (2012) Material und Energieflussanalyse—Methodik zur Optimierung von Prozessen und Prozessketten. In: Neugebauer R (ed) Proceedings of the ICMC 2012/2nd international colloquium of the cluster of excellence eniPROD. Wissenschaftliche Scripten, Auerbach, pp 99–128

Horngren CT, Datar SM, Rajan MV (2012) Cost accounting: a managerial emphasis, 14th edn. Prentice Hall, Boston

Hyršlová J, Bednaříkova M, Hájek M (2008) Material flow cost accounting, only a tool of environmental management or a tool for the optimization of corporate production processes? Sci Pap Univ Pardubice Ser A 14:131–145

ISO (2006) ISO 14040 Environmental management—life cycle assessment—principles and framework

ISO (2010) Draft International Standard ISO/DIS 14051 Environmental management—material flow cost accounting—general framework

ISO (2011) ISO 14051 Environmental management—material flow cost accounting—general framework

Jasch C (2009) Environmental and material flow cost accounting, principles and procedures. Springer, Dordrecht. doi: 10.1007/978-1-4020-9028-8

Kokubu K, Campos MKS, Furukawa Y, Tachikawa H (2009) Material flow cost accounting with ISO 14051. ISO Manag Sys 5(2):15–18

Kokubu K, Kitada H (2010) Conflicts and solutions between material flow cost accounting and conventional management thinking. In: Proceedings of the 6th APIRA. http://apira2010.econ.usyd.edu.au/conference_proceedings. Accessed Nov 12 2012

Loew T (2003) Environmental cost accounting, classifying and comparing selected approaches. In: Bennett M, Rikhardsson P, Schaltegger S (eds) Environmental management accounting—purpose and progress, Kluwer, Dordrecht pp 41–56

Möller A (2010) Material and energy flow-based cost accounting. Chem Eng Technol 33(4):567–572. doi:10.1002/ceat.200900491

Nakajima M (2006) The New Management Accounting Field Established by Material Flow Cost Accounting (MFCA). Kansai Univ Rev Bus Commer 8:1–22

Nakajima M (2008) The new development of material flow cost accounting (MFCA): MFCA analysis in power company and comparison between MFCA and TPM (Total Productive Maintenance). Kansai Univ Rev Bus Commer 10:57–86

Nakajima M (2009) Evolution of material flow cost accounting (MFCA): characteristics on development of MFCA companies and significance of relevance of MFCA. Kansai Univ Rev Bus Commer 11:27–46

Nakajima M (2010) Environmental management accounting for sustainable manufacturing, establishing management system of material flow cost accounting (MFCA). Kansai Univ Rev Bus Commer 12:41–58

Nakajima M (2011) Environmental management accounting for cleaner production: systemization of material flow cost accounting (MFCA) into corporate management system. Kansai Univ Rev Bus Commer 13:17–39

Prammer HK (2009) Integriertes Umweltkostenmanagement Bezugsrahmen und Konzeption für eine ökologisch nachhaltige Unternehmensführung. Habilitation University of Linz, Linz

Schmid U (2001) Schwachstelle Datenqualität: Flusskostenrechnung in der Praxis bei Sortimo International. Ökol Wirtsch 2001(6):16–17

Schmidt M (2012) Material flow cost accounting in der produzierenden Industrie. In: von Hauff M, Isenmann R, Müller-Christ G (eds) Industrial ecology management. Gabler, Wiesbaden, pp 241–255. doi:10.1007/978-3-8349-6638-4_15

Schubert A, Goller S, Sonntag D, Nestler A (2011) Implementation of energy-related aspects into model-based design of processes and process chains. In: Proceedings of the 2011 international symposium on assembly and manufacturing. doi:10.1109/ISAM.2011.5942344

Schweitzer M, Küpper H-U (2011) Systeme der Kosten- und Erlösrechnung, 10th edn. Vahlen, Munich

Statistisches Bundesamt (ed) (2011) Statistical yearbook 2011 for the Federal Republic of Germany including "International tables", Wiesbaden

Strobel M, Redmann C (2002) Flow cost accounting, an accounting approach based on the actual flows of materials. In: Bennett M, Bouma JJ, Wolters T (eds) Environmental management accounting: informational and institutional developments. Kluwer, Dordrecht, pp 67–82

Strobel M, Wagner F (1999) Flußkostenrechnung als Instrument des Materialflußmanagements. uwf 7(4):26–28

Sygulla R, Bierer A, Götze U (2011) Material flow cost accounting, proposals for improving the evaluation of monetary effects of resource saving process designs. In: Proceedings of the 44th CIRP conference on manufacturing systems. http://www.tu-chemnitz.de/wirtschaft/bwl3/DownloadAllgemeinOffen/44thCIRP_MFCA.pdf. Accessed 12 Nov 2012

Viehmann S (2012) Winterkorn will VW grüner machen. zeit online. www.zeit.de/auto/2012-03/volkswagen-umweltschutz. Accessed 12 Nov 2012

Viere T, Prox M, Möller A, Schmidt M et al (2011) Implications of material flow cost accounting for life cycle engineering. In: Hesselbach J, Herrmann C (eds) Globalized solutions for sustainability in manufacturing. Proceedings of the 18th CIRP international conference on life cycle engineering. Springer, Berlin pp 652–656. doi:10.1007/978-3-642-19692-8_113

Wagner B, Nakajima M, Prox M (2010) Materialflusskostenrechnung—die internationale Karriere einer Methode zu Identifikation von Ineffizienzen in Produktionssystemen. uwf 18:197–202. doi:10.1007/s00550-010-0189-1

Life Cycle Based Evaluation and Interpretation of Technology Chains in Manufacturing

A Methodology for Industrial Use

F. Klocke, B. Döbbeler, M. Binder, R. Schlosser and D. Lung

Abstract This chapter describes a methodology to gather, assess and interpret the ecological impact of technology chains within industrial manufacturing. The explained methodology leads to significant information about high consuming processes and important energy and material flows. Industrial companies cannot allocate the exact consumptions in the manufacturing processes. Especially costs and consumptions for media like compressed air or centrally provided lubricants are mostly distributed by means of the number of machines rather than by actual consumption figures. By utilising the presented methodology not only information about real consumptions, but furthermore ecological data can be generated for various purposes such as ecological product declarations and evaluation of alternative production chains. The methodology is exemplarily applied in two industrial case studies and results of these studies are shown in this chapter.

1 Introduction

Due to the global trends of this century rising world population, the striving for prosperity in the emerging nations and the globalization—adequate strategies to face associated problems need to be found. The framework of a sustainable

F. Klocke · B. Döbbeler (✉) · M. Binder · R. Schlosser · D. Lung
Laboratory of Machine Tools and Production Engineering (WZL) of RWTH, Aachen University, Steinbachstr. 19, 52074 Aachen, Germany
e-mail: b.doebbeler@wzl.rwth-aachen.de

F. Klocke
e-mail: f.klocke@wzl.rwth-aachen.de

M. Binder
e-mail: m.binder@wzl.rwth-aachen.de

D. Lung
e-mail: d.lung@wzl.rwth-aachen.de

E. Henriques et al. (eds.), *Technology and Manufacturing Process Selection*, Springer Series in Advanced Manufacturing, DOI: 10.1007/978-1-4471-5544-7_7,

development aims at meeting the human needs nowadays while preserving the needs of next generations. A sustainable development is based on the equal consideration of three aims: the ecological, economic and social dimension (Hauff 1987).

In order to assess the environmental performance of a product, the Life Cycle Assessment (LCA) has become a widely accepted methodology (Guinee and Heijungs 1993, p. 1925; Reich-Weiser et al. 2010, p. 2). Within the LCA, a systematic approach of gathering the required data and assessing the environmental impacts on various environmental issues of concern is guaranteed. The LCA does not only focus on the usage but embraces the assessment of the environmental impact during the production, usage, transportation, maintenance and disposal in line with the term life cycle (Guinee and Heijungs 1993, p. 1925).

This chapter focuses on the assessment and evaluation of technology chains in industrial manufacturing and thus primarily on the production phase of the life cycle. At first, the procedure of a LCA according to standards is described. Following, its application to manufacturing systems is discussed. Specifically for the manufacturing sector, some suitable evaluation schemes are presented subsequently, followed by two industrial case studies of the automotive sector. Concluding, the article is summarized and an outlook is given.

2 Life Cycle Assessment Within Industrial Manufacturing

In this section the application of the Life Cycle Assessment methodology to industrial manufacturing is described. The LCA is documented and standardized within DIN EN ISO 14040/14044. The LCA approach is relative. It refers to a functional unit, assessing technology chains to a manufactured product. All the material and energy flows and the effects in various impact categories are linked to the functional unit. The procedure, illustrated in Fig. 1, is subdivided into four main steps (DIN EN 14040 2009; DIN EN 14044 2010):

1. Goal and scope definition
2. Inventory analysis
3. Impact assessment
4. Interpretation.

According to this procedure, this section is structured into four parts. First, the appropriate definition of the balance shell as a critical prerequisite for a distinctive analysis is discussed. This part covers the first step of the LCA, the goal and scope definition. Following, the data acquisition including the adequate extent and level of detail is discussed. The third part focuses on the evaluation of the gathered data. The LCA covers impact categories which assess the potential impact of the considered object representing environmental issues of interest like acidification or global warming. In the fourth and last step the results are discussed as a basis for conclusions.

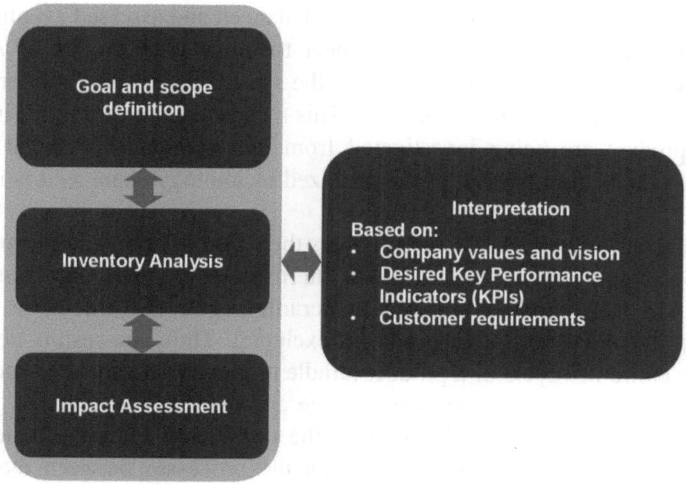

Fig. 1 Methodology of life cycle assessment according to DIN 14040/14044

For every part of the four step procedure the specifications of the standard are summarized and the application to industrial manufacturing is described. Results of Life Cycle Assessments of technology chains are presented as industrial case studies in the following section.

2.1 Definition of the Goal and Balance Shell

The goal definition of the LCA embraces the definition of the assessed object and thus the functional unit to which the entire analysis is related. As the LCA is a relative procedure, the impacts have to be related to a specified product or function. In order to achieve comparability of studies, the functional unit is a quantitative measure of the functions that the good or the service provides (Finnveden et al. 2009, p. 2).

Furthermore the impact category or categories to be used have to be determined. The impact category specifies the environmental category of concern such as climate change or the scarcity of resources. Completing the goal of the LCA, the reasons for conducting the LCA and the addressees have to be named. The entire further approach has to be matched and constantly questioned in order to reach the defined goal. Thus the step of goal definition is of crucial importance.

The scope of a LCA including the boundaries and the level of detail depends on the aim and the addressees of the study. Both depth and breadth of the LCA can vary extensively with the intention (DIN EN ISO 14044, p. 6).

For the assessment of technology chains in manufacturing the definition of the balance shell comprehends the constitution of the process steps to be investigated, the types of energy and material flows to be balanced and assessed eventually. The level of detail needs to be chosen subject to the aim of the LCA. Assessing manufacturing systems or process chains, the system boundaries are normally set according to a cradle to gate assessment. This means that the ecological impacts of a given product are being investigated from the creation of material or semi-finished products until the end of the analyzed technology chain as it is illustrated in Fig. 2.

The analysis of technology chains covers the phase of raw materials production by the consideration of the materials and intermediate products and focuses on the investigation of the production phase. In a cradle to gate assessment, the phases of utilization and recycling or disposal are excluded. Thus, the results do not represent the entire life cycle of a product (cradle to grave) but can be linked with the results of an assessment of the phases usage and disposal/recycling.

An alternative balance shell excludes the consideration of the materials and intermediate products and thus focuses on the processes between the gates of delivery. According to that scope this variant is called gate-to-gate assessment.

Technology chains in manufacturing consist of several process steps. The process steps, illustrated in Fig. 2, cause material and energy consumptions as well as emissions to air, water and soil. The attribution to the produced functional unit is made on the basis of the functional unit's proportional share of the consumptions and emissions from each process (Finnveden et al. 2009, p. 2).

The scope of the study needs to be documented and the assumptions, restrictions and system boundaries need to be transparent. The defined settings in this

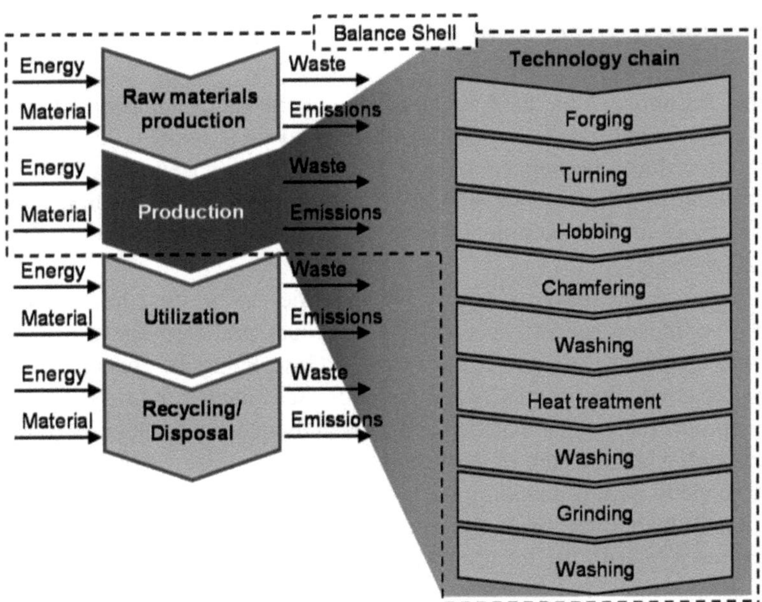

Fig. 2 The balance shell in LCA of manufacturing systems (based on Schiefer et al. 2009)

first step of the procedure are not immutable proceeding with the inventory analysis, the impact assessment and interpretation. If the further procedure reveals the need to adjust the scope of the study, the definition of the first step may be changed in iteration.

2.2 Measurements Within the Balance Shell

In the first step of the LCA, the initial goal and scope of the study has been defined. Based on the definition, the energy and material flows of the system have to be quantified. The quantification of the so called inventory data is carried out in the second step of the LCA, the life cycle inventory analysis (LCI).

On the one hand, the results of a LCA depend on the quality of the acquired data and thus on the completeness, representativeness and the precision of the data. On the other hand, the data acquisition is a major effort (time and labor) (Finnveden et al. 2009, p. 6) of the LCA procedure and therefore should not be conducted more detailed then required. The process of data collection is not necessarily limited to measurements. Calculation or estimations are explicitly allowed by the standard as suitable methods of quantifying elementary input and output streams of a technology chain (DIN EN 14044 2010, p. 23). The methodology of the data quantification needs to be transparent for every data source and thus thoroughly documented. During the inventory analysis, new requirements and restrictions may be identified while learning more about the analyzed system. The data collection process is iterative.

The data collection within the systems boundaries embraces energy inputs, raw material inputs, ancillary inputs, products, co-products, waste as well as emissions to air, water and soil (DIN EN 14040 2009, p. 26). The in- and outputs need to be counted, estimated, calculated or measured by sensors. Independent of the form of data acquisition, the input and output streams need to be related to the throughput of the functional unit. For example, if a stream is measured in a volume flow (m^3/h), it has to be linked with the process cycle-time (functional unit/h) in order to obtain feasible information on work piece detail level (m^3/functional unit).

Reducing the effort of assessing the LCI data, there are Life Cycle Inventory Databases available on the market that contain data sets for a variety of industries and impact categories. The databases cover inventory data on products and basic services that are required in every Life Cycle Assessment, such as raw materials, electricity generation, transport processes and waste services (Finnveden et al. 2009, p. 6). Furthermore there is software available on the market that helps users to conduct LCA. These tools like Umberto, GaBi and Ecoinvent contain data from previous studies and thus reduce the effort by providing standard features like material creation for instance (Reich-Weiser et al. 2010, p. 2). As the analyzed system may be complex, flow diagrams help to give a proper overview about the inventory data. Sankey diagrams for instance offer a good opportunity to visualize the analyzed system and help to identify further need of data acquisition.

The energy and material flows in technology chains are complex and interdependent (Reich-Weiser et al. 2010, p. 1 f.). For the LCI the collection of data within the system boundaries is required, which is usually not yet available in manufacturing companies.

The relevant input and output streams of a manufacturing system depend on the assessed object and the designated level of detail as well as the impact category of interest. If the output of CO_2-equivalent is the goal of the study, the negligence of detergents in a washing process might be tolerable. If the acidification potential of the technology chain is of interest, the impact of detergent usage might be of high relevance. Prior to the acquisition of data an estimation of the relevance of material and energy streams is a reasonable step in order to reduce the effort in acquiring data. Common data collections comprehend energy inputs in the form of electrical energy enabling the machinery and equipment or natural gas and district heating used for heat treatments. In addition, raw material inputs, intermediate products as well as recycled material (e.g. degraded material, chips) need to be considered. Furthermore manufacturing supplies like water, tools, oil, detergents, paint and glue on the input side as well as waste, exhaust, effluents and emissions on the output side play a role in assessing manufacturing systems. Determining the flows of relevance is a part of the LCI. An example for a process chain in manufacturing and relevant, exemplary input and output streams are displayed in Fig. 3. The flows, illustrated as arrows, need to be quantified by input and output per functional unit, e.g. Wh per manufactured piece. As mentioned before, the selection and the level of detail of the gathered data depend on the aim and scope of the study.

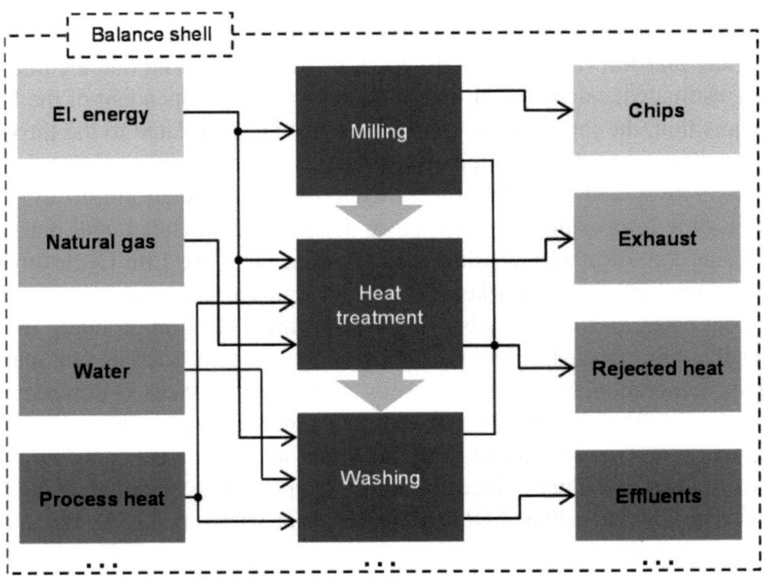

Fig. 3 The inventory analysis of an exemplary technology chain

In manufacturing systems central units cause material and energy flows that are not directly accountable to a certain product and thus the functional unit. Examples for central units in manufacturing systems are centralized lubricoolant facilities or compressors for pressurized air which supply the entire site or various divisions. In order to attribute the caused material and energy flows to the functional unit, the energy and material flows have to be captured and related to the provided product (e.g. lubricoolant, pressurized air). The energy and material flows caused by the central unit are attributed to the processes and eventually the functional unit by the proportion of the use of the provided product, see Fig. 4. The effect of the inputs and outputs caused by the central unit is related to the process step proportionately to the used share of the provided flow. Hence, the central unit has to be considered during the LCI.

Furthermore, the technical facility equipment causes energy and material flows, predominantly energy expenditure, that cannot be attributed to the functional unit on cause basis. Lighting or air conditioning of the plant can be named exemplarily. The energy and material flows can be related to the processes by the required floor space of the technical equipment used for the specific process steps.

In order to reduce the effort to conduct the inventory analysis, life cycle inventory databases are established in various industries. Within manufacturing industries the CO_2PE!-initiative collects data to foster availability of required data and thus, reduce the constraints of conducting a LCA (Kellens et al. 2012; Reich-Weiser et al. 2010, p. 4). Investigating manufacturing systems, a combined use of databases and own data acquisitions seems to be sensible. However, as technology chains are complex and may differ although the identical product is manufactured, the use of aggregated data needs to be questioned and tailored inventories for specific technologies should be preferred (Finnveden et al. 2009, p. 7).

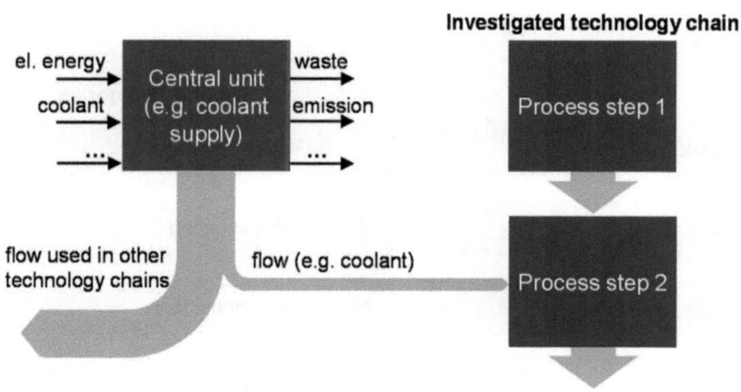

Fig. 4 The attribution of central units to the investigated technology chain

2.3 Estimation and Assessment of the Gathered Data

The result of the inventory analysis is the quantified balance of the entire system within its boundaries defined in the first step. The information does not contain any assessment so far. The third step of the LCA is the impact assessment (LCIA) which evaluates the environmental significance of a certain flow against the background of an impact category or several categories. Thus the LCA perspective changes from an exclusively quantitative to a qualitative perspective. The inventory data is linked with specific environmental impact categories determined in the first step of the LCA (see Fig. 5).

Impact categories are a set of scientifically elaborated factors. They quantify the theoretical impact of every material and energy flow gathered within the inventory analysis. The factors attribute different types of data described by different units (e.g. kWh/functional unit, kg material/functional unit, Nm^3/functional unit) to one single impact category in order to achieve comparability. The sources of the chosen impact categories and parameters need to be referenced. A set of widely accepted impact categories and data sets already exists in the literature, of which some are listed in Table 1. A survey of the existing impact categories is given in ISO/TR 14047. The majority of existing impact categories can be attributed to three main "areas of protection": human health, natural environment and natural resources (Finnveden et al. 2009, p. 8).

Although for most LCA studies existing impact categories are selected, it is possible to define new impact categories and data sets in order to fulfill the defined goal and scope of the LCA (DIN EN ISO 14044 2010, p. 34 f). The correlation between the inventory data and the impact, mathematically described by a set of factors, is deduced by characterization models. These models describe the relation between the LCI results and the impact categories and thus need to be referenced and documented for a transparent assessment. For instance, the gases contributing to the greenhouse effect (CO_2, CH_4, N_2O, CF_4, C_2F_6 etc.) are assessed by the degree of

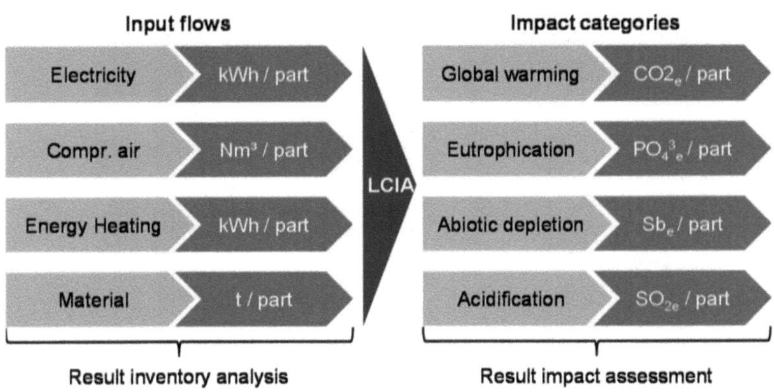

Fig. 5 Assessing the quantitative information of the inventory analysis in the LCIA

Table 1 Examples for ecological impact categories (Guinée and Heijungs 2002, p. 56 ff.)

Impact category	Indicator
Global warming	Global warming potential (GWP) [kg CO_2-Equivalents]
Eutrophication	Eutrophication potential (EP) [kg PO_4^{3-}-Equivalents]
Acidification	Acidification potential (AP) [kg SO_2-Equivalents]
Resources (abiotic)	Abiotic depletion potential (ADP) [kg Sb-Equivalents]
Stratospheric ozone layer	Ozone depletion potential (ODP) [kg CFC-11-Equivalents]
Human toxicity	Human-toxicity potential (HTP) [kg p-DCB-Equivalents]

affecting the atmosphere's ability to absorb infrared radiation and normalized to the effect of CO_2 (unit [kg CO_2-Equivalents]) (Finnveden et al. 2009, p. 8).

The LCIA simplifies the enormous complexity of a nearly arbitrary amount of flows into a couple of operating figures. Hence, the LCIA is a very powerful step to ease the assessment on the one hand but also implies the risk to conceal the reality and to derive wrong conclusions.

Within manufacturing industries many companies have committed themselves to reduce their CO_2-emissions. To evaluate the initial situation, the savings and the drivers of emission, the assessment of technology chains using the Global Warming Potential is an option. Although global warming is in the focus of the public interest, the global warming potential (GWP) expressed in kilogram of carbon dioxide equivalents does not serve as a representative indicator for other types of impacts and hence the overall environmental impact (Laurent et al. 2010).

2.4 Interpretation

The interpretation of the results should meet the definition of the goal and the purpose of the study. Possible foci of the interpretation may be the support of decisions, the identification of the levers to reduce the environmental impact of a product or the preparation of a report. The interpreter should be aware of the fact that the results of a LCA are not actual effects but potential effects that are modeled.

In line with the iterative approach of a LCA the completeness, sensitivity and consistency of the results as well as the decisions made in the prior steps need to be critically reflected. If an adjustment of the assumptions, definitions and decisions made in the prior steps seems inevitable, the approach has to be adapted.

Interpreting the results of the assessment of technology chains the environmental performance of the usage phase and disposal needs to be taken into account. Within the assessment of manufacturing, it is typical that sophisticated manufacturing technologies, that worsen the environmental performance of the production phase through additional efforts, help to reduce the impact on the environment in the usage and disposal phase by prolonging the life, reducing weight or enhancing modular use for instance.

When the LCA is methodically completed, the reporting about the results is an integral part of the study (DIN EN ISO 14040 2009, p. 32). Depending on the addressee, various aspects may be highlighted, but the reporting on a LCA needs to embrace the relationship with the LCI results, a description of the data quality, the category endpoints to be protected, the selected impact categories and the characterization models, the impact factors and the profile of the indicator results (DIN EN ISO 14040 2009, p. 32). The more information is published, the more transparency is created and thus credibility may be achieved.

Interpreting the results of a LCA conducted in the field of manufacturing systems, the key drivers of environmental performance may be identified. Assessing complex technology chains the impact can be attributed to the process steps in order to provide information about the environmental relevance. The transparent allocation of the potential effects may already foster measures of improvement. Furthermore the results of the inventory analysis help to identify real production costs as indirect costs are normally not attributed to products on cause basis.

3 Suitable Evaluation Schemes for Manufacturing

In order to evaluate the ecological impact of products within the manufacturing phase, it is necessary not only to assess the consumptions during the whole life cycle, but also to use a suitable evaluation scheme. This evaluation scheme needs to build up on the life cycle assessment and its results. In the literature a few schemes have been discussed and used within life cycle based product and service evaluations:

- Embodied Energy (cumulative energy consumption)
- Eco-Indicator 99
- Ecological Footprint.

There are several other possibilities to assess the ecological impact of manufacturing processes, but these mentioned methodologies are the most suitable within manufacturing. The evaluation will be described and discussed below.

3.1 Cumulative Energy Demand (CED)

A clear methodology for the embodied energy of products is provided by the cumulative energy demand. The cumulative energy demand represents the sum of all primary energy consumption which is necessary to create (CED_M), use (CED_U) and recycle or dispose (CED_{RD}) materials used within the balance shell (VDI 4600):

$$CED = CED_M + CED_U + CED_{RD} \quad (1)$$

The cumulative energy demand during CED_M includes all efforts for extraction, processing and transport of materials during manufacturing. In this energy share also the consumptions for consumables and their recycling or disposal are included in order to consider all consumptions during this phase. Consequently, the primary energy demand during the usage and the recycling phase, including consumables, electrical energy and maintenance, is summarized in CED_U and CED_{RD}, respectively. These efforts include all cumulative energy demands of the used consumables, intermediate products as well as transport consumptions.

Since Life Cycle Inventory (LCI) databases usually include primary energy consumptions of many materials, processes and services, the assembly of the cumulative energy demand is easily determined within the life cycle assessment. During this assessment all cumulative energy demands of single materials and processes are aggregated up to the considered final product, process chain or service. Hence, not every consumable or material within the balance shell has CED_M, CED_U and CED_{RD} at the same time. For example, consumables during the material creation will be considered within the CED_M of the final product and therefore have no CED_U. This differentiation is necessary in order to prevent the assessment from double counting of consumptions (VDI 4600).

Naturally, the use of a cumulative energy demand has limits regarding the accessibility of CED data on lower levels of evaluation. For instance, not every screw in a machine can be assessed and considered. Therefore qualified assumptions regarding materials and processes with low influences due to small energy and material consumptions are necessary. With the growing number of life cycle analyses and product declarations in respect to environmental data, the estimation of inaccuracy becomes manageable.

Recent studies show that the cumulative energy demand using primary energy or fossil energy consumptions leads to outcomes regarding the derived measures which are comparable to extensive impact assessment methodologies such as the ecological footprint or the eco-indicator 99. Still, there are many single exceptions that do not lead to comparable results, but for various reasons. Due to different calculation methodologies each procedure has main impact factors that influence the outcome more than in other methodologies. Nevertheless, the cumulative energy demand represents a first and important step towards general evaluation of ecological impacts (Huijbregts et al. 2006, 2010).

The CED is a useful tool for assessing most ecological impacts of manufacturing. It is no methodology for assessing specific ecological impact categories. Nevertheless, it provides an evaluation criterion which can be tracked, documented and used for improvement measures. It also provides the possibility to combine data from different companies which contributes to the life cycle perspective.

The cumulative energy demand will exemplarily be highlighted in one of the following case studies.

3.2 Eco-Indicator 99

The eco-indicator 99 is a methodology to assess the ecological impact of products and services. On different levels of aggregation it is able to provide ecological data. A single digit evaluation for simple use as well as impact categories are possible outcomes depending on necessity. Three steps have to be performed: Inventory analysis, damage prediction and weighting (Fig. 6) (Goedkoop 2000).

During the *inventory analysis* all relevant emissions, resource extractions and land-use of the life cycle processes have to be determined. This analysis is a common component of a LCA. The second step, the *damage model*, aims at calculating the damage of the inventory analysis results to resources, human health and the ecosystem quality. The last step, that is only necessary to present a single-score, the eco-indicator, is the *weighting of the damage categories*. The weighting is the most critical and controversial step within the methodology, because of the weighting factors that cannot be determined absolutely. For each category a special unit is used to describe the damage. For damage to human health this is DALY (Disability Adjusted Life Years), which is an indicator for the number of years of disabled living. The damage to the ecosystems can be described by the loss of species over an area and time (% m^2 year). Finally, the damage to resources is measured in surplus energy needed for future extractions of minerals and fuels (Goedkoop 2000).

The eco-indicator 99 represents one of the few methodologies which aim at quantifying the inflicted ecological damage. It targets not only on providing the results of impact categories, but on calculating a unique indicator derived from the impact category results. This increases the usability due to simplicity, but also raises the possibility of false conclusions.

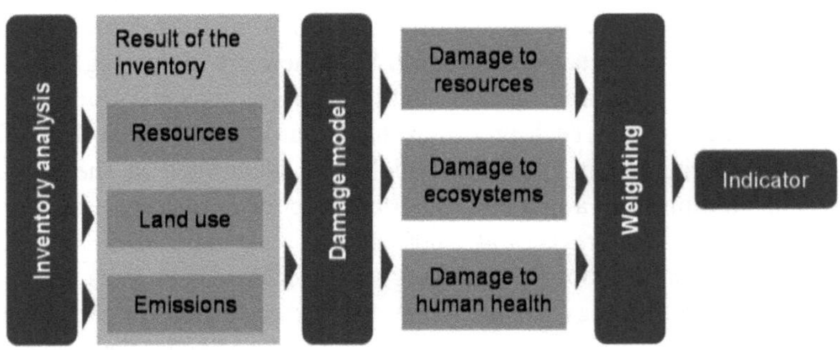

Fig. 6 General procedure of eco-indicator calculations (based on Goedkoop 2000)

3.3 Carbon Footprint

Not only in the automotive industry, but also in legislative regulations the carbon footprint is common. Aims to reduce the carbon dioxide emissions of companies, consumers and countries are widely spread and anchored in directives and visions. Due to the high impact of energy resources on the carbon dioxide equivalent emissions (CO_{2e}), this methodology of impact assessment creates results similar to the cumulative energy demand. Still, in various industries such as chemicals or pharmaceutics the use of certain materials leads to significantly different results than CED procedures.

The carbon footprint typically considers the following greenhouse gases: CO_2, CH_4, N_2O, SF_6, hydrofluorocarbons (HFC) and perfluorocarbons (PFC) analogue to the impact category global warming (Laurent et al. 2010). Yet, there is an ongoing discussion whether to include other greenhouse gases into the carbon footprint and about a clear definition. This definition does not exist in detail, but the typical use of the term carbon footprint includes the given gases as well as CO_2 itself (Wiedmann and Minx 2007).

The representativeness of the carbon footprint especially regarding the human toxicity impacts is controversially discussed. Recent results show that the carbon footprint is heavily depending on the energy production technologies. Since renewable energy production such as wind or hydro emit less greenhouse gases than usual fossil energy generation, this dependency is logical. Not only the carbon footprint, but also other impact calculations are influenced basically by the energy sources considered. Energy is usually the basis for many processes ranging from material creation, manufacturing to the recycling (Laurent et al. 2010).

Finally, the carbon footprint is a simple way to assess the global warming impact and also might be used for ecological impact estimation. For more appropriate and specific impact evaluations a whole life cycle assessment which addresses various impact categories is necessary. The main disadvantage of a whole LCA is still the difficult and unique evaluation and the bigger number of indicators which leads to an evaluation dilemma. The continuous life cycle assessment within manufacturing is not yet implemented, but the orientation of companies towards ecological evaluation procedures already began. Therefore, it is a good starting point to address the actual legislation and European focus on the carbon emission equivalents in manufacturing. With this first approach it is possible to provide a carbon footprint for products. If and when a broader and more extensive life cycle-based approach will be used cannot be foreseen, but the number of methodologies is continuously increasing.

4 Industrial Case Studies

In order to show the importance and applicability of the shown evaluation methodologies as well as the life cycle assessment, two different case studies will be presented in this section. Both evaluations are part of public funded projects in Germany and show that the ecological impact is an additional tool for evaluation the performance of technology chains.

4.1 Life Cycle Assessment in the Automotive Sector

Within the project BEAT (Evaluating energy efficiency of alternative processes and technology chains) funded by the BMBF (Federal Ministry of Education and Research), the technology chains of two reference products have been investigated and assessed in detail (Schlosser et al. 2011; ICMC 2010). The consortium of the project consists of the industrial partners Robert Bosch GmbH and Daimler AG, the SME consultant Effizienz Agentur NRW, the software company and consultant PE International AG and the research institutes IEM (institute for electrical machines) and WZL (Laboratory for Machine Tools and Production Engineering) of RWTH Aachen University.

During the project the technology chains of two demonstrator products for the automotive sector have been analyzed. On the one hand a gear-wheel, produced for the A-Class of Daimler has been analyzed. It is made of the case-hardened steel 20MoCr4 (Mat. 1.7321) at the plant in Rastatt, Germany. On the other hand a diesel injection nozzle produced by Robert Bosch GmbH has been investigated. The nozzle is manufactured of low-alloy steel 100Cr6 (Mat. 1.3505) in the Bosch plant in Bamberg, Germany. The products as well as the technology chains are shown in Fig. 7.

Within the framework of this research project, a comprehensive analysis of the technology chains was conducted. Extensive measurements on the shop floor and process level have been carried out and a level of detail in the LCI has been acquired, that extends the accuracy which needs to be achieved in order to conduct a sound LCA. Within the data acquisition electrical energy, water, lubricoolants, detergents, technical heat and coolants as well as natural gas and pressurized air have been measured in central units for providing usable media as well as in the processes themselves. Also the lighting, ventilation, heating and handling processes of waste materials such as metal scrap and waste water in the considered part of the plant have been assessed. For all the processes the consumptions for one work piece have been determined and included in the modeling afterwards. Among other goals, the aim of the research project was to identify the input and output flows that are highly significant and those being negligible. Detailed analyses regarding the necessary electrical energy in cutting processes have been carried out in order to deduct the influence of the cutting parameters, such as cutting velocity, feed and depth of cut (Schlosser et al. 2011).

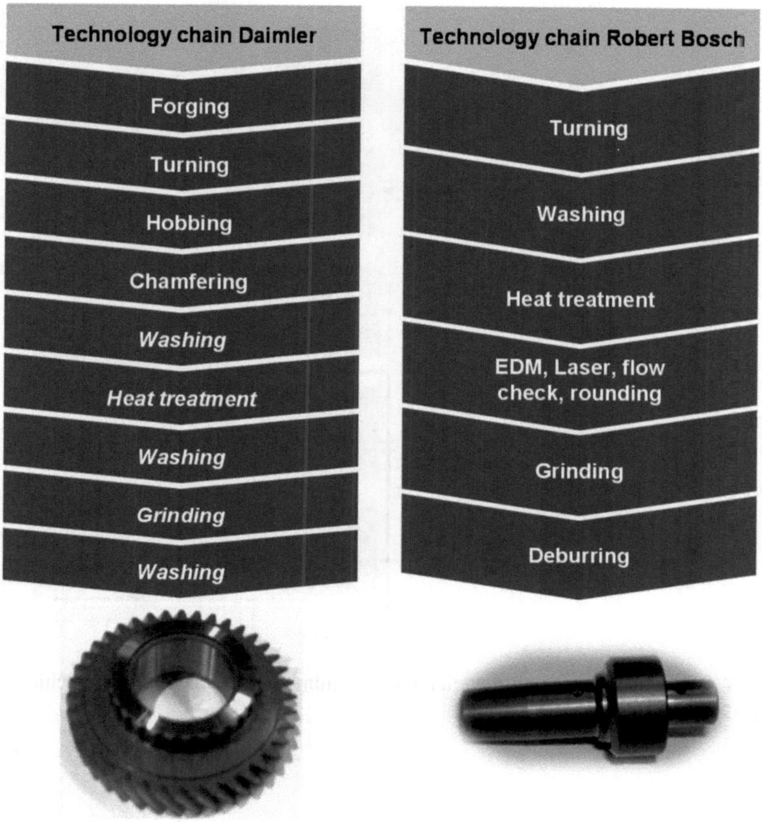

Fig. 7 The technology chains of the demonstrator products within the automotive industry (Schlosser et al. 2011, p. 2 f.)

Figure 8 illustrates an excerpt of the interdependencies and complexity of the energy and material flows which were quantified within the first process steps of the technology chain of the Daimler work piece (see Fig. 7). Due to the fact that the acquisition of the data is a major effort of the study, the level of detail needs to be adjusted to the defined aim of the LCA.

The technology chains have been modeled in the software GaBi. Using the connected databases and the inventory data gathered within the project, a unique data set was created. The model enables to analyze the technology chains against the background of various impact categories.

Figure 9 displays the share of the average electrical power consumption as well as the global warming potential per process step of the considered technology chain of the injection nozzle by Bosch. One-third of the process load of the entire technology chain can be attributed to the furnace of the heat treatment, as expected a high proportion. Due to the relative approach of the LCA, the environmental impact on global warming per functional unit—significantly determined by the

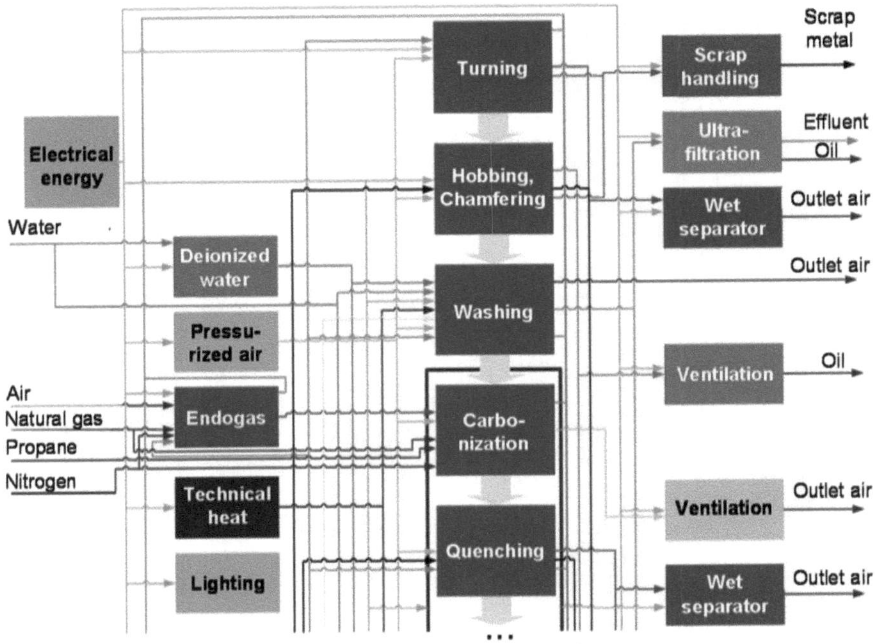

Fig. 8 Overview of energy and material flows within the technology chain (Schlosser et al. 2011, p. 3)

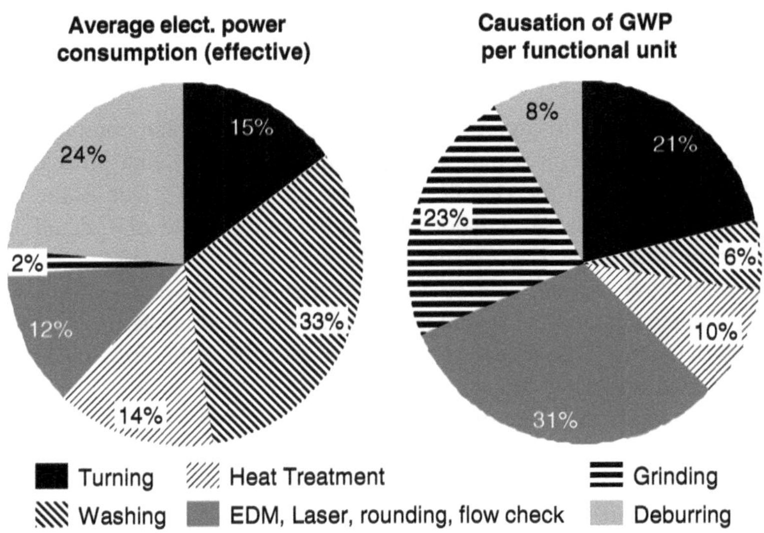

Fig. 9 Comparison of the average power consumption with the contribution to GWP of every process step in the technology chain

consumption of electrical energy—can be identified as comparatively low. Even though the furnace has a high process load, the impact per unit is small in consequence of the big batch size that can be processed simultaneously. Hence, the environmental impacts, caused by the process can be allocated to many products. These degression effects do not occur in process steps like turning or grinding. The economies of scale affect the relative impacts negatively in an analog way. Therefore the impact on GWP in relation to the entire technology chain of turning (21 %) and grinding (23 %) are high. The same argument can be applied to the step of EDM, Laser, rounding and flow check (31 %) as the throughput is low.

The described example illustrates the benefit of the relative LCA approach. The analysis of the results reveals the true drivers of a certain impact category. In this specific case, the significance of process steps that do not attract attention by their high installed electrical power is proven. For example the washing process has a very high installed electrical power in comparison to other processes, but in respect to the global warming potential the other processes dominate. Nevertheless, especially in SME-environments, the installed electrical power is a criterion that is often addressed in order to identify the levers of improving energy efficiency and eventually the environmental impact of a production system. An additional approach is provided by the results of the LCA that can be integrated into existing evaluation methods of manufacturing process chains. This will be the biggest benefit of life cycle assessment approaches. Especially changes within the technology chains in respect to process substitutions, removal or additional process steps can be assessed and estimated.

4.2 Life Cycle of a Forming Tool

Within the sub-project 3.1.1 of the Innovation Alliance Green Carbody Technologies (InnoCat), which is funded by the BMBF (Federal Ministry of Education and Research), the life cycle of a deep-drawing forming tool used for car body parts such as doors has been assessed. The InnoCat project concentrates on energy and resource efficiency in major processes within the car manufacturing aligned to the following topics:

- Topic 1: Planning of low-energy production
- Topic 2: Performance of the press plant
- Topic 3: Resource efficient tool and die industry
- Topic 4: Energy and resource efficient body construction
- Topic 5: Energy efficient painting.

The life cycle has been assessed at three partners, Römheld & Moelle GmbH, AUDI AG and Volkswagen AG, from the foundry, the machining and build up to the usage and maintenance, see Fig. 10.

For the recycling phase assumptions have been made to consider this phase according to the others. Phases of manufacturing at the single partners are illustrated with varying colors indicated below.

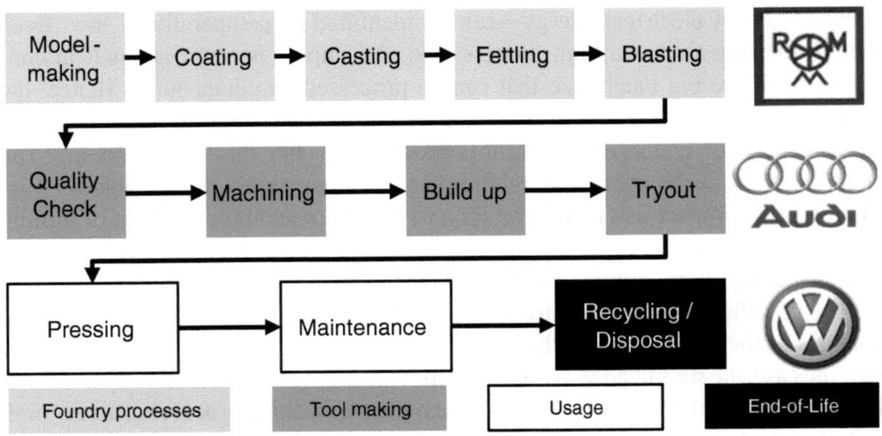

Fig. 10 Overview of the assessed technology of the forming tool

In this project it was important to define the functional unit in order to use the results afterwards in the respective branches of the partners. Whereas Audi and Volkswagen build cars, Römheld & Moelle does not only supply the automotive industry. Within the foundries basically all energy and material consumptions are associated to the weight of usable cast. Therefore all expenses for creating one usable ton of cast material have been assessed and used as the basis for further investigations and evaluations. The link between the foundry processes and the tool making consequently could be created by the weight of the tool parts used at Audi: matrices, die, upper part, lower part and blank holders. The functional unit at tool making and usage therefore was one forming tool. During the usage phase also the maintenance for keeping the forming tool running throughout the life cycle has been considered. The whole life cycle was determined by the number of car body parts which are produced by one forming tool. In this case a production number of one car body part for 1, 500, 000 cars has been assumed. This leads to a usage phase of 6 years (Klocke et al. 2012).

During the entire production phase extensive measurements, calculations and estimations in line with DIN ISO 14040/44 have been performed. Energy and material consumptions considering the process level as well as plant installations such as air conditioning, heating and lighting were considered. For the material consumption a similar approach as in the BEAT project was used. For each material and energy form extensive datasets from the GaBi Software were used for the assessment and evaluation on the work piece level (Klocke et al. 2012).

From this extensive assessment approach some interesting results could be derived. The approach included the cumulative energy demand on the basis of primary energy (see previous sections). On this approach a holistic evaluation was performed. During the production phase of the forming tool (see Fig. 11) the majority of the energy is consumed during the forming, melting and blasting. Within this process step the melting energy as well as the material itself is considered. Due to the high influence of material creation this finding is logical.

A very similar result can be derived from the maintenance and tryout share. Due to sample parts that are pressed during these process steps, the energy consumption is significantly higher than during model making, coating, fettling or build up. During the machining step a large amount of electrical energy is consumed per part. This is due to high base loads and long production times during machining. Furthermore, no scale effects caused by a high number of produced parts at the same time can be achieved.

One of the most interesting findings of the project was based on the life cycle assessment. It was obvious that energy and material consumptions during the usage phase of the forming tool were significantly dominant, see Fig. 11.

The overwhelming influence of the usage phase can be explained by the final parts that were produced. Within the primary energy consumption during usage of the forming tool also the energy difference between coil and waste material has been accounted for. Although this methodology of evaluation mixes the work piece and the forming tool life cycle, it provides a good overview of both life cycles. Furthermore it highlights the possibility to include measures that will decrease the energy consumption during usage, even though they will result in higher expenses during the manufacturing of the forming tool. A good way to achieve both targets is to decrease the forming tool weight. During manufacturing

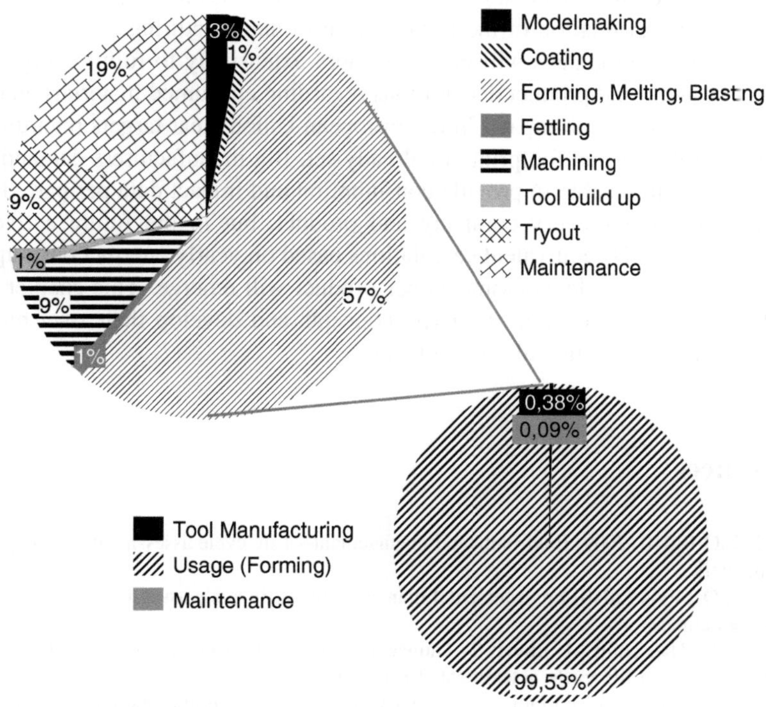

Fig. 11 Shares of primary energy consumption during manufacturing of the forming tool (Klocke et al. 2012)

this will lead to lower material consumptions on the one hand, on the other hand higher efforts are needed because of the complexity of the forming tool. During usage the power necessary to move the tool will be reduced (Klocke et al. 2012).

5 Summary and Outlook

As shown in this chapter there are several possibilities to determine the ecological impact of manufacturing systems, especially technology chains. The basic requirement always is a life cycle based assessment of the industrial environment. Measurements, estimations and calculations are necessary prerequisites for further evaluations. These evaluations are usually based on the life cycle assessment data which is provided by suitable software solutions. This life cycle assessment data has to be interpreted. This usually is a crucial step in industrial companies. The number of different impact categories leads to the difficulty of deriving clear and understandable conclusions of the presented data. Therefore three different methodologies have been introduced in this chapter: the cumulative energy demand, the eco-indicator and the carbon footprint. A suitable methodology for companies has to be in line with the vision and orientation of the company. If the right metrics are used the ecological impact assessment is an additional tool for the description and evaluation of the performance in manufacturing.

In the future companies will be urged more extensively to fulfill energy management systems and product declarations regarding consumed materials, energies and even the ecological impact. In the European Union clear targets regarding the reduction of the carbon footprint are already set. Big manufacturers demand from their suppliers to be able to provide consumption and impact data per part. This is not yet possible for every company and product, but in the future a trade-off between simple CO_2-equivalent calculations based on yearly energy consumptions and full-scale life cycle assessments per product will emerge. This chapter helps possible users to understand the basic concepts and develop their own methodology to use in an industrial environment.

References

DIN EN ISO 14040 (2009) Environmental management—Life cycle assessment—Principles and framework

DIN EN ISO 14044 (2010) Environmental management—Life cycle assessment—Requirements and guidelines

Finnveden G, Hauschild M, Ekvall T, Guinée J, Heijungs R, Hellweg S et al (2009) Recent developments in life cycle assessment. J Environ Manag 91:1–21

Goedkoop M (2000) The eco-indicator 99 LCIA methodology—an introduction. In: Eidgenössische technische Hochschule Zürich (ed) eco-indicator 99—eine schadensorientierte Bewertungsmethode. Zürich

Guinée J (2002) Handbook on life cycle assessment. Operational guide to the ISO standards. Int J LCA 7:311–313

Guinee J, Heijungs R (1993) A proposal for the classification of toxic substances within the framework of life cycle assessment of products. Chemosphere 26:1925–1944

Hauff V (1987) Unsere gemeinsame Zukunft: [der Brundtland-Bericht der] Weltkommission für Umwelt und Entwicklung. Eggenkamp, Greven

Huijbregts M, Hellweg S, Frischknecht R, Hendriks H, Hungerbühler K, Hendriks A (2010) Cumulative energy demand as predictor for the environmental burden of commodity production. Environ Sci Technol 44:2189–2196

Huijbregts M, Rombouts L, Hellweg S, Frischknecht R, Hendriks A, van de Meent D et al (2006) Is cumulative fossil energy demand a useful indicator for the environmental performance of products? Environ Sci Technol 40:641–648

ICMC (ed) (2010) Sustainable production for resource efficiency and ecomobility International chemnitz manufacturing colloquium (ICMC)

Kellens K, Dewulf W, Overcash M, Hauschild M, Duflou J (2012) Methodology for systematic analysis and improvement of manufacturing unit process life-cycle inventory (UPLCI)-CO2PE! initiative (cooperative effort on process emissions in manufacturing) Part 1: Methodology description. Int J LCA 17:69–78

Klocke F, Schuh G, Döbbeler B, Pitsch M, Schlosser R, Lung D et al (2012) Simplified life cycle analysis of a forming tool in the automotive industry. In: Dornfeld D, Linke B (eds) Leveraging technology for a sustainable world. Proceedings of the 19th CIRP conference on life cycle engineering, University of California at Berkeley, Berkeley, USA. Springer, Heidelberg

Laurent A, Olsen S, Hauschild M (2010) Carbon footprint as environmental performance indicator for the manufacturing industry. CIRP Ann 59:37–40

Reich-Weiser C, Vijayaraghavan A, Dornfeld D (2010) Appropriate use of green manufacturing frameworks. In: CIRP (ed) 17th CIRP International conference on life cycle engineering (LCE 2010), Anhui, China

Schiefer E, Hora M, Hermenau U (2009) Entwicklung umweltgerechter produkte. Integration von EcoDesign in die betriebliche praxis. Hessen-Umwelttech News 2:15–16

Schlosser R, Klocke F, Döbbeler B, Riemer B, Hameyer K, Herold T et al (2011) Assessment of energy and resource consumption of processes and process chains within the automotive sector. In: Hesselbach J (ed): Globalized solutions for sustainability in manufacturing. Proceedings of the 18th CIRP International conference on life cycle engineering, Technische Universität Braunschweig, Braunschweig, Germany. Springer, Heidelberg

VDI 4600: Cumulative energy demand—Terms, definitions, methods of calculation

Wiedmann T, Minx J (2007) A definition of 'Carbon Footprint'. In: Pertsova C (ed) Ecological economics research trends. Nova Science Publishers, Hauppauge

Selecting Manufacturing Process Chains in the Early Stage of the Product Engineering Process with Focus on Energy Consumption

Martin Swat, Horst Brünnet and Dirk Bähre

Abstract Manufacturing process chains describe the concept of how the transformation of a raw material into a finished product is achieved. Within the planning phase of the process chains not only technical and economic requirements must be met but also ecological aspects need to be considered, e.g. the energy consumption during the production phase. The aim of this chapter is to illustrate how the energy consumption of process chains can be considered in the early stage of the planning phase. It provides an overview of the methods that are available to describe and predict the energy demand of consumers in process chains. The presented method is based on planning data like characteristic power consumption parameters of manufacturing equipment and related time parameters. It aims at predicting the energy consumption per product. The data is needed for predictive assessment of alternative process chains and to assess the impact of energy consumption during the production phase in life cycle considerations. Finally, this chapter presents an example for the energy-aware design and selection of a preferred process chain from several alternatives. By this it is illustrated how the presented heuristic approach can be applied.

1 Introduction

Over the last years, the consumption and cost of energy have become ever more important issues for governments, consumers and industrial companies. Electrical energy is one of the most valuable energy forms and it is mainly used in the industrial sector. Among energy carriers, electrical energy is the most commonly used to operate manufacturing processes. In Germany, for example, the industrial

M. Swat (✉) · H. Brünnet · D. Bähre
Institute of Production Engineering, Saarland University,
Campus A4.2 66123 Saarbrücken, Germany
e-mail: m.swat@mx.uni-saarland.de

E. Henriques et al. (eds.), *Technology and Manufacturing Process Selection*,
Springer Series in Advanced Manufacturing, DOI: 10.1007/978-1-4471-5544-7_8,

sector accounts for 43 % of the overall electrical energy consumption (Tzscheutschler et al. 2009). In addition, a survey of manufacturing processes indicates a trend towards more precise processes that consume more electrical energy per processed material volume (Gutowski et al. 2009). That is why reducing the energy consumption becomes a vital aspect for manufacturers during the development of industrial products. In this context the product life cycle perspective provides valuable insight: the most significant impact for energy reduction results from preventive action (Schulz et al. 1999). In other words, methodologies for the prediction of energy consumption are needed, e.g. for the procurement of manufacturing equipment (Kuhrke 2011) or for the optimization of energy costs by intelligent production scheduling (Fang et al. 2011; Pechmann and Schöler 2011).

Manufacturing companies have the chance to exploit this potential during the product engineering process when manufacturing technologies are selected and combined to manufacturing process chains. If companies manage to assess and optimize the energy consumption of manufacturing process chains during the design phase, they will be able to sustainably reduce energy consumption during the entire production phase. This applies specifically to the high-volume production of high-quality products. Likewise, the high demand concerning the quality and reliability of precision parts leads to complex process chains and poses the challenge to find the one with the lowest energy consumption per produced part. Also the product design and its influence on the energy consumption during the use phase must not be neglected.

This chapter presents an approach for the prediction of energy consumption of manufacturing process chains during the product engineering process. In Sect. 2 the understanding of the product life cycle and the product engineering process are specified for this chapter. Further, it is explained how the goal of reducing the energy consumption can be brought into line with other objectives for the selection of manufacturing process chains. The proposed approach for predicting the energy consumption is presented in Sect. 3. It describes the applied system boundaries and includes corresponding measurement results that characterize manufacturing equipment with two power consumption parameters. An exemplary application in Sect. 4 demonstrates the principles of the proposed method. Finally, it is shown how the results for the prediction of energy consumption can be included in product life cycle considerations.

2 Product Engineering as Part of the Product Life Cycle

2.1 The Product Life Cycle Perspective

In order to specify this chapter, the definition of the product life cycle (PLC) and the investigated product will be briefly determined. The product life cycle can be defined in two different ways. The first describes the life cycle from an economic viewpoint as the evolution of a product. In this definition of the *economic life cycle*, the product passes the following phases: introduction, growth, maturity and the decline stage. These phases are characterized by the development of the product sales over time (Kotler et al. 2009). The second definition represents the physical progress of the product from its creation through the distribution and use phase to its final disposal. This is called the *physical product life cycle* (Sundin 2010). In manufacturing systems the life cycle of the manufacturing processes with the regarding manufacturing equipment and the manufactured product intersect. Hence, it has to be clarified which product is under investigation. This chapter refers to the physical product life cycle of the manufactured product. Special focus is given to the energy consumption during the product creation which is largely predetermined by the definition of the manufacturing process chain.

2.2 The Product Engineering Process

The physical life cycle can be described by the phases: creation, distribution, use and the recycling or final disposal of the product. The focus in this chapter is on the creation phase which can be depicted in detail by the product engineering process. The product engineering process comprises the entire creation phase, starting with the initial product idea, followed by the product and process design to the actual manufacturing of the product. Figure 1 shows the product engineering process as part of the life cycle.

Before a new product engineering project is approved, estimations of costs and potential sales are evaluated and a project plan is created. The aim is to eliminate inappropriate and just pursue promising product ideas. Subsequently, product designs that fulfill the features of the pursued product idea are worked out. Following, the challenge for the process design is to find and select manufacturing processes that are capable to realize the defined product design with the desired features. Product and process design strongly influence each other. On the one hand, the demand for e.g. a lightweight design can make it necessary to apply certain manufacturing processes. On the other hand, if a manufacturing process is preset, e.g. due to vast investment in special manufacturing equipment, this may impose restrictions to the product design. In addition, single manufacturing processes have to be combined to manufacturing process chains in order to transform a raw material into the defined product. This leads to a variety of alternative process chains that are capable of creating the product. Before the start of production, much effort is put into the design, assessment and selection of process chains in order to ensure cost efficient and high quality production (VDA 1998; Klocke et al. 2000; Ehrlenspiel et al. 2007; Chrysler et al. 2008). This planning effort is justified especially for the series production of high-tech products where high unit costs combined with large quantities offer significant potential for cost reduction. The same relation applies to the energy consumption in manufacturing:

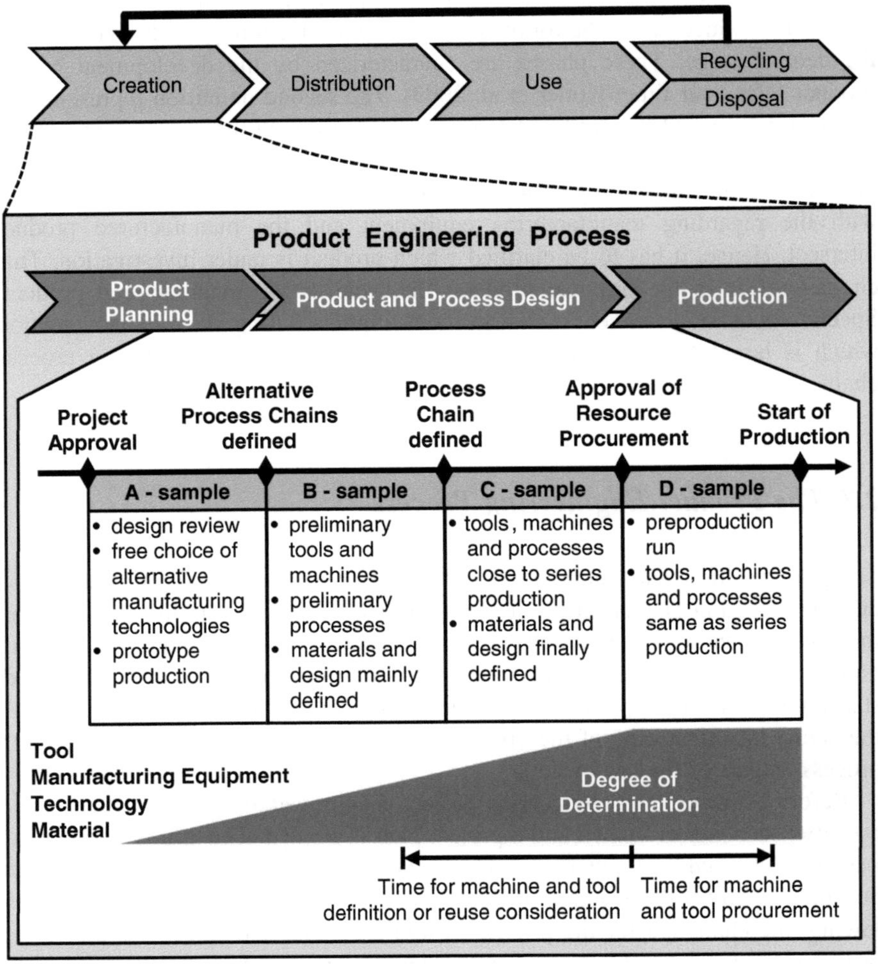

Fig. 1 The product engineering process within the life cycle, following VDA (1998) and Chrysler et al. (2008)

large shares of energy consumption and costs can be influenced in the planning stage. In order to consider energy consumption as one important decision criterion, a method is required that allows predicting the energy consumption of alternative manufacturing process chains. This method must take into account the existing planning process and the limited energy-relevant information that is available in the planning process. A closer look at the product-oriented description of the engineering process provides understanding of which energy-relevant information is available and at what point in time the assessment must be available for decision-making.

The description in Fig. 1 is based on a concept used in automotive industries. It comprises the product and process design. The four phases A to D represent the particular status of the product sample. Within each sample phase, certain attributes of the product and the process design are defined (Bähre et al. 2011). In sample phase A, the product is produced as a prototype. A selection of alternative processes is considered to manufacture the product. Materials, machines and tools are not finally defined yet. Subsequently, the B-sample phase is characterized by the use of preliminary machines, tools and processes for manufacturing the samples. In this phase, usually a number of alternative process chains is still considered for evaluation. The selection of one preferred process chain is made at the latest with the end of the B-sample phase. Afterwards, machines, tools and processes close to series production are utilized to manufacture the C-samples. The definition of tools and machines for series production is completed. As a result, the approval of resource procurement is given. Finally, in the D-sample phase, the product is manufactured in the same way as the series product. The degree of determination of the final process chain increases along the sample phases. Thus, the energy consumption of the process chain is becoming increasingly determined. According to the above remarks, the definition of the process chain for the series production is finished and a decision is made within the B-sample phase. It is evident that this is the latest point in time for assessing the energy consumption of alternative process chains.

The definition of manufacturing equipment, i.e. machine tools, starts within the B-sample phase. Two scenarios are possible: either new machine tools will be defined and purchased or existing machines can be considered for reuse. The energy consumption of the applied manufacturing equipment contributes to the energy consumption of the manufacturing process chain. Hence it is important to consider the energy consumption as one aspect when purchasing new manufacturing equipment. Kuhrke (2011) presents a procedure for cutting machine tools that enables the purchasing department to inquire reliable and comparable consumption data from machine tool manufacturers. Thereby it is possible to assess the energy efficiency and the investment costs of cutting machine tools. Due to high investment costs for new equipment, it can be advantageous to take into account existing manufacturing equipment for reuse in a new process chain. In this case, advantages and possible disadvantages of the reuse have to be balanced. An unacceptable energy consumption can for example lead to an exclusion of old manufacturing equipment from further reuse considerations (Weyand et al. 2011). In both cases no final definition of the machine tools is available before the process chain design is fixed. This indicates that the planning data are not finally determined when the assessment method has to be applied. However, a general specification of both the manufacturing equipment and the process parameters is still possible. For instance, technically and economically reasonable process parameters for manufacturing can be derived from the product description and the manufacturing processes considered for the process chain (Klocke et al. 2000; Esawi and Ashby 2003; Peças et al. 2013). Companies may maintain such data as they represent their manufacturing experience and know-how.

2.3 Selection of Manufacturing Process Chains

The selection of manufacturing process chains is a challenging task to perform as the planner has to deal with a variety of alternatives and objectives. Due to this complexity, it is crucial to start the selection based on a clear structure of the pursued objectives and decision criteria. It is evident that the main objective is to select the best available process chain. This comprises a number of sub-objectives like the achievement of technological requirements as well as the reduction of costs and environmental impact (Müller et al. 2008; Abele et al. 2012; Reinhardt et al. 2012). In order to find out how the reduction of energy consumption can be included in the selection of manufacturing process chains, it is important to distinguish fundamental and means objectives for this decision situation. A fundamental objective is pursued for its own sake. In contrast, a means objective is only pursued to achieve one or more other fundamental objectives (Eisenführ et al. 2010). The sub-objectives shown in Fig. 2 are fundamental for the selection of the best process chain among several alternatives. Whereas the reduction of energy consumption is not an end in itself here, i.e. it is not a fundamental objective for the selection of the best process chain.

The reduction of energy consumption is rather pursued to reduce energy costs and the CO_2-emission related to a manufacturing process chain. In other words it is a means objective that is striven to reduce the manufacturing costs and the environmental impact. This distinction is often neglected but important to clarify the goals of the process chain selection.

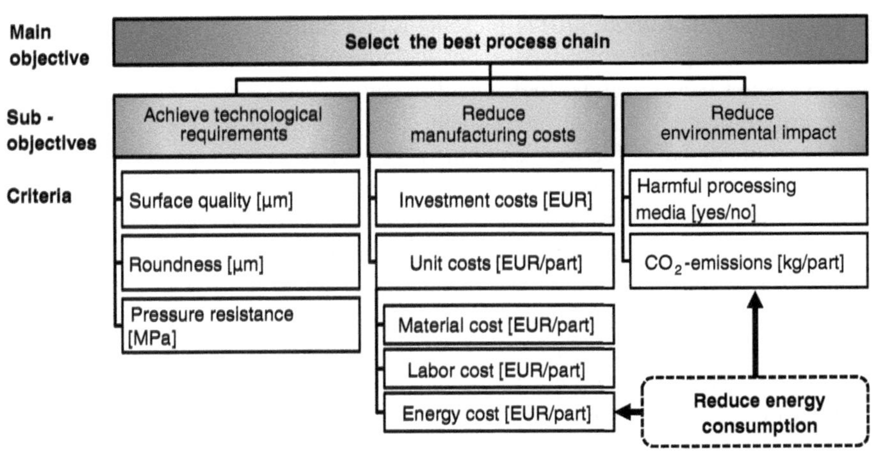

Fig. 2 Objectives and decision criteria for the selection of manufacturing process chains

3 Methodology

3.1 Energy Consumption of Manufacturing Process Chains

Electrical energy is used for a wide range of applications in industrial production systems, e.g. processing of material, heating, lighting and transportation. Investigating and comparing the energy consumption of manufacturing process chains requires a structured and applicable description of relevant energy consumers. This description must also provide a reasonable definition of the considered system boundaries as illustrated in Fig. 3.

A process chain contains manufacturing processes that enable the transition of a product from an initial state (A), e.g. a round or prismatic bar material, to the final state (B). The manufacturing processes are interlinked by supporting processes such as transport, handling or storage operations. Supporting processes are planned subsequently to the technologically induced manufacturing processes (Klocke et al. 2000) and are not considered for the energy assessment.

For the manufacturing process chains three important scopes of energy consumption can be identified (Schulz et al. 1999). The energy applied to perform the manufacturing technology, e.g. cutting or molding material, is a subset of the energy that is consumed by the used manufacturing equipment, e.g. a turning or press machine. The energy consumption of the manufacturing equipment, including the energy consumption for the manufacturing technology, can be measured at the electric main connection of the system. Such measurements require certain effort for the installation of metering equipment and the conduction of the measurement itself. Developments in the machine controls of major machine controller manufacturers show a trend towards integrated functionalities for energy consumption measurements (Siemens 2012; Bosch Rexroth 2012). Those functionalities will help to facilitate the determination of the energy consumption for manufacturing equipment. In contrast, the energy consumed by peripheral systems, e.g. compressed air or coolant units, cannot be directly measured for single processes. Those peripheral systems are often central units that supply several processes in the process chain. The energy consumption of the peripherals must be assigned to each process by allocation rules.

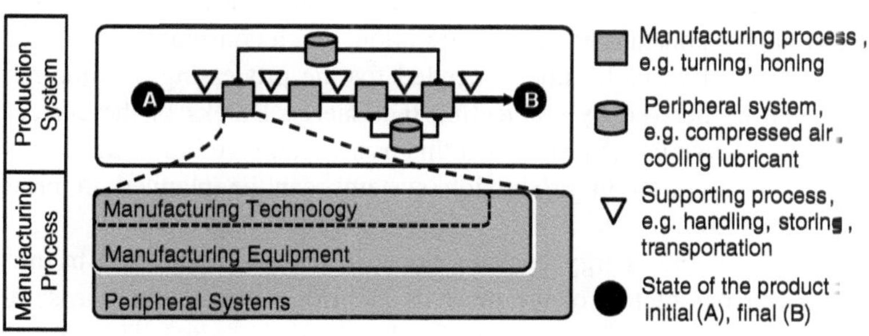

Fig. 3 System boundaries for the energy consumption of manufacturing process chains

3.2 Prediction of the Energy Consumption

The major challenge in predicting the energy consumption of manufacturing process chains is to provide energy-relevant planning data. For each process the direct energy consumption of the manufacturing equipment and the amount of indirect consumption of peripheral systems have to be considered. Therefore, the energy consumption as the integral of electrical power over time needs to be predicted with sufficient effort and accuracy.

3.2.1 Manufacturing Equipment

Several approaches have been presented in order to describe and predict the power consumption profile of manufacturing equipment (Schulz et al. 1999; Kordonowy 2002; Dietmair and Verl 2008; Weinert et al. 2009). Kordonowy (2002) distinguished fixed and variable shares for the energy consumption of cutting machine tools. He presented a qualitative model of power consumption depending on the work load for a selection of cutting machine tools. While this model gives insight into the relation between workload and energy requirement of cutting machine tools, it lacks the capability of forecasting the energy consumption.

A calculation model for cutting machine tools was developed by Dietmair and Verl (2008) in order to foster the energy efficient optimization of machine tools. The machine tool is therefore divided into functional modules, e.g. main spindle and drive control. Each module is assigned to nine defined operational states of the machine and subsequently the power consumption of the machine tool in each operational state is provided by measurements. With these parameters and with the knowledge of the usage profile that describes the order and duration of the operational states, the energy consumption of the machine tool can be calculated.

Weinert et al. (2009) also used operational states, like turned-off, warm-up, standby, processing or stopping, to describe the power consumption of manufacturing equipment. Their so-called Energy Blocks represent the time-dependent behavior of the power consumption during these operational states. Therefore, the behavior of an existing system is measured. Subsequently, the behavior is approximated by linear mathematical functions in order to represent the measured power consumption. According to the operational state they represent, the Energy Blocks can have a constant or variable duration, e.g. a constant time is needed for the warm-up and a variable time is needed for the processing. A time-resolving prediction can be derived by connecting the Energy Blocks in the correct order according to a known usage profile. On this basis, load leveling or peak shaving of the energy consumption in production systems can be planned in production planning and control.

A prediction of the energy consumption can also be conducted by means of only two power parameters of the most frequent operational states in series production and the related time parameters for the planned product. Therefore, Schulz et al. (1999) proposed to determine two averaged power consumption parameters that are multiplied with the product related time parameters of the manufacturing equipment (VDI3423 2002). Equation 1 shows the calculation adapted from Schulz et al. (1999).

$$E_{manufacturingequipment} = P_{process\emptyset} \cdot t_{process} + P_{basic\emptyset} \cdot (t_{occupied} - t_{process}) \quad (1)$$

$P_{process\emptyset}$ is the averaged power consumption during the performed machining task, e.g. drilling a hole. It is multiplied with the regarding processing time $t_{process}$ · $P_{basic\emptyset}$ is the averaged power consumption of the manufacturing equipment when it is not processing. It is multiplied with the occupied time that exceeds the processing time. Such periods can occur due to technical or organizational problems like a lack of parts and tools or the setting-up. All mentioned time parameters refer to one produced part. The manufacturing equipment can also be switched off during weekends and holidays. In that case, the power consumption is zero.

In order to verify the applicability of the two averaged power consumption parameters for the description of the energy consumption of manufacturing equipment in series production, a long-term measurement was conducted. The results of the investigation of two comparably equipped honing machines (A and B) are shown in Fig. 4. The measurement was carried out at the machine's main connection by using a meter for 3-phase AC that was extended by additional current transformers with a split core for quick and easy assembly (Christ 2012; Hobut 2012). The measurement period was set to four hours during the normal series production to get a reliable picture of the relevant machine states, their frequency of occurrence and the related power consumption. Whether the machines were in the processing or in the basic state was tracked by a signal taken

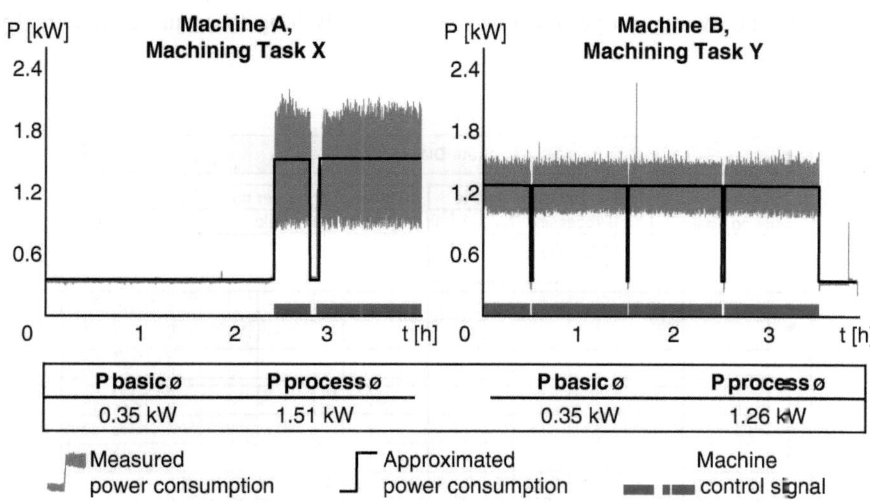

Fig. 4 Power consumption profiles of comparably equipped honing machines, machining bores of different size in series production

from the machine control. The signal was zero when no machining task was carried out and the machine was in basic state.

The results show short regular periods where the system switches to the basic state. These periods are due to an exchange of pallets for the supply of raw materials. Long and aperiodic periods in the basic state occurred due to a lack of raw materials. The investigation indicates that the averaged power consumption parameters $P_{basic\emptyset}$ and $P_{process\emptyset}$ can be determined and used for the approximation of the consumed energy. Further parameters with significantly differing power consumption and significant magnitude were not found. However, the investigation also shows an influence of the performed machining task and the applied process parameters on the power consumption of the machine. While $P_{basic\emptyset}$ is equal because of the comparability of the manufacturing equipment, a difference in $P_{process\emptyset}$ can be observed. In case of machining task X, a larger surface was machined than in case of machining task Y. This results in a higher power consumption of machine A compared to machine B.

In order to predict the energy consumption of the manufacturing equipment, also the time parameters must be predicted. The processing time can be calculated by using technically and economically reasonable process parameters (e.g. cutting speed, depth of cut, feed rate) and planning procedures for the determination of manual operations (Schiefer 2000). For the calculation of the occupied time per part it is required to know how frequent the manufacturing equipment operates in the processing state ($f_{process}$), the basic state (f_{basic}) and how frequent it is switched off (f_{off}) in longer breaks. Therefore, shift protocols of two weeks for machine A were exemplarily evaluated. The results are given in Fig. 5.

The evaluation shows that machine A was operated on average to 74 % in the processing and to 26 % in the basic state. Note that the machine was not switched off during longer breaks. In this example the distribution is equal for both weeks. However, the evaluation period must be sufficiently long to ensure a statistically

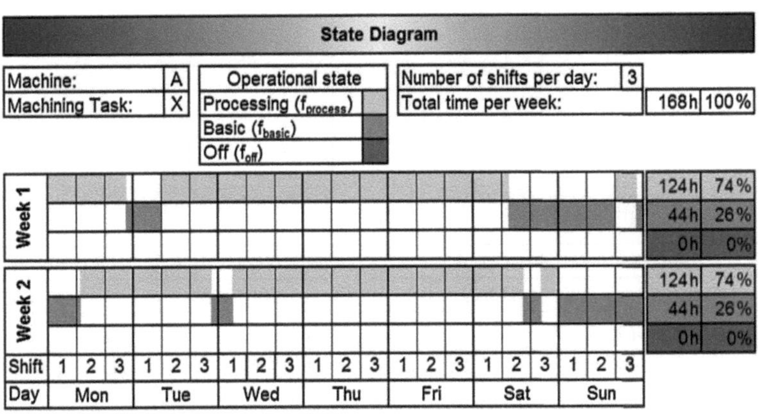

Fig. 5 Time parameters based on the evaluation of shift protocols for machine A

reliable distribution among the operational states. Based on the knowledge of $t_{process}$ and $f_{process}$, the occupied time can be calculated according to Eq. 2.

$$t_{occupied} = t_{process} \cdot \frac{f_{basic} + f_{process}}{f_{process}} \tag{2}$$

3.2.2 Peripheral System

In series production, peripheral systems usually not only provide one process but a number of processes with the needed process media. Peripheral systems usually have an electric connection that is separate from the manufacturing equipment. Hence, they must be measured and allocated to the product separately. The peripherals provide manufacturing processes with process media like compressed air and lubricant. Today, the energy consumption of peripheral systems is mostly allocated on shop floor level but neither to individual manufacturing equipment nor to the products manufactured by the manufacturing equipment. However, a sufficiently exact assessment of alternative manufacturing process chains includes such a detailed allocation to the manufactured products. The peripheral energy consumption cannot be directly allocated to a product but there are two parameters that can be applied for the allocation: the media consumption rate of the process ($MCR_{Process}$) and the energy consumption rate of the peripheral system ($ECR_{Periphery}$) as shown in Fig. 6.

The $MCR_{Process}$ is the input of media into the process related to the production rate that describes the output of the process. To supply the adjusted media consumption of the process, the peripheral system consumes electric energy. This energy consumption (input) related to the media volume (output) of the peripheral

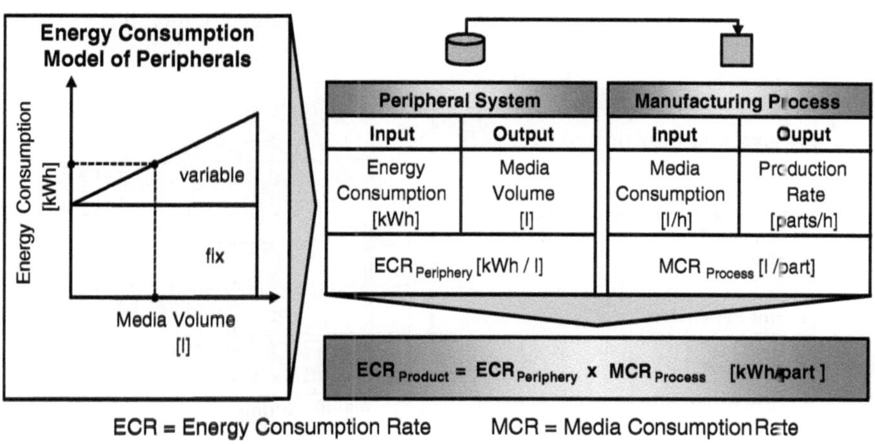

Fig. 6 Allocation procedure for energy consumption of a peripheral system that provides media to the connected processes, energy consumption model following Gutowski et al. (2006)

system displays the energy consumption rate of the periphery ($ECR_{Periphery}$). The multiplication of the two rates yields the energy consumption rate per product ($ECR_{Product}$).

This allocation scheme is based on an energy consumption model of the peripherals. The model takes into account the number of connected manufacturing processes and the adjusted media consumption of the processes that influence the workload and energy consumption of the peripherals. The workload-dependent energy consumption data must be gained by measurements, gathered from suppliers or approximated by data of comparable peripherals.

3.3 Database for Energy-Relevant Planning Data

Regardless of the methods for determining and predicting the power consumption, a database is needed to manage and provide the power consumption data for the planning process. The investigation of the honing machines showed an influence of the machining task on the power consumption. The manufacturing equipment and the applied process parameters also influence the power consumption. Hence, these three parameters define a case of application. Due to the effort related to the measurement of existing machines or the parameterization of simulation models it will not be possible to determine the energy-relevant planning data for every case of application. However, the definition of parameter classes can help to limit the number of combinations and to provide sufficiently exact power consumption data for the planning case. The classes for the manufacturing equipment can be distinguished by the construction type or the type of a main component, e.g. the main spindle. Furthermore, the machining tasks for the honing process can be classed by the machined surface area and the process parameters by the rotational or lifting speed. The cubes of the data model for the honing process shown in Fig. 7 can be identified by a class of each parameter. The highlighted cube, for example,

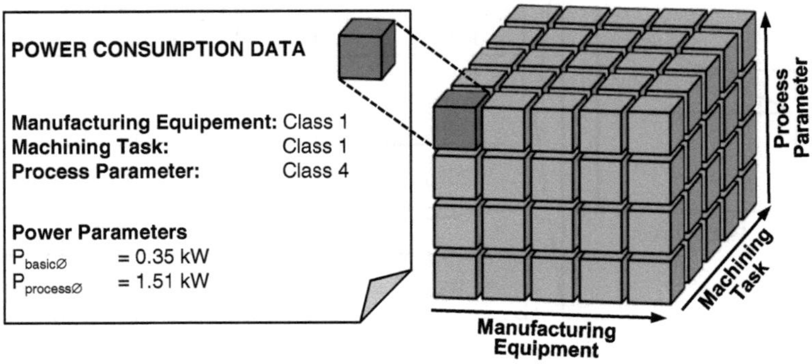

Fig. 7 Database with data-sets for the energy-relevant planning data for a manufacturing process, with power consumption data taken from the example in Fig. 4

contains the power consumption parameters of machine A performing machining task X. Once this data was determined, it can be applied for the planning of a comparable honing process. In case no cube is available for a new planning situation, the data must be determined or an available cube with a differing but comparable class of manufacturing equipment, machining task or process parameters might be used for approximation. The definition of parameter classes is also reasonable as no detailed information about the manufacturing equipment, the machining task and the process parameters will be available. However, the definition of a respective parameter class is still possible.

Maintaining such a database is possible for big companies. In addition, manufacturers of production equipment who offer their products as an industrial product-service system (IPS^2) can also maintain databases with energy consumption parameters. In an IPS^2, the manufacturer does not sell the production equipment but the usage of the equipment. Due to this business model, the provider of the equipment can gain access to energy consumption data from the use phase of its products (Weinert 2009). This data can be used to supply power consumption parameters for the planning of manufacturing process chains. In consequence, the manufacturer of production equipment can provide the planning data as an additional service. Moreover, a cooperation of production equipment manufacturers in an IPS^2 consortium also allows to provide data for a wide range of processes.

4 Exemplary Application

4.1 Product Design

In order to illustrate the product engineering process and the previously presented method for considering the energy consumption, a commonly used product geometry for high pressure applications is used as an example: a hydraulic tube or thick-walled cylinder. Fuel injection lines of diesel injection systems in automobiles are popular examples for the application of such components. They are subjected to high mechanical stresses during operation. In case of dynamic pressure profiles, this may lead to fatigue and hence a limited lifetime. Over the last decades, conventional design approaches for internally pressurized components dominated the process chain layouts in many application areas. Here, high strength steels with large wall thicknesses were used for hollow components in order to reach the desired strength (Kendall 2000). Additionally, highly sophisticated surface finishes are necessary in order to avoid crack initiation at imperfections. However, the challenge in product and process chain design nowadays is not only to achieve a certain component strength but instead to reduce the weight at the same time. Especially in car manufacturing, this trend is obvious. One promising approach is to exploit the full material potential (Brünnet et al. 2011) For this purpose, selectively induced compressive residual stresses are gaining importance. By considering the same wall thickness and inducing pronounced compressive

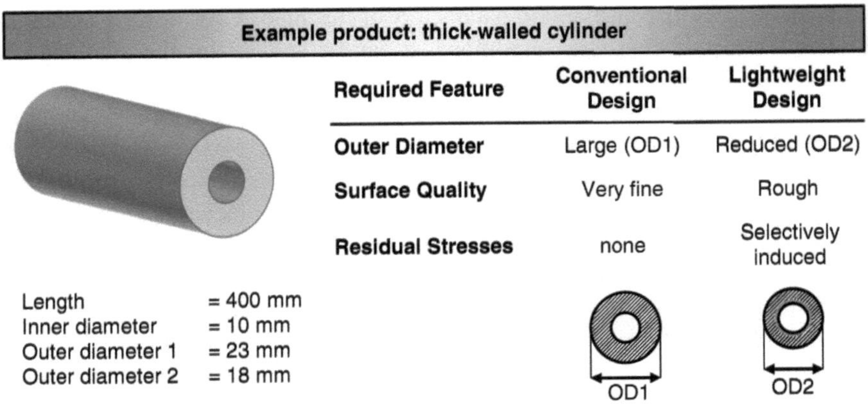

Fig. 8 Alternative product design approaches

residual stresses, the fatigue strength may be significantly increased. One process to selectively induce compressive residual stresses is hydraulic Autofrettage (AF). Figure 8 shows the two principle design approaches, which result in alternative manufacturing process chains.

In case of the considered tube, both designs include an inner bore diameter of 10 mm. The conventional design asks for a sufficient wall thickness (OD1 = 23 mm) and manufacturing processes that generate very good surface finishes in the bore. By inducing selectively induced compressive residual stresses, a reduced wall thickness (OD2 = 18 mm) is sufficient and the bore is less vulnerable to machining imperfections. Hence, less effort has to be spent in surface finishing.

4.2 Considered Processes and Process Chain Design

The process chain design starts based on a semi-finished product that is a high strength steel solid cylinder with finished external machining. Subsequently, the required processes for manufacturing the part must be defined. The following five processes are considered to manufacture the example product: Drilling, Reaming, Honing, Abrasive Flow Machining and Autofrettage. While Drilling and Reaming are traditional and well-known cutting processes, the remaining processes are advanced manufacturing processes. Therefore, they are briefly introduced.

Honing, in particular long-stroke honing of bores, is an abrasive machining process with undefined cutting edges. It is mainly used as a finishing process (DIN8589 2003). The abrasives are bound in a so-called honing stone. The honing tool with the stone rotates and executes a linear oscillating movement along the bore while the honing stone is forced against the bore surface. Honing can produce precision surfaces with a roughness up to 1 μm (Schmitt et al. 2011). Typical applications for long-stroke honing can be found at hydraulic and pneumatic parts.

Abrasive Flow Machining (AFM) is an abrasive machining process where a polymer-based pliable carrier medium with abrasive particles is forced through internal geometries. Surface qualities with a roughness below 2 μm can be achieved (Sankar et al. 2007). AFM is typically applied for finishing internal passages of pressurized parts.

Autofrettage (AF) uses a low viscosity medium to over-pressurize the inner diameter of a hydraulic component. This single over-pressure exceeds the material's yield strength to introduce plastic deformation and as a consequence pronounced compressive residual stresses (Gibson 2008; Seeger et al. 1994). The compressive residual stresses superimpose the later applied load stresses during the operation and effectively increase the fatigue strength of the components (Löhe and Lang 2002). Additionally, less wall thickness is necessary compared to a conventional design without the use of compressive residual stresses.

The application of the manufacturing processes and their combination in a process chain are determined by the required features included in the product design and by the ability of the manufacturing processes themselves. The processes need a certain input and generate a certain output regarding these features. The difference between the needed input and the generated output defines the ability of the process (Fallböhmer 2000). However, a process can also leave certain features unaffected. In Fig. 9 the feature index depicts the ability of the five

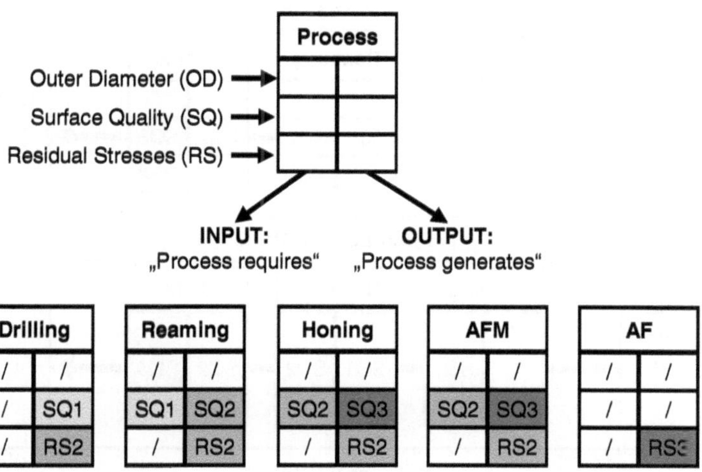

Feature Index	Outer Diameter (OD)	Surface Quality (SQ)	Residual Stresses (RS)
/	None / no influence	None / no influence	None / no influence
1	Large	Rough	–
2	Reduced	Fine	Unintentionally induced
3	–	Very fine	Selectively induced

Fig. 9 Ability of the considered manufacturing processes

considered manufacturing processes regarding the outer diameter (OD), the surface quality (SQ) and the residual stresses (RS).

The outer diameter in connection with the set inner diameter of the bore yields the wall thickness of the tube. This is a feature that is inherent to the product design and is not affected by the considered manufacturing processes. The outer diameter is large in the conventional design (OD1) and it is reduced in the lightweight design approach (OD2). In contrast, all processes except from AF influence the surface quality of the bore. Drilling allows the manufacturing of the bore and generates a rough surface (SQ1). This surface quality can be improved to a fine surface (SQ2) by reaming the bore. The abrasive cutting processes of honing and AFM require a fine surface and generate a very fine surface quality (SQ3). AF does not affect the surface quality but selectively induces residual stresses into the surface of the bore (RS3). Unintentionally induced residual stresses (RS2) can also be a result of all other processes. However, these residual stresses are supposed to be not significant.

Based on the two design approaches and the introduced processes, three alternative process chains with the ability to generate the example part can be derived, as shown in Fig. 10. Process chains A and B consider the conventional product design with sufficient wall thickness and a very fine surface finish of the bore to prevent crack initiation. In contrast, process chain C is based on the lightweight design with reduced wall thickness and selectively induced

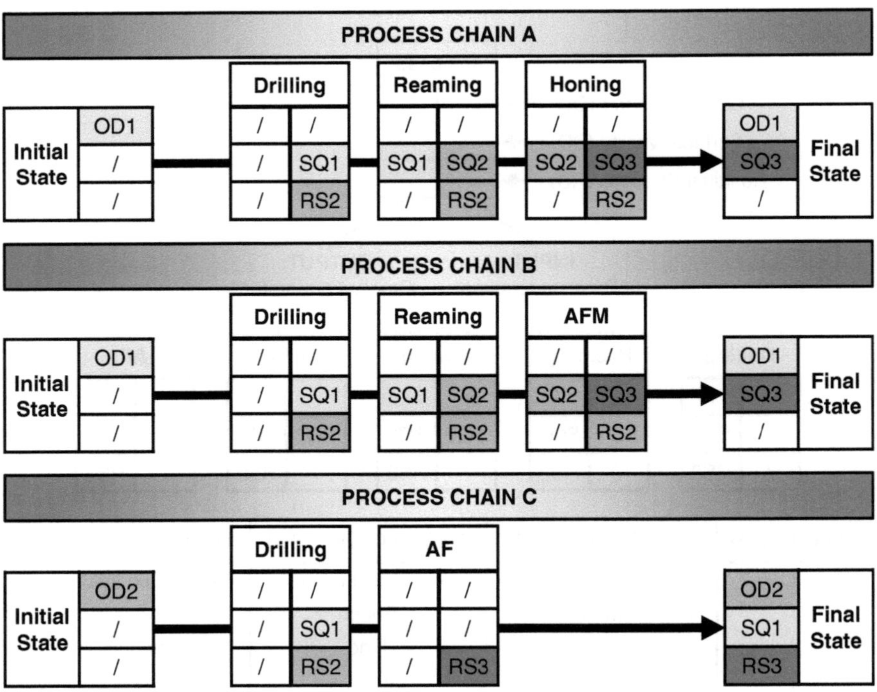

Fig. 10 Alternative process chain designs

compressed residual stresses. Note that the initial and the final state refer to the feature index that is required by the product design. However, the required index can also be exceeded by the process chain, e.g. process chain A and B induce some residual stresses (RS2) to the final state where no residual stresses are required by the product design.

The process chains A and B are equal except for the applied finishing process: the honing process for A and the AFM process for B. Due to the induced residual stresses by the AF process in process chain C, the bore is less vulnerable to machining imperfections. As a consequence, the surface quality generated by the drilling process is sufficient with the result that the reaming process can be eliminated.

4.3 Energy Balancing of the Alternative Process Chains

In order to predict the energy consumption of the alternative process chains, an assignment of the energy consuming manufacturing equipment to the processes is necessary. This assignment can be done similarly to the previous selection of the manufacturing processes. Therefore, the characteristics of the product, e.g. geometrical characteristics and required surface quality, must be assigned with the ability of the manufacturing equipment, e.g. number of axes and achievable working pressure. The conduction of the alignment leads to an exclusion of inappropriate manufacturing equipment from further consideration and to the selection of equipment that is capable of manufacturing the product (Trommer 2001). This assignment combined with the description of the machining task and the process parameters allows the selection of the consumption parameters of the equipment as described in Sect. 3.3. Figure 11 summarizes the procedure for the energy balancing procedure of the manufacturing process chains.

The energy consumption of the alternative process chain designs with the assignment of the production equipment is the basis for the assessment. Power consumption data together with the time data of the processes are applied to predict the energy consumption of the manufacturing equipment. Accordingly, the allocation scheme is used to distribute the energy consumption of the peripheral systems to the manufacturing processes. Note that the energy consumption of single processes should be carefully compared to each other. The energy consumption of one process depends on the design of the process chain and the contribution of the process to manufacture the final product design. AF for example might have a higher energy consumption compared to honing and AFM. At the same time, it allows to eliminate the reaming process. This indicates that a comparison should be conducted in terms of the entire process chain and not in terms of single processes. On the other hand, the comparison of the total consumption of processes in a process chain provides evidence for further improvement in the detailed planning and realization once a preferred process chain has been selected. Finally, a ranking of the alternative process chains concerning their energy consumption per produced part can be derived. That is the input data for the final assessment of the process chains that

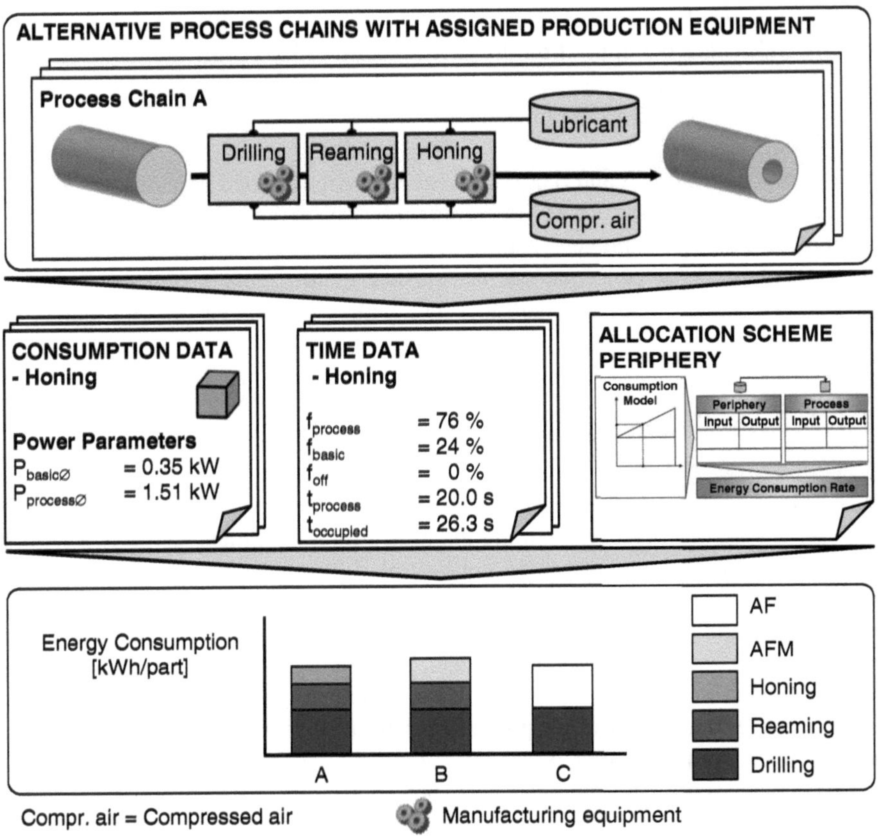

Fig. 11 Energy balancing procedure for manufacturing process chains

considers technological, economic and ecological criteria, e.g. energy cost and CO_2-emissions. Further, the CO_2-emissions are the reference for a comparison of the consideration of ecological life cycle aspects.

4.4 Life Cycle Considerations

The alternative manufacturing process chains for the example part allow for two different product designs with different material input and weight of the produced parts. In order to give a complete picture and to avoid ecological burden shifting between the life cycle phases, also the CO_2-emissions for the creation of the material input and the CO_2-emissions during the use phase of the part must be considered. The application of the AF process in process chain C allows reduced material input compared to the process chains A and B. The reduced weight also results in reduced fuel consumption during the use phase. The weight of the parts

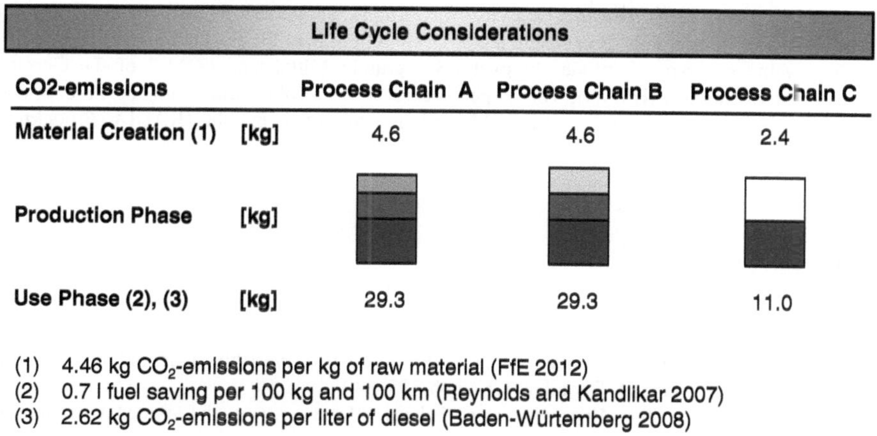

Fig. 12 CO_2-emissions related to the parts produced by the alternative process chains

can be calculated using the dimensions of the parts and the density of steel. The calculated CO_2-emissions for the material creation and during the use phase of the parts are summarized in Fig. 12

The results show that the effects of different product and process chain designs on the ecological performance for the material creation and during the use phase of the part must not be neglected. The application of the presented method will allow to also quantify the amount of energy consumption and CO_2-emissions of the process chains during the production phase.

5 Conclusion and Outlook

Energy Consumption and the related costs and CO_2-emissions become ever more important issues for manufacturing companies. The design and selection of manufacturing process chains have significant influence on the energy consumption during the entire production phase. In order to sustainably reduce the energy consumption, this issue must be considered already in the design of manufacturing process chains. In this chapter, methods to predict the energy consumption of manufacturing equipment and a methodology to predictively balance the energy consumption per produced part were presented. The methodology was developed with regard to the requirements during the product engineering process. The application of the method using a comprehensive example for the selection of a preferred process chain illustrated the method. In order to foster the proposed method and its practical application, a software tool is beneficial and will be developed. If companies manage to determine and select the process chain with the lowest energy consumption per part, they have the basis to sustainably exploit the full energy saving potentials of their production systems. Life cycle aspects must not be neglected here.

Acknowledgments This research activity was kindly supported by the research project "Ganzheitliche Gestaltung energieeffizienter technologischer Prozessketten (enPROchain)" funded by the program "Zentrales Innovationsprogramm Mittelstand (ZIM)" of the German Ministry of Economics (BMWi). This paper is a result of a joint effort with the Process and Software Engineering Company ENCOM and is supported by the KADIA Produktion GmbH + Co.

References

Abele E, Schrems S, Eisele C, Schraml P (2012) Simulation-based assessment of the energy consumption of manufacturing processes. In: Proceedings of the 19th CIRP conference on life cycle engineering (LCE), Berkeley, USA, pp 375–380

Baden-Würtemberg (2008) Energiesparend Fahren. Broschure of the Baden-Würtemberg ministry of internal affairs

Bähre D, Swat M, Steuer P, Trapp K (2011) Energy consumption: One criterion of the sustainable design of process chains. In: Proceedings of the 9th global conference on sustainable manufacturing, St. Petersburg, Russia, pp 164–170

Bosch Rexroth (2012) MTX ega—energy analysis tool. www.boschrexroth.com. Accessed 12 Apr 2012

Brünnet H, Yi I, Bähre D (2011) Modeling of residual stresses and shape deviations along the process chain of autofrettaged components. J Mater Sci Eng A, 1/7A:915–936

Christ (2012) Data sheet power transmitter CLT 313. http://www.christ-elektronik.de/elements/products/files/CLT313_DatEN_Rev06.pdf. Accessed 14 Apr 2012

Chrysler Corporation, Ford Motor Company, General Motors Corporation (2008) Advanced product quality planning (APQP) and control plan. Reference Manual, 2nd edn

Dietmair A, Verl A (2008) A generic energy consumption model for decision making and energy efficiency optimisation in manufacturing. In: Proceedings of the 18th international conference on flexible automation and intelligent manufacturing (FAIM), Skövde, Sweden, pp 123–133

DIN8589 (2003) Manufacturing processes chip removal—Part 14: Honing and superfinishing. Classification, subdivision, terms and definitions. Beuth, Berlin

Ehrlenspiel K, Kiewert A, Lindemann U (2007) Cost-efficient design. Springer, Berlin

Eisenführ F, Weber M, Langer T (2010) Rational decision making. Springer, London

Esawi AMK, Ashby MF (2003) Cost estimates to guide pre-selection of processes. Mater Des 24:605–616

Fallböhmer M (2000) Generieren alternativer Technologieketten in frühen Phasen der Produktentwicklung. Ph.D. thesis, RWTH Aachen, Aachen

Fang K, Uhan N, Zhao F, Sutherland JW (2011) A new shop scheduling approach in support of sustainable manufacturing. In: Proceedings of the 18th CIRP international conference on life cycle engineering (LCE), Braunschweig, Germany, pp 305–310

FfE—Forschungsstelle für Energiewirtschaft e.V. (2012) Ganzheitliche Bilanzierung von Grundstoffen und Halbzeugen. http://www.ffe.de/images/stories/Berichte/Gabie/metall.htm. Accessed 30 Apr 2012

Gibson MC (2008) Determination of residual stress distributions in autofrettaged thick-walled cylinders. Dissertation, Cranfield University

Gutowski G, Branham M, Dahmus J, Jones A, Thirez A (2009) Thermodynamic analysis of resources used in manufacturing processes. Environ Sci Technol 43:1584–1590

Gutowski T, Dahmus J, Thiriez A (2006) Electrical energy requirements for manufacturing processes. In: Proceedings of the 13th CIRP international conference on life cycle engineering (LCE), Leuven, Belgium

Hobut (2012) Split-core current transformers. http://www.hobut.co.uk/images/ files/Split-Core%20Clip%20Together%20range%20-%20new.pdf. Accessed 14 Apr 2012

Kendall DP (2000) A short history of high pressure technology from bridgman to division 3. J Press Vessel Technol 122:229–233

Klocke F, Fallböhmer M, Kopner A, Trommer G (2000) Methods and tools supporting modular process design. Robot Comput Integr Manuf 16:411–423

Kordonowy D (2002) A power assessment of machining tools. Bachelor of science thesis in mechanical engineering, Massachusetts Institute of Technology, Cambridge, Massachusetts

Kotler P, Keller K, Brady M, Goodman M, Hansen T (2009) Marketing managment. Pearson, Harlow

Kuhrke B (2011) Methode zur Energie- und Medienbedarfsbewertung spanender Werkzeugmaschinen. Ph.D. thesis, Technical University of Darmstadt, Darmstadt

Löhe D, Lang KH (2002) Residual stresses and fatigue behavior. In: Totten GE et al. (eds) Handbook of residual stress and deformation of steel. ASM International, Materials Park, Ohio

Müller E, Engelmann J, Strauch J (2008) Energieeffizienz als Zielgröße in der Fabrikplanung. wt Werkstattstechnik online 98:634–639

Peças P, Ribeiro I, Silva A, Henriques E (2013) Comprehensive approach for informed life cycle-based materials selection. Mater Des 43:220–232

Pechmann A, Schöler I (2011) Optimizing energy costs by intelligent production scheduling. In: Proceedings of the 18th CIRP international conference on life cycle engineering (LCE), Braunschweig, Germany, pp 293–298

Reinhardt S, Fischl M, Reinhart G (2012) Characterization and weighting scheme to assess the resource efficiency of manufacturing process chains. In: Proceedings of the 19th CIRP conference on life cycle engineering (LCE), Berkeley, USA, pp 509–514

Reynolds C, Kandlikar M (2007) How hybrid-electric vehicles are different from conventional vehicles: the effect of weight and power on fuel consumption. Environ Res Lett 2:014003

Sankar MR, Jain VK, Ramkumar J (2007) Abrasive flow machining (AFM): An overview. http://www.psgtech.edu/ psgias/smart_machine_tools/V.K.Jain.pdf. Accessed 22 Nov 2011

Schiefer E (2000) Ökologische Bilanzierung von Bauteilen für die Entwicklung umweltgerechter Produkte am Beispiel spanender Fertigungsverfahren. Ph.D. thesis, Technical University of Darmstadt, Darmstadt

Schmitt C, Bähre D, Forsch K, Klein H (2011) Feinstbearbeiten hochgenauer Bohrungen durch Honen. In: Hoffmeister HW, Denkena B (eds) Jahrbuch Schleifen, Honen, Läppen und Polieren, 65th edn. Vulkan, Essen

Schulz H, Atik A, Schiefer E (1999) Product development must consider environmental aspects. In: Kuljanic E (ed) AMST'99: Advanced manufacturing systems and technology Springer, Wien, New York, 39–52

Seeger T, Schön M et al (1994) Autofrettage I—Dauerfestigkeitssteigerung durch Autofrettage. Forschungsvereinigung Verbrennungskraftmaschinen, Vorhaben Nr 478

Siemens (2012) SINUMERIK Ctrl-energy. www.siemens.com. Accessed 12 Apr 2012

Sundin E (2010) Life-cycle perspectives of product/service-systems: in design theory. In: Sakao T, Lindahl M (eds) Introduction to product/service-system design. Springer, London

Trommer G (2001) Methodik zur konstruktionsbegleitenden Generierung und Bewertung alternativer Fertigungsfolgen. Ph.D. thesis, RWTH Aachen, Aachen

Tzscheutschler P, Nickel M, Buttermann HG (2009) Energieverbrauch in Deutschland: Stand 2007, Daten, Fakten, Kommentare. BWK—Das Energie-Fachmagazin 61/6: 6–14

VDA—Verband der Automobilindustrie e.V. (1998) Quality assurance prior to serial application–project planning, vol 4 Part 3, 1st edn.

VDI 2423 (2002) Technical availability of machines and production lines—terms, definitions, determination of time periods and calculation. Beuth, Berlin

Weinert N, Chiotellis S, Seliger G (2009) Concept for energy-aware production planning based on energy blocks. In: Proceedings of the 7th global conference on sustainable manufacturing, Madras, India, pp 75–80

Weyand L, Bley H, Swat M, Trapp K, Bähre D (2011) Energy consumption as one possible exclusion criterion for the reuse of old equipment in new production lines. In: Proceedings of the 18th CIRP international conference on life cycle engineering (LCE), Braunschweig, Germany, pp 287–292

Manufacturing with Minimal Energy Consumption: A Product Perspective

Alexandra Pehlken, Alexandra Kirchner and Klaus-Dieter Thoben

Abstract The aim of this paper is to highlight energy intensive process steps in compound feed production and their importance for the overall LCA in feed processing. The carbon footprint has become a relevant measure among the animal and feed experts for comparing product or process performances. Our research focuses on the energy intensive process steps in the feed production line. Due to the fact that there is a high pressure on feed and food quality, there are very limited possibilities to change the process since a certain energy input is needed to reach the requested quality. The solution can be provided through an intelligent network process control. The network needs access to automatic sensor control devices that are installed inline to measure varying product parameters like the changing water content in grains for example (due to rainy or dry seasons). The knowledge of certain product parameters will influence the process control like for example steam addition to the process. An intelligent network supports the best energy performance for producing the requested compound feed quality by the customer. The paper summarizes the information along the compound feed production that is necessary for an energy efficient process control and explains how an intelligent network can contribute to the latter.

1 Introduction

The supply of safe and healthy food plays a vital role in Europe and the feed industry is an integral part of the production of animal products. The European compound feed industry is an important economic sector with more than 4,000

A. Pehlken · K.-D. Thoben
Bremen University, Badgasteinerstr.1, 28359 Bremen, Germany

A. Kirchner (✉)
International Research Association of Feed Technology, 38110 Braunschweig, Germany
e-mail: a.kirchner@iff-braunschweig.de

E. Henriques et al. (eds.), *Technology and Manufacturing Process Selection*, Springer Series in Advanced Manufacturing, DOI: 10.1007/978-1-4471-5544-7_9,

production units (FEFAC 2010). Animal feeding stuffs, including feed materials and compound feed, are the main input to livestock production with approx. 450 million tons in total consumed by European livestock each year (FEFAC 2010).

Feeding stuffs are subdivided into: feeding stuff grown and used on the farm of origin, industrially produced supplements to be used on the farm of origin and grain based industrial compound feed (Jeroch et al. 1993). In the year 2010 the amount of industrial compound feed in Europe was about 150 million tons (FEFAC 2010), which relates to one-third of the total animal feeding.

The food and beverage consumption is known as one of the consumer activities with the highest environmental impact (FEFAC 2009; Tukker et al. 2006). European feed and food production faces a number of ongoing trends that are understood to be untenable and unsustainable. Driven by the ever increasing global demand for meat, edible oils and biofuels, the establishment of soy or palm oil plantations in South-America and Indonesia has resulted in the conversion of large areas of forests with high conservation value and threatening the rich biodiversity in these ecosystems. But also transport and transformation of the harvested products generate adverse effects on the environment, in particular caused by climate relevant emissions and the use of fossil energy (Steinfeld et al. 2006; Thomassen et al. 2008; Dalgaard et al. 2007). These are exemplary aspects that urge towards a more sustainable feed and food production.

The European compound feed producers have a role to play in minimizing environmental impacts while meeting the demand for high quality products (FEFAC 2009). The challenging combination of changing raw materials, increasing raw material prices and electricity costs as well as modified consumer preferences is forcing European feed producers to maximize efficiency and to minimize environmental impacts. Opportunities to save energy in feed production are limited but there is scope in the manufacturing operations and the rationalization of logistics (FEFAC 2009). Actually, reliable results on the necessary energy consumption for feed processing are missing altogether. The specific energy demand per ton may vary between 45 and 120 kWh due to various refinement steps. This might be misinterpreted e.g. by LCA practitioners who might use data that may represent only one-third of the actual energy consumption. How do non-feed-processing experts know which data to use? For this reason it is important to highlight energy demands in the compound feed production in which feed characteristics as well as the nutritional value of feed stuffs are linked to the process treatment intensity. There is a strong relation between the feed quality and the process treatment. Some animals prefer very hard pellets (cows) and some prefer very soft pellets (pigs).

To identify and quantify the energy needs as well as the materials and to evaluate the environmental burden it is common to use the services of the Life Cycle Assessment (LCA). But current LCA and environmental impact studies e.g. the calculation of the Carbon Footprint of pork meat are often undifferentiated due to varying system boundaries and the neglect of various technological production methods of animal feeding stuffs. There is hardly any information available for the feed processing and its impact on the whole LCA. This chapter is going to provide some additional information on the complexity of feed compound processing and highlights energy efficiency potentials along the manufacturing process.

To provide further insight into the potential for energy savings of industrial feed compounding, an intelligent network process control will be established to help identify the carbon footprint and to optimize the energy input into the process without negatively influencing the product quality.

2 State of the Art in Feed Processing

Industrial feed compounding is focusing on the refining of agricultural raw materials like grains. Most of the technological processes in compound feed manufacturing have an impact on the nutritive value and hygienic status of the feed and sometimes these processes act synergistically.

Feed mixtures are composed of up to 40 different agricultural macro components, e.g. wheat, corn, barley, soy and limestone, and effective micro components like vitamins and enzymes which are added in ppm-amounts. In addition to conventional feed components by-products, for instance, from bio-fuel production like Dried Distillers Grains with Solubles (DDGS) are increasingly used as feedstuffs (Lardy 2007). The composition of feed mixtures primarily depends on the period of growth of raw materials and the recipient animal species. Currently it is also adjusted by e.g. availability and market price of raw materials.

As mentioned above, in 2010 the amount of industrial compound feed in Europe was about 150 million tons (FEFAC 2010). In the same year, a total amount of compound feed of approx. 22 million tons were produced in Germany: 9.5 million tons feed for fattening pigs, 6.2 million tons feed for cattle, 3.5 million tons feed for broilers and two million tons feed for layers (Noras 2012). Most European feed compounders—which are typically small and medium-sized (SME)—are producing diet formulations on customers' demand in single-line production plants just in time. Despite the complexity of this ambitious objective, customer preferences tend towards an ever increasing number of formulations. For a medium-sized production plant with an annual production average of 200,000 tons this may result in a short-term supply of, for instance, more than 200 different compositions.

The feed processing—shown in a simplified flow chart in Fig. 1—starts with the incoming raw materials, which are delivered by trucks, discharged into collecting vessels and stored short-term in raw material silos.

Then components are automatically dosed and weighed based on the respective feed formulation. All formulations are recorded in the internal process control information system. The refinement starts with the grinding of the specific grainy raw materials, which is commonly conducted as a compound grinding by the use of hammer mills. Additionally, roller mills as well as disc mills can be used, depending on the desired grinding product. A feed-flour which is produced by the use of hammer mills generally shows a high amount of fine particles whereas a

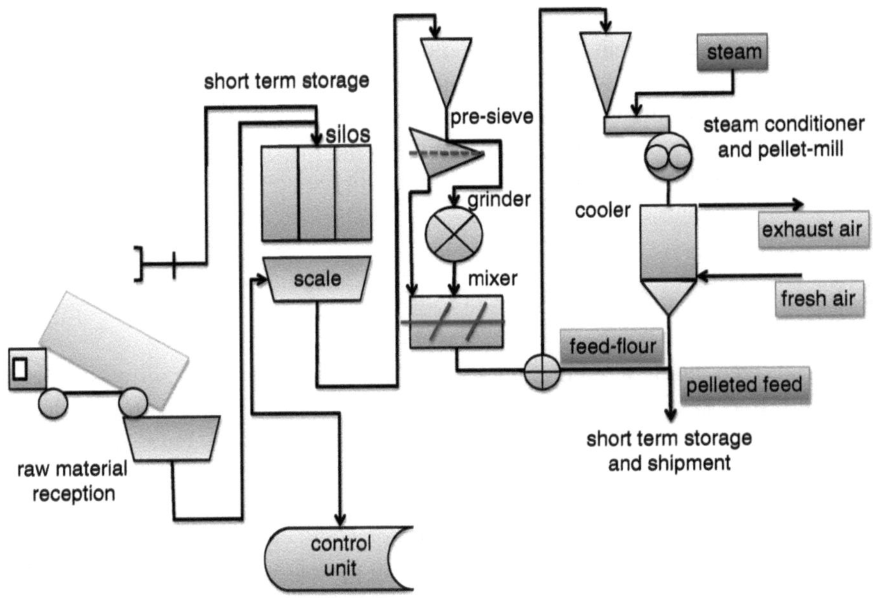

Fig. 1 Flow sheet of a single-line compound feed production plant

feed-flour produced by the use of a roller mill has less fine particles and shows characteristically a narrow particle size distribution range. The optimum particle size for the best result has been a matter of controversy for almost as long as feeds have been pelleted (Behnke 2006). Moreover, technological properties of feeding stuff powders e.g. miscibility, flow ability and segregation are influenced by the properties of their components especially by the particle size distribution. A coarser particle size may be linked to segregation effects on the one hand and reduced pellet durability (PDI) on the other.

However, a pre-sieve should be installed to separate fines before grinding in order to unload the milling process and reduce its energy consumption. Fines are returned to the material flow after the mill. Different raw materials will cause different grinding characteristics and will reveal different particle size distributions (Löwe 2011). An exemplary composition of a fattening pig feed is given in Table 1.

Depending on the recipient animal species different particle sizes and respective feed structures need to be achieved. This makes it difficult to provide mean values for feed. Even cow feed is varying according to animals' age and lactation phase. Latest results in animal nutrition research on the influence of the particle size of fattening pig feed show that a coarser structure of the feed decreases the passage rate in the stomach, which leads to an increase of the dry substance matter of the stomach content. This panders the growth of lactobacilli, which reduces the pH-value in the stomach. In turn, a low pH-value minimizes the survivability of ubiquitary Salmonella or other bacteria and therefore results in a better livestock

Table 1 Composition of a fattening pig feed

	Component	Amount in %
1	Wheat	40.70
2	Barley	15.00
3	Rye	13.00
4	Soy extraction meal	8.20
5	Wheat bran	8.00
6	Rape seed expeller	6.50
7	Wheat gluten	6.20
8	Calcium carbonate	1.35
9	Vegetable oil	0.50
10	L-Lysin-Monohydrochloride	0.37
11	Sodium chloride	0.18

hygiene (Kamphues et al. 2007a, b; Große Liesner 2008). At the moment these results tend to promote a coarser grinding of feeding stuffs. It can be seen that due to the animals' digestion the grinding process does not solely result in smaller particles. It turns out that it is even more complex since the overall particle size distribution is important.

An adequate feeding of animals requires—besides grain-based components—an addition of essential mineral components (e.g. salt, limestone), trace elements and additives, with the use of high potential additives being regulated by law (European Commission and Council 2005). Ensuring equitable feeding of all animals in stock, animal health as well as quality demands for food of animal origin, micro components in ppm amounts must be mixed homogeneously with macro components. The mixing of feed compounds is generally conducted as discontinuous processes whereas several discontinuous mixers are available. Following the mixing step small amounts of liquids—e.g. molasses and vegetable oils may be applied to avoid segregation of flour-like feeding stuffs. The desired mixture homogeneity as well as the segregation tendency depends on the particle size distribution of the feeding stuffs and the agglomeration rate of the fines (Kirchner 2010).

To improve digestibility, the preparation of feedstuffs became a major focus in animal nutrition research. Pelleting may generate an improved animal digestion performance. The integrated thermal process modifies starch content and improves tastiness of feeding stuffs. Other advantages are the avoidance of selective feeding and an improvement of technological characteristics like flow ability as well as livestock management. Pelleting in combination with short-term conditioning is most common as hydrothermal treatment in feed mills; about 80 % of the European compound feed productions are pelleted. Conditioning by adding saturated steam to the mash feed is used to prepare the flour for compaction. The material moisture as well as the temperature is increased depending on the amount of steam as well as the dissipative heating by mechanical energy input (mixing and kneading). The steam will condense uniformly on the particle surfaces and thus

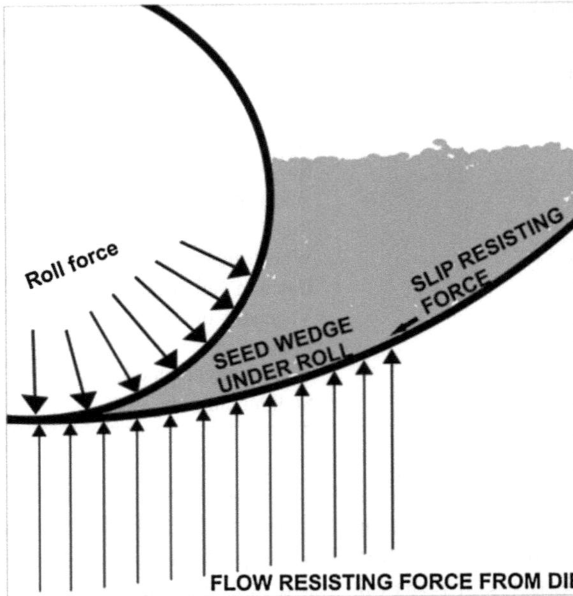

Fig. 2 Roll and die relationship (Behnke)

provide lubrication of the pellet, reducing the wear and increasing the production rate. Liquid bridges between the particles will ensure sufficient pellet durability. The feed mixture is uniformly applied as a layer to a die ahead of the rollers. The formation of the pellets actually occurs at the nip between the rolls and the die (Behnke 2006; Friedrich et al. 1978) as illustrated in Fig. 2.

The die provides not only the final diameter of the pellet but also the resistance force on the feed. Furthermore it has a direct influence on the throughput rate, the pellet quality (Behnke 2006) and on the energy consumption. The main characteristic product property is the pellet durability, which is itself characterized by different test methods as Pellet Durability Index (PDI) and directly correlated to the energy input while pelleting. The energy demand of the pellet mill is between 10 and 25 kWh/t depending on machinery and process parameters as well as on material properties (Friedrich et al. 1978, Glandorf and Genschel 2010). If a value of 25 kWh/t is assumed for the pellet mill, the whole plant energy consumption cannot result in 45 kWh/t and should therefore be closer to the 120 kWh/t as mentioned in Sect. 1. Even on the technical level there are efforts identified: latest machinery developments in the field of pellet mill make use of a direct drive with the motor directly connected to the main drive shaft and the dies, thus reducing transmission losses. The direct drive also allows that the circumferential die speed can be adjusted during the production process. This means that the mill can be adjusted to suit any feed formulation, in many cases without requiring a change of the die. A reduction in the energy consumption of 15–25 % of the pellet mill may be possible (Ziggers 2011). In order to ensure the storage and transport ability, the moisture and heat added to the processed material have to be adjusted to a maximum moisture content of 14 % and a material temperature of about 5–10° above surrounding conditions by cooling and drying the pellets in counter flow or belt coolers.

Feed safety is increasingly recognized as an important topic in the feed-to-food value chain and subject to EU legislation (European Commission and Council 2003, 2005; FEFAC 2009). However, the production units for livestock have become larger driven by economies of scale. In this concern, hygienic regimes—including feed hygiene—are essential to prevent disease outbreaks and secure hygiene and feed ingredients of animal origin. Heat treatment (hydrothermal and hydrothermal-mechanical treatment) offers an opportunity to eliminate bacteria from feed components and is common in feed mills especially with respect to the nutritional value and digestibility (Heidenreich 1999). Hydrothermal-mechanical treatment of feeding stuffs can be done by the use of an annular gap expander conducting a high-temperature short-term process (HTST) to reduce bacteria count as well as Salmonella on the one hand and increase starch digestibility on the other. A new development is the crown-expander, an adaption of the traditional annular gap expander, which additionally holds a die and a cutting device. The die is designed as a ring-shaped extension of the horizontal tubular case of the expander (Fig. 3) that could either be designed with longitudinal slots or holes. This ring-shaped extension allows the shaping of the feed e.g. in pellet-form (Graf Reichenbach 2011). These "crown-pellets" show a significantly coarser structure in comparison with conventional pelleted feed since additional no grinding effect occurs.

3 Life Cycle Approach

As mentioned in the beginning of the chapter there is a lack of data concerning LCA in animal feed. Information on feed processing can hardly be found nor is it accessible in databases (European Commission—DG Joint Research Centre—Institute for Environment and Sustainability).

Fig. 3 Crown expander after shutdown

This section is intended to highlight some information on the feed production for further inclusion in LCAs performed by LCA experts. As mentioned in the previous section the feed processing is a complex system with many variables and a mean value for feed processing in general is not possible.

The situation of feed compound production is different to those in standard serial production, like car manufacturing for example. Input resource parameters into the plant are always related to natural resources that vary in their composition. The composition is dependent on weather conditions, the land use and other influencing factors that we are not aware of. Nevertheless, these factors cannot be controlled and therefore the varying quality parameters have to be handled. At the end of the production process the processor guarantees a defined output quality which is strictly controlled. Today the process control within the plant can only be influenced by experts who know whether the production runs smooth or not and they may interfere with the process.

Additionally, the process is run according to customer demands and not to the product itself. Various customers for pig feed for example are demanding a different recipe for pig feed. Also, the demand for feed is provided to the feed processor on short notice, which leads to feed processing within 24 h with no time for reasonable logistical planning.

The life cycle approach in this paper only applies to the manufacturing plant. Therefore the system borders embrace the storage silo for incoming products, the processing steps and finally the storage of the finished compound feed, as highlighted in Fig. 4.

An optimally conditioned compound feed relates to positive animal growths due to optimal feed efficiency with regard to digestion. We consider this matter as an optimal life cycle approach of feed intake. This is only possible through efficient

Fig. 4 Life cycle of animal products and system border of this paper

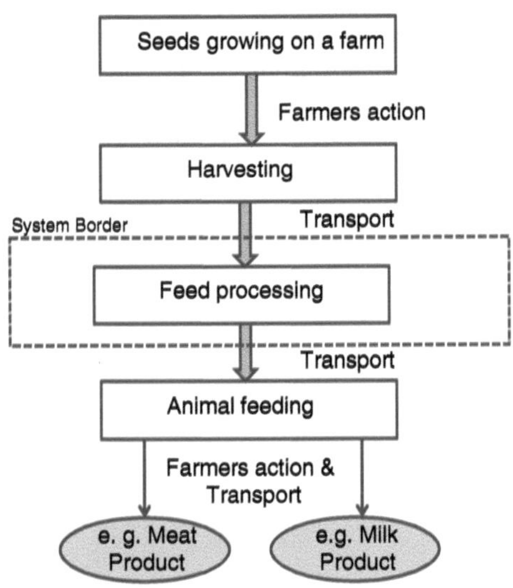

product control and measuring. The next section will therefore describe an attempt at achieving an efficient feed processing by taking into account the previous and the following life cycle chain.

4 Energy Intensive Processes

4.1 Energy Saving Potential on Technological Level

There is only little information available on the energy demand of the compound feed processing and if available—in most cases—the numbers are not scientifically proven. However, thermal and hydrothermal treatment of the feed mixtures is correlated with an energy demand that may vary between 45 and 120 kWh per ton whereas variations are caused by differing refinement steps. The total yearly electricity consumption of a medium-sized compound feed plant (240,000 t/a) is approx. 6 GWh.

The energy demand of the processing steps of compound feed production is exemplarily illustrated in Fig. 5. The data was conceived by an industrial feed processing plant in 2011. The compacting (= pelleting) has the highest share in electrical consumption with no conditioning included since hot steam is provided by fossil fuel usage. Pelleting in combination with short-term conditioning in general shows the highest energy demand of about 60 % of the total energy demand (Friedrich 1983).

With regard to energy saving potentials specific conditions of the industrial compound feed production need to be taken into account. The expenditures for raw materials amount to 80 % of the retail price of the feeding stuffs. A cost optimized production is therefore essential for competitiveness of small and medium sized compound feed producers since energy consumption is the main driver in cost

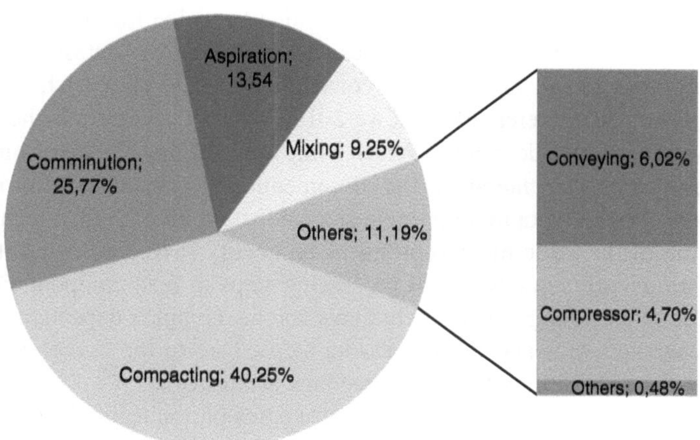

Fig. 5 Energy demand distribution in percentage for compound feed processing by example

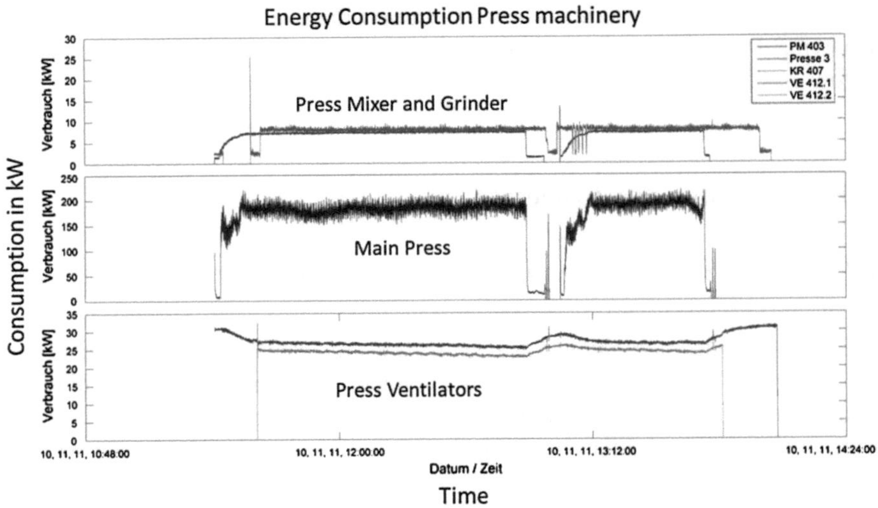

Fig. 6 Energy consumption of all machinery related to the compacting process in one batch (Kottowski 2012)

reduction. The feed production by customer preferences is linked to the production of small batch sizes causing a high number of recipe changes combined with idle times as can be seen in Fig. 6. The figure demonstrates relevant machinery involved in the pressing step: the main process in compound feed production. The main press (in the middle of Fig. 6) is showing the batch size with constant energy consumption and an interruption of operation followed by a decline of energy consumption in the middle and an increase of energy consumption after starting the process again. Other machinery like press ventilators are running continuously. Moreover, short-term commissioning is state of the art in feed processing which makes it difficult to optimize job-order. Customer habits are difficult to change but they have a high potential in energy savings due to recipe logistics.

Long-standing research work focuses especially on an optimization of the pelleting process as the main energy consuming process (Friedrich et al. 1978; Friedrich 1983; Heidenreich 2001; Löwe 2010). Of major concern is the reduction of the specific energy demand (kWh/t) while maintaining the feed quality and safety parameters. Depending on the aforementioned research results in animal nutrition concerning an optimal particle size and structure of feed, a technological optimization of the grinding step promises an energy saving potential, too. Nevertheless, an overall strategy for all processing steps in feed compounding is still missing due to the wide range of ingredients and the complex dependencies of raw material characteristics and the processing steps. Due to the complexity of feed processing the possibilities for energy savings are restricted.

At the moment, several energy saving strategies on the technological level are conceivable and currently investigated in several applied research projects. The highest potential for energy savings are described as follows:

- Step grinding

Step grinding is a size reduction accomplished in steps or stages, usually incorporating two grinders. The first grinder is performing a pre-grinding e.g. of single components and the second mill is grinding the total mixed portion. Additionally, combined sieving may enhance the energy efficiency, too (Heimann 2006). Advantages of step grinding are a lower specific energy demand in comparison with a conventional grinding and a better process control concerning the aspired particle size distribution. The energetically optimized configuration of the grinding equipment to be used with respect to a minimal specific energy demand is one of the interests to be investigated in a current research project at the International Feed Association (IFF) in Brunswick, Germany. In this concern, the applicability of a disc wheel grinder will be tested.

- Pelleting aids with surfactant or lubrication characteristics

Fat as a simple lubricant reduces friction in the die but leads to softer pellets with a higher PDI. The addition of pelleting aids with surfactant characteristics shows an energy saving potential especially when pelleting becomes difficult to process formulas (Cooksley 2010). Pelleting aids may bind for example fatty components while increasing the reactive surface area and creating a higher number of contact points between particles (Löwe 2010), resulting in a high PDI. However, the correlation of cause and effect using surfactants as pelleting aids is still unclear. Further research is needed.

- Improvement of heat and mass transfer efficiency

A currently started research work at the IFF Research Institute is focusing on the optimization of the heat and mass transfer efficiency while pelleting in combination with short-term conditioning and cooling. The feasibility of waste heat recovery is an important aspect, too. Experimental investigation by systematic variation of process conditions and variations of heat and mass flows shall lead to energy functions for the conditioning step, the pelleting step and the cooling step. Their discussion is expected to show energetically optimized operating points.

- Simulation-based job-order

The impact of customer specific production of small batch sizes on the specific energy demand is illustrated in Fig. 7 as resulting from experiments. These results have already been confirmed by cooperating feed processing companies. Main energy consuming factors are identified as idle times while recipes are changed and rinsing batches, which may be produced to avoid cross-contamination. This is a drawback of batch production that cannot be avoided in feed processing theses days.

Producing larger lots from a number of batches will reduce idle times to a minimum. A strategy development for an optimized job-order by the use of material flow simulation may help to reduce the total energy consumption and

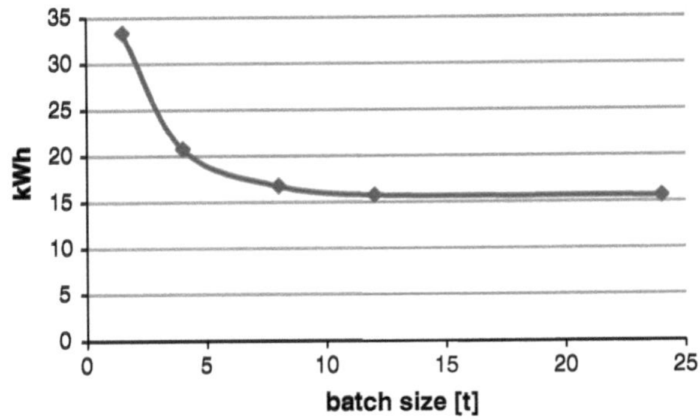

Fig. 7 Influence of the batch size [t] on the specific energy demand [kWh/t] (Glandorf and Genschel 2010)

improve the process efficiency of the compound feed production on common single-line production plants.

In perspective, the enforcement of cooperative and interdisciplinary research work in the global sustainability context is necessary for an overall sustainability approach and strategy in the feed and food industry.

4.2 Energy Saving Potential on Process Control Level

After some technical options for energy savings had been identified in the previous sections there is also potential on the process control level. Most of the process is run automatically with a quality control for the end product. Often the process is overdriven to exceed the requested product quality. A low quality product is not only rejected by the customer but also the animals themselves are refusing to eat. Therefore, the feed producer intends to deliver a product above the desired quality.

A first step towards an optimized process control that relates to varying input and process parameters is the use of online measuring devices. Online measuring devices may represent analyzers based on NIR technology (for immediate analysis of e.g. water, starch and protein) or automatic particle analyzers for immediate particle size analysis. By connecting these devices to the process control they are able to influence the control directly and immediately. The particle analyzer will provide information on the actual particle size distribution and could give a signal if the material was getting either too coarse or too fine. The process control will then correct the hammer mill for example by adjusting the speed. The question to be solved is how the devices will control the processing. Experts have to define rules on how the moisture content of the input materials influences the subsequent process steps. By adding an optimum of steam to the conditioning step the following pelleting step can be run at optimal energy consumption. A similar relation regarding the particle size has already been discussed in Sect. 4.1. The rules for the process and material property dependencies can be implemented in an expert system that communicates with the process control. The Bremen University is currently running a project that is considering the implementation of an expert system in a compound feed process (Redecker and Pehlken 2012).

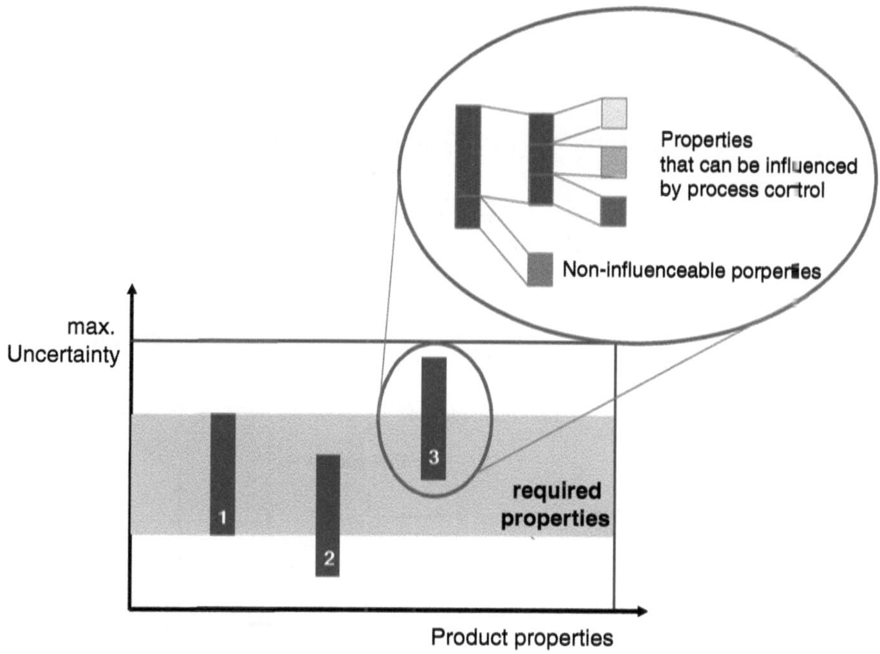

Fig. 8 Relation between product properties and their uncertainties

The aim is to relate the energy consumption not only to the process steps but also to the various recipes. By identifying the carbon footprint for each recipe the energy input into the process can be optimized without influencing the product quality. This is the highest challenge, since the quality of the end product cannot be changed due to animal demands. For example industrial products (like automobile, aircraft, etc.) can be constructed by reducing the weight without loosing quality properties. In the food and feed processing industry reducing weight means that the animals or we (humans) are put on diet. The research focus lies on the investigation of energy efficiency of the production process without changing the end product quality. This can be achieved by monitoring the process through online measuring systems and by developing an expert system that is able to control the process using fuzzy rules.

The definition of the fuzzy rules depends on the identified uncertainties within the process. During the handling of renewable materials uncertainties can be found in their composition due to the fact that natural resources with varying properties, like for example their water content, are handled. The varying input quality affects the process control because of different conditions during the process. A higher water content in the input results in adding less fluids during the processing for example. In addition to water there are other parameters like protein, ash and starch content, etc.

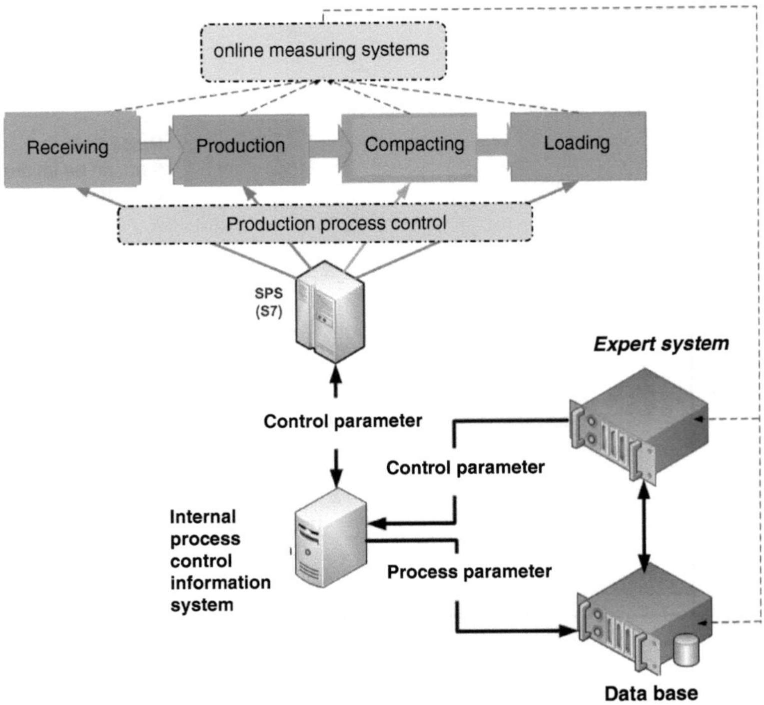

Fig. 9 Expert System in feed compound production (Redecker, Pehlken 2012)

The uncertainties in the feed process can be divided into (see also Fig. 8):

- uncertainties that can be influenced by the process control (adjusting the water content by adding water)
- and uncertainties that cannot be influenced by the process control (chemical composition).

The relation of an expert system with the process control under a quality assurance approach is depicted in Fig. 9. The development of the expert system of the production process is performed according to uncertainty management. Uncertainties are present at any time in the process and therefore have to be taken into account during the decision making process. The uncertainties are formed into fuzzy rules and are implemented as a new software tool (expert system) that supports the process control. The process control can be optimised receiving product properties (like water, protein, and starch content for example) that are delivered by online measuring devices. The combination of fuzzy rules and the product parameters measured online will lead to an efficient process. This information can even help assessing the process set up and logistics (what is the best cyclic order of the recipes to consume the least energy?) (Forgy 1982; Beierle and Kern-Isberner 2008).

The system will handle uncertainties combining the defined rules with the natural variation of product properties that cannot be influenced. As a result the system is handling data ranges instead of data points. Due to changing product properties (e.g. water content in the input) the system is able to correct the water content during the process and adjusts the conditioning process before pelleting.

5 Conclusions and Further Research

The global demand for agricultural products, meat and milk is still increasing. This results in a higher demand for feeding stuffs whereby the importance of the feed industry in the sustainability of livestock production and consumption is unambiguous. To provide high quality and cost effective products feed producers and machinery suppliers focus on the energy efficiency of machinery and the efficiency of feed processing.

The complexity of feed processing for various customer and animals has been shown in this chapter. Recipes are changing many times during the day and the input product quality is changing according to weather conditions. One mean value for feed processing is therefore impossible.

The implementation of an intelligent network process control is an important step to include experts knowledge in the process control and to face upcoming sustainability topics like changing raw materials in diet compositions and the improvement of feed efficiency exemplarily by the reduction of nutrient losses. An intelligent network is not yet state of the art in the compound feed production but online measuring systems are increasingly involved in assisting quality control. Further research in finding rules on process dependencies and product quality is still necessary.

References

Behnke KC (2006) The art (science) of pelleting. In: American Soybean Association (ed) ASA technical report series-feed technology, Singapore, pp 5–9

Beierle C, Kern-Isberner G (2008) Methoden wissensbasierter systeme. Grundlagen, algorithmen, anwendungen. Springer, New York

Cooksley J (2010) Processing aid boosts feed mill productivity and efficiency. Feed Compounder (May 2010):28–31

Dalgaard R, Halberg N, Hermansen JE (2007) Danish pork production. An environmental assessment. http://www.lcafood.dk/djfhus82ny.pdf. Accessed 29 May 2012

European Commission—DG Joint Research Centre—Institute for Environment and Sustainability EUROPA-Site on LCA Tools, Services and Data—LCI dataset area. http://lca.jrc.ec.europa.eu/lcainfohub/datasetArea.vm. Accessed 31 May 2012

European Commission and Council (2003) Directive 2003/99/EC of the European parliament and of the council of 17 November 2003 on the monitoring of zoonoses and zoonotic agents. Zoonosen-Verordnung

European Commission and Council (2005) Regulation (EC) no 183/2005 of the European parliament and of the council of 12 January 2005 laying down requirements for feed hygiene. Futtermittel-Hygieneverordnung
FEFAC (2009) Environment report. Accessed 29 May 2012
FEFAC (2010) Feed and food statistical yearbook 2010. Accessed 29 May 2012
Forgy CL (1982) Rete: a fast algorithm for the many pattern/many object pattern match problem. Artif Intell 19(1):17–37. doi:10.1016/0004-3702(82)90020-0
Friedrich, Jansen, Robohm (1978) Der Verdichtungsvorgang bem Pelletieren von Mischfutter. Untersuchungen der Einflussgrößen mit dem Ziel, den Verschleiß zu reduzieren und den Energiebedarf zu lenken. Schlussbericht zum AiF-Vorhaben 3154, Braunschweig-Thune
Friedrich W (1983) Warum ist dampfzugabe beim pelletieren wirtschaftlicher als wasser? die mühle + mischfuttertechnik 120(14):173–178
Glandorf H, Genschel AHJ (2010) Energieeinsparpotenziale in einem Mischfutterwerk
von Graf Reichenbach H (2011) The KAHL Crown Expander—A Way Out of the Feed Structure Dilemma. http://www.akahl.de/akahl/files/Prospekte/Prospekte_englisch/en_SD-MM-Der-Kronenexpander-E-0411.pdf. Accessed 10 Apr 2012
Große Liesner (2008) Zur Gesundheit der Magenschleimhaut von Absetzferkeln unter dem Einfluss der Futterstruktur und—konfektionierung. Dissertation, Tierärztliche Hochschule
Heidenreich E (1999) Benefits and side effects of expanding. Feed Tech 3(1):21–23
Heidenreich E (2001) Untersuchungen zur alternativen Mischfutterherstellung mit optimiertem Energieverbrauch. Studies on alternative compound feed production with optimised specific energy consumption. Kraftfutter/Feed Mag 11:394–408
Heimann M (2006) Step grinding for improved efficiency of grain and meal products. In: American Soybean Association (ed) ASA Technical report series-feed technology, Singapore, pp 53–55
Jeroch H, Flachowsky G, Weissbach F (eds) (1993) Futtermittelkunde. Gustav Fischer Verlag, Jena
Kamphues J, Brüning I, Papenbrock S, Mößeler A, Wolf P, Verspohl J (2007a) Lower grinding intensity of cereals for dietetic effects in piglets. Livestock Sci (109):132–134
Kamphues J, Papenbrock S, Visscher C, Offenberg C, Neu N, Verspohl J, Westfahl C, Häbich A (2007b) Bedeutung von Futter und Fütterung für das Vorkommen von Salmonellen bei Schweinen. Übers Tierernährg (35):233–279
Kirchner A (2010) Liquid addition to feeding stuff powders as an aspect of feed quality and safety. Archiva Zootechnica 13:55–61
Kottowski C (2012) Analyse von unscharfen Parametern in Materialflüssen mit Erstellung eines Carbon Footprints. Diploma thesis, Bremen University
Lardy G (2007) Feeding coproducts of the ethanol industry to beef cattle. http://www.ag.ndsu.edu/pubs/ansci/beef/as1242.pdf. Accessed 09 Nov 2012
Löwe R (2010) Factors affecting pelleting and energy consumption. Grain Feed Milling Technol 121:14–17
Löwe R (2011) Technological aspects of feedstuff comminuting and compacting. In: DLG-Verlag (ed) Proceedings of the society of nutrition physiology band 20. Berichte der Gesellschaft für Ernährungsphysiologie. DLG-Verlag, Frankfurt am Main, pp 147–149
Noras B (2012) Absatz- und Umsatzwachstum für die Futtermittelindustrie. Ursachen: hohe Rohstoffkosten und positive Entwicklung der Tierhaltung, Deutscher Verband Tiernahrung, Bonn
Redecker M, Pehlken A (2012) The impact of processing natural resources on uncertainties in life cycle assessment, SETAC World Congress 2012, Berlin
Steinfeld H, Gerber P, Wassenaar T, Castel V, Rosales M, de Haan C (2006) Livestock's long shadow: environmental issues and options. Food and Agriculture Organization of the United Nation, Rome
Thomassen M, van Calker K, Smits M, Iepema G, de Boer I (2008) Life cycle assessment of conventional and organic milk production in the Netherlands. Agric Syst 96(1–3):95–107. doi:10.1016/j.agsy.2007.06.001

Tukker A, Huppes G, Guinée J, Heijungs R, Koning A de, van Oers L, Suh SGT, van Holderbeke M, Jansen B, Nielsen P (2006) Environmental impact of products (EIPRO): analysis of the life cycle environmental impacts related to the final consumption of the EU-25

Ziggers D (2011) AllAboutFeed—processing: power in motion—New Kubex T pellet mill. http://www.allaboutfeed.net/processing/pelleting/power-in-motionnew-kubex-t-pellet-mill-12595.html. Accessed 10 Apr 2012

Integrated Framework for Life Cycle-Oriented Evaluation of Product and Process Technologies: Conceptual Design and Case Study

Uwe Götze, Andrea Hertel, Anja Schmidt, Erik Päßler and Jörg Kaufmann

Abstract In order to capture long-term effects of technology decisions regarding technical, economic or ecological targets of a company and its supply chain partners, life cycle-oriented evaluation of product and process technologies is necessary. For enabling a systematic evaluation, methodological support is needed. Thus, this chapter presents a framework for life cycle-oriented evaluation of product and process technologies consisting of three pillars: generic product, life cycle and process models, a decision theory-based procedure model allowing a structured and integrated evaluation, and a method and tool set. The framework will be illustrated by a case study of life cycle-oriented evaluation of alternative technologies for manufacturing mountain bike-frame components.

1 Introduction

Due to the dynamics of markets and the increasing customer needs concerning quality and unique selling propositions of products, manufacturing companies have to develop and apply innovative product and process technologies to gain and sustain long-term competitive advantages. The related challenges are obvious: the potential of technologies has to be identified and evaluated, taking short- and long-term effects regarding economic, technical or ecological targets of the company itself and its supply chain (SC) partners (customers, suppliers etc.) into account. This comprises manufacturing costs but also other life cycle (LC) costs as well as

U. Götze (✉) · A. Hertel · A. Schmidt
Chair of Management Accounting and Control, Chemnitz University of Technology, Chemnitz, Germany
e-mail: uwe.goetze@wirtschaft.tu-chemnitz.de

E. Päßler · J. Kaufmann
Department of Lightweight Structures and Polymer Technology, Chemnitz University of Technology, Chemnitz, Germany

E. Henriques et al. (eds.), *Technology and Manufacturing Process Selection*, Springer Series in Advanced Manufacturing, DOI: 10.1007/978-1-4471-5544-7_10,

revenues and the underlying functional performance of technologies and environmental criteria. At the same time, technologies typically imply a lot of complex and interdependent decisions concerning product and process alternatives (about product functions and working principles, materials, manufacturing processes etc.) which determine the economic competitiveness of the technology and its success or failure. Because costs and performances are strongly influenced by decisions made in early stages of the design process, the evaluation of technology alternatives should be applied even in these stages—mostly with little information available.

Therefore, instruments which support a highly significant, structured and economic decision-making in due consideration of the variety of SC partners, targets, alternatives, influencing factors, and effects over time are required. Despite literature provides various approaches which may be used to solve specific partial problems of technology assessment and decision-making, such holistic or integrated instruments do not exist as far as we know. Furthermore, most of the methods and tools available for product designers focus on production costs and barely include LC cost aspects (Janz et al. 2005). Thus, this chapter intends to contribute to close these gaps by presenting a theory-based integrated framework for LC-oriented evaluation of product and process technologies which is applicable by manufacturing companies in different stages of product and process design—when bigger part of the product and process performance is determined.

2 The Framework and its Pillars

2.1 Overview

The proposed framework for LC-oriented evaluation of product and process technologies consists of three pillars:

- generic product, life cycle and process models,
- a decision theory-based procedure model which allows a structured and integrated evaluation of technologies and the corresponding decision-making, considering targets, (sub-)alternatives, influencing factors, and consequences,
- a method and tool set which can be applied in the different steps of the procedure model.

In the following sections these pillars are discussed in more detail. Moreover, LC-oriented evaluation of alternative technologies is shown in a case study using the example of mountain bike (MTB)-frame components.

2.2 Product, Life Cycle and Process Models

In order to enable the handling of the complexity coming along with technology evaluation we recommend using generic product, life cycle and process models. The term "product" denotes a physical object which is created by a manufacturing enterprise (Gielingh and Suhm 1993). This can be a complex end product which is made up of various individual components like parts or groups of parts (DIN 6789 1990) or a component itself. Since technology decisions are usually made about components (Reuter 2007), even the requirements, analysis and evaluation tasks concerning alternative materials and manufacturing processes have to be performed on component level.

For any kind of design task it is necessary to have an idea of the intended structure of the product which, thus, has to be described. Therefore, basic **product models** in the form of function-oriented and production-oriented product breakdown structures are useful. The function structure of a product represents its functions and the logical interrelations between them (Janz et al. 2005); it defines the decomposition of the required overall functioning of the product (derived from customer requirements) into more detailed sub-functions. By means of these functional structures, essential physical components, which implement functions and sub-functions, can be identified and described from another, production-oriented modeling perspective (Ulrich 1995; Reuter 2007; for an example of assigning machine components to machine tool functions for a metal cutting machine tool see ISO 2012). The resulting production-oriented structure (shown in Fig. 1 by the example of a car) represents the hierarchical physical composition of the components—of parts, components (in a strict sense), (function-oriented) systems, and modules. Thus, it predetermines the logical structure of assembly processes (Göpfert 2009) as well as the structure of product information documents such as lists of parts and engineering drawings (DIN 6789 1990) which may also be interpreted as product models in a wide sense.

The product models described above are essential for engineering processes and therefore are embedded in design methodologies like functional analysis (VDI 2221). Furthermore, they are useful for generating and managing architectural knowledge, which is required to understand the linkages between customer requirements, system parameters, and component performance specifications in order to support product and process technology as well as make or buy decision-making (Venkatesan 1992). It should be noted that various other types of product models may be used for supporting product (technology) engineering by visualization, simulation, optimization etc. (Cameron and Gani 2011)—at least if the term product model is understood in a wide sense. Especially in the field of computer aided design, numerous distinctive modeling approaches do exist (such as geometrical modeling, kinematic modeling, finite element analysis, product dynamic behavior modeling, ergonomic and esthetic modeling (Yan 2003)). However, these modeling approaches and the product models generated by them

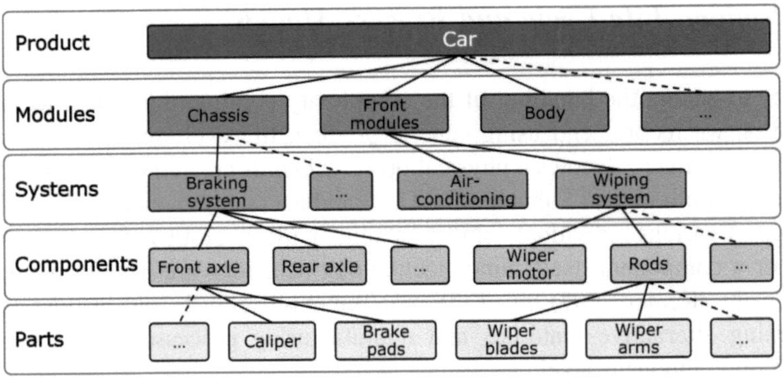

Fig. 1 Decomposition levels of a car (according to Pfaffmann 2001)

are methods and tools facilitating product (technology) design in more detail (for methods and tools see Sect. 2.4).

Since components typically affect the performance of the end product with respect to technical, environmental and economic criteria (e.g., weight, process time, energy consumption, or costs) over their individual entire LCs, a product analysis has to cover these different criteria as well as the LCs of components. Furthermore, the definition of LC phases of components and products is of particular importance for LC evaluation. This kind of structuring of LCs may be based on existing product LC models. However, to show the interrelatedness between product and component decisions, a generic model of an **integrated product and component LC** consisting of four LC phases is presented here:

- **design**—includes pre-production activities like product and component design, process planning as well as prototyping of components (most of technology-related decisions have to be made in this phase),
- **manufacturing and assembly**—comprises the manufacturing of components like parts and their aggregation to more complex components or complete end products (assembly often constitutes the last stages of production),
- **usage and service**—starts after end products have been sold to end customers who use them and, thereby, consume resources including energy; additionally maintenance and repairing tasks have to be fulfilled,
- **end-of-life**—after the end of usage, end-of-life options like part reuse (after disassembly), product reuse, material recycling, disposal, and incineration with or without energy recovery have to be chosen (Mazhar et al. 2005).

Figure 2 demonstrates that LCs of components are embedded in products' LCs. Like the product as a whole, components have to be designed and manufactured. Furthermore, they influence the usage properties after having been assembled to an end product, have to be repaired and/or replaced, recycled etc. Enabled by a modular product structure—different actors (symbolized by the small boxes) can be involved in the LC: the focal company and its suppliers with design and

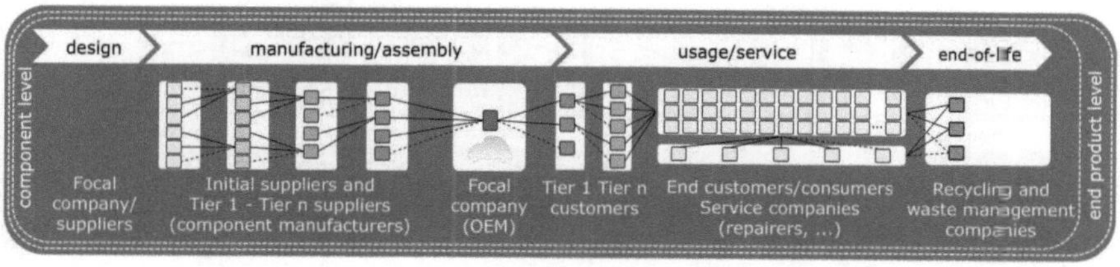

Fig. 2 Integrated product and component LC and actors involved (similar to Lambert et al. 1998)

production activities, Tier 1—Tier n customers including retailers as well as consumers, service companies, and companies fulfilling recycling and waste management tasks.

For LC-oriented evaluation of product and process technologies, it is not sufficient to consider LC phases of products and components—the processes running during the phases influence the technology performance as well and, thus, have to be included into the analysis. In order to handle these processes, it seems to be reasonable to use **process models**. Process models exist in different interpretations and types (an overview is given by Gadatsch 2010). The type of process models which is focused here aims at the systematization of processes (for a different type of more detailed process models, the ITO-models, see Sect. 2.4). A prominent example is the SC Operations Reference-model that pre-structures the processes in SCs and, thus, supports communication among SC partners, helps to manage processes, and facilitates SC integration by providing common process definitions (Supply Chain Council 2010).

Following such approaches, we propose to use **generic process models** for a structured and systematic describing and visualizing of processes in the different **LC phases**. These models are expected to pre-structure the relevant processes which may occur in the integrated product and component LC as well as their relationships and to define reference processes with standard descriptions. They support a consistent presentation of processes and the setting of boundaries in a way that makes sense to the evaluation tasks. Companies can use them as a framework for their process modeling which, thus, is simplified: the process models solely have to be adapted to a specific context.

Figure 3 shows that the processes, on the one hand, have to be structured in a horizontal direction, following the value chain and the LC. For example, the SCOR-model classifies plan, source, make, deliver and return processes. On the other hand, a (vertical) decomposition into hierarchy levels enables a more detailed description and analysis of the LC-related processes. Thus, like product structures, process structures can be hierarchical. Then, processes are decomposed into more detailed sub-processes on a lower level, e.g., production activities into process groups, processes, operations (e.g., all activities done with one machine), and passes (e.g., single moves of a tool or an employee) enabling a finer detail view (Cameron and Gani 2011; Gielingh and Suhm 1993). As a prerequisite, processes have to be

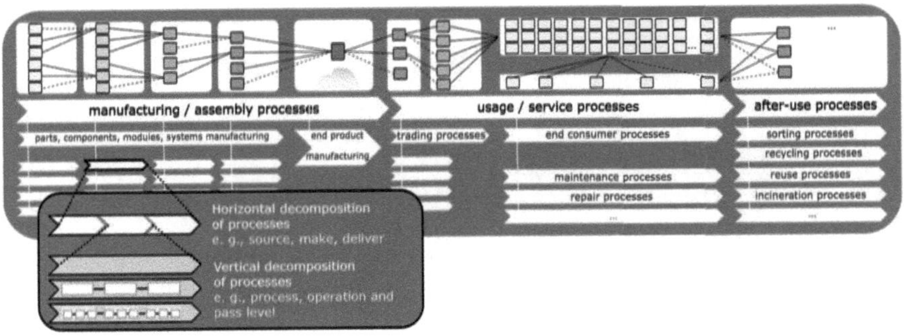

Fig. 3 LC-oriented process model

understood as a linked sequence of operations and those as a sequence of passes (Gaitanides 1996). The level of detail should be defined according to the underlying problem. For example, it can be useful to concentrate on a special type of process which is expected to be important for evaluation tasks (e.g., on make processes) and to analyze it on a high level of detail (e.g., on pass level).

Generally, reference models can be developed by the scientific community, by other company-independent institutions (as the Supply Chain Council), or by single companies. Besides the SCOR-model, some other structuring approaches are provided by literature and by standards that can be used for horizontal process decomposition. According to DIN 8580 (2003), for example, transformation of materials into outputs can be realized by manufacturing processes of six groups: primary shaping, forming, cutting, joining, coating, and changing of material properties. Besides, production also involves additional activities which enable material flows but do not directly impact the conversion of materials. According to VDI 2860, these activities can be divided into three groups: handling as well as stocking and transporting. In addition, there are activities which, though they support production, are only indirectly associated with production activities (e.g., maintenance and repair of machinery or production engineering). From this classification a quite detailed structure of the underlying production (or make) processes can be derived Fig. 4.

Which processes and how detailed they are mapped, is dependent on both the purpose of the models and the modeling point of time. Depending on whether

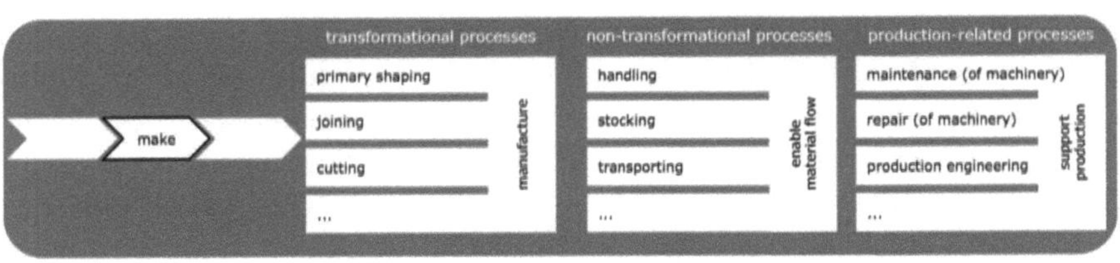

Fig. 4 Horizontal decomposition of production (make) processes

modeling takes place within early design stages or operations scheduling, the planning team avails itself of various data including product information documents such as material data, list of parts, engineering drawings, or prototypes (DIN 6789 1990) as well as machine data sheets, company-specific data collections describing production processes, technical literature, or process-specific data provided by technology suppliers (Eversheim et al. 1997).

2.3 Integrated Procedure Model for Life Cycle-Oriented Evaluation of Technologies and Decision-Making

Since LC-oriented design or selection as well as evaluation of product and process technologies is typically characterized by a high complexity, it is necessary to decompose these tasks with regard to both products and processes (Nahm and Ishikawa 2004). Thus, the technology design or selection problem should be systematically separated into different partial problems in order to enable solving these problems and synthesizing the results to a total solution (see VDI 2221). This principle can be adapted to the evaluation tasks which are inherent in technology design or selection processes: Fig. 5 shows the decomposition of evaluation tasks referring to products as well as processes and the aggregation of evaluation results to overall performance measures.

Taking the principle of decomposition and linking of problems/solutions into account, the **integrated procedure model for LC-oriented technology decision-making** (Fig. 6) consists of two levels: the level of LC model, concerning technology selection as a whole and fostering the global comparison of the candidate technologies, and the level of sub-models, taking up the product and process-related partial problems. In order to enable defining the scope of analysis and evaluation, the initial step of the **procedure model (at the LC level)** comprises the

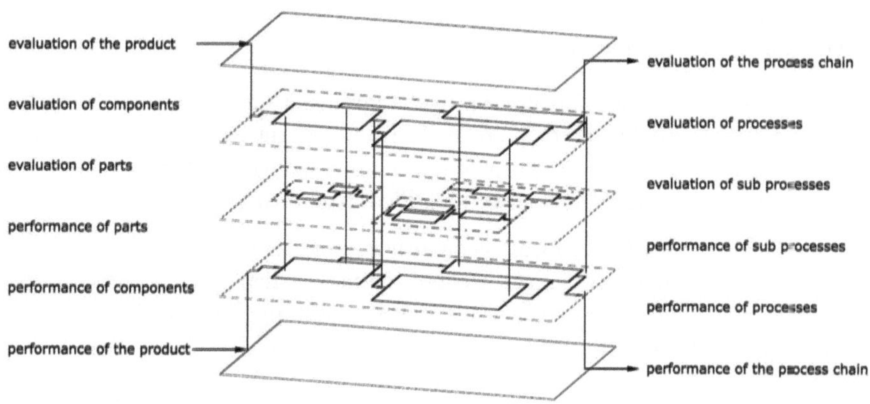

Fig. 5 Principle of decomposition and linking of problems/solutions (based on VDI 2221)

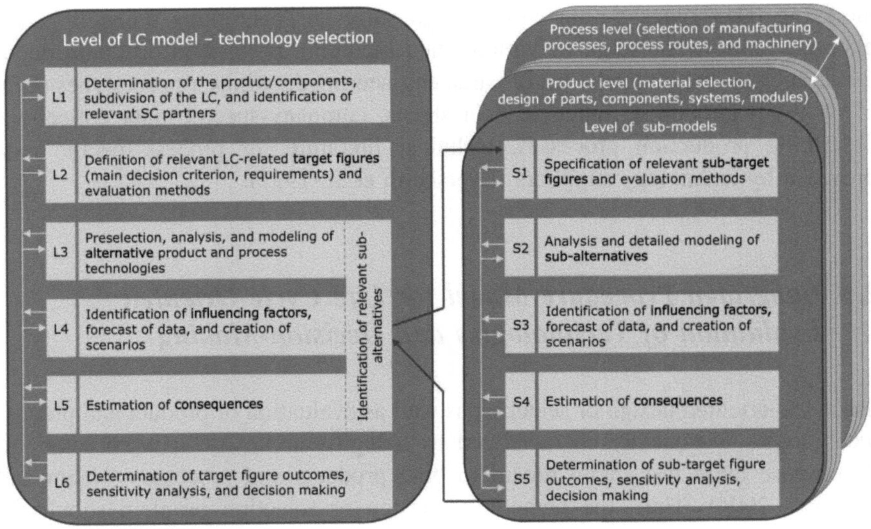

Fig. 6 Integrated procedure model for LC-oriented technology decision-making (similar to Götze et al. 2010 and Hertel et al. 2011)

preliminary characterization of the product and its components (establishing a first product breakdown structure), the definition and subdivision of the LC, and the identification of relevant SC partners (L1).

Afterwards, it is possible to define all relevant LC-related **target figures** (L2). Those can, in principle, refer to various dimensions such as:

- technical/customer-oriented—e.g., conforming to requirements, ranging from basic functional to esthetic requirements,
- social/employee-oriented—e.g., reduction of health risks,
- political/legal—e.g., adhering to safety-related norms or other regulations and legislations,
- ecological—e.g., recyclability of components and/or
- economical—e.g., reduction of LC costs.

For systematization and integration of the targets, a preliminary target system should be developed by defining main decision criteria as well as subordinate targets/restrictions. Therefore, the potential non-monetary or monetary effects of the candidate technologies have to be identified. *Non-monetary effects* can refer to

- mechanical, physical, chemical, geometrical, optical and tactile properties of products and components (e.g., strength, stiffness, shape, complexity, surface finish, weight, corrosion and heat resistance or energy absorption capacity, or suitability for manufacturing and assembly) as well as

- process characteristics with respect to time (e.g., manufacturing and assembly time or holding time), quality (e.g., susceptibility to damage or scrap rate), or flexibility.

On the one hand, by these primary effects secondary effects may be caused, e.g., image effects (resulting from performance criteria or recyclability etc.) or systemic effects (e.g., a lower overall weight of products resulting from downsizing of components or from reducing the number of parts by integrating functions). On the other hand, both primary and secondary effects lead to *monetary effects* (on *costs and revenues*, and other monetary items that are not focused here). For example, manufacturing process time has an impact on manufacturing costs; the weight of components influences the fuel costs of a car; technical features, additional value adding services or ecological aspects (Meffert et al. 2008; Busch and Beucker 2004; DIN EN 60300-3-3 2005) affect customer's benefits, their purchasing decisions and, thus, revenues (Fig. 7).

From an economic point of view, LC profits of technologies should be chosen as main decision criterion and other multi-dimensional criteria as subordinate targets/restrictions—due to the fact that most companies primarily pursue monetary targets. Additionally, non-monetary effects (e.g., process characteristics, weight of components) are influencing factors of costs and revenues and, thus, reflected by the results of profit calculation (their significant consideration in the calculation presupposed).

For enabling the calculation of LC profits, relevant LC cost and revenue items have to be identified from the decision-makers′ point of view. For example, the cost items may comprise costs of product and process design and prototyping, manufacturing and assembly costs, service and usage costs including warranty and liability costs, costs of energy consumption, maintenance and repair costs, insurance costs, as well as end-of-life costs, e.g., for disassembly, product reuse,

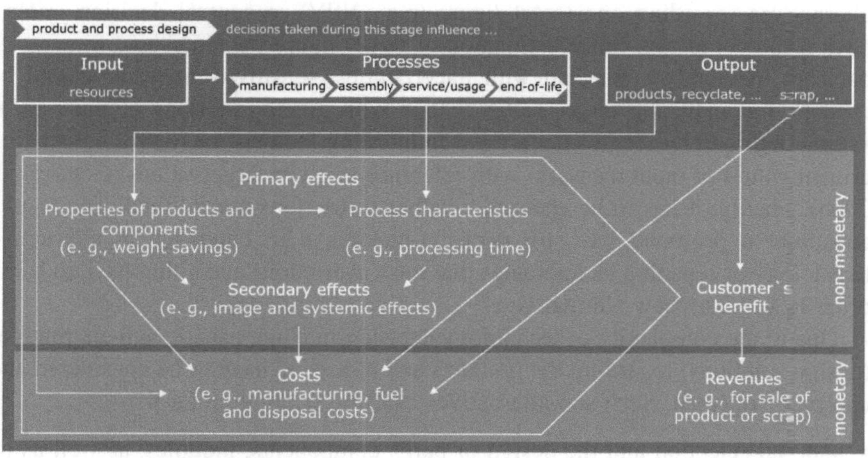

Fig. 7 Pattern of non-monetary and monetary effects within LC

material recycling, disposal, or incineration. Furthermore, it has to be analyzed how the decision-maker and other relevant actors of a SC participate in LC costs and revenues taking into account the relationships between them (the revenues of one actor become part of the costs of the subsequent member of SC, the concrete revenues (and costs) are influenced by competitive and legal aspects, contractual arrangements etc.). This analysis is particularly important against the background that companies will only (be able to) engage themselves in SCs if they expect (and realize) economical success.

After relevant target figures are defined, adequate evaluation methods have to be chosen. For calculating the LC profits of alternative technologies, methods of investment appraisal, especially the net present value (NPV) method, are predestined (see Sect. 2.4). Non-monetary effects may be aggregated to an utility value (for utility value analysis see Götze et al. 2008).

In the next step, **alternative product and process technologies** (L3) have to be preselected as far as possible, usually based on product requirements (Eversheim et al. 1997). For the evaluation of the candidate technologies it is essential to describe them with the help of product and process models (see Sect. 2.2).

Success of technology decisions depends on a multitude of internal and external **influencing factors** such as condition of production facilities, competitive environment, demand, legislation and politics, social values, technological trends, prices etc. Therefore, crucial influencing factors as well as their (in-)direct impacts on each other should be identified and data have to be forecast by using internal and external information systems. Additionally, the forecasts of single influencing factors are aggregated to consistent scenarios (L4). Afterwards, the **consequences** caused by preselected technology alternatives and influencing factors have to be forecast as well, using adequate analytical or subjective methods of prognosis (L5).

After all relevant consequences have been estimated, the target figure outcomes can be calculated (L6). Dependent on the concrete target system, the outcomes of alternatives regarding one target figure (e.g., NPV) or several decision criteria have to be compared. Since estimation of target figure outcomes and forecast of data are inevitably connected with uncertainties, sensitivity analysis should be used (i) for analyzing target value change with given variations of input measures—and thereby identifying critical influencing factors—or (ii) for calculating critical values of input measures with reference to given target values (Götze et al. 2008). Finally, it has to be checked whether restrictions are fulfilled. The results constitute a profound base for decision-making, e.g., choosing an alternative technology or rejecting technologies that miss restrictions and going back to L3 in order to generate new alternatives.

Due to the complexity of technology decision-making, usually sub-alternatives (design solutions) regarding the products and processes have to be considered and chosen or rejected. Such sub-alternatives can, for example, refer to

- material selection and the design of parts, components, modules, as well as to
- manufacturing processes, process routes, and machinery.

An exemplary task is the selection of prepregs with different matrix or fibres materials with respect to the preselected technology (e.g., thermoforming) and limitations (e.g., regarding space or weight) resulting from the higher level systems. In general, sub-alternatives are interrelated in different ways: interdependencies exist between products and processes (e.g., between materials, functions, design, and manufacturing processes) as well as between different process and product levels (e.g., the design of a component often depends on the design of its lower level components (except for the case of product modularity when components can be designed concurrently (Nahm and Ishikawa 2004))) Again, product and process models may be helpful to structure the sub-alternatives (see 2.2).

The corresponding decision problems are identified within the steps L3–L5 (primarily in L3). In order to enable systematic and consistent handling of sub-alternatives, they should be evaluated and compared on a separate subordinate level, the **level of sub-models**. At this level, the solution of (partial) decision problems is possible following steps S1–S5 (which have the same inner logic as steps L2–L6 of LC model level).

By reducing complexity with the help of sub-models, technology decision-making can be tackled more effectively. Additionally, the procedure model can be used on different levels of detail in very early design phases as well as in later phases.

2.4 Methods and Tools

LC-oriented evaluation of product and process technologies has to be supported by different methods and tools (Almeida et al. 2011). Simplifying, it is differentiated between methods and tools for (quality) engineering, economic evaluation, and life cycle assessment of environmental impacts.

Methods and tools for **(quality) engineering** can be integrated into multidimensional LC-oriented engineering approaches such as product design for manufacture, assembly, service, and recycling (design for X) (Kuo and Zhang 1995; Klein 2010) or life cycle engineering (Peças et al. 2009; Hesselbach and Herrmann 2011). They include (Pahl et al. 2007):

- quality function deployment (QFD),
- theory of inventive problem solving (TIPS),
- design of experiments (DOE),
- failure mode and effects analysis (FMEA),
- computer aided engineering (CAE) including computer aided design (CAD),
- simultaneous/concurrent engineering and rapid prototyping,
- various simulation tools (e.g., finite element analysis (FEA) including multi-body, process or structure simulation), and
- materials data bases.

These methods and tools are used to evaluate and to design high quality products/components in a relatively short time and a cost-efficient way (Klein 2010).

Methods and tools for **economic evaluation**, which are focused here, comprise LC-oriented approaches that can be subsumed under the term *life cycle costing (LCC)*. Assuming that objects like products and components or the inherent technologies follow a certain path of (organic) life, LCC is intended to evaluate total LC costs and revenues considering trade-offs between phases and periods. For structuring and describing LCs, generic LC models can be used (see 2.2). These LC models might also show typical courses of costs and revenues (or cash flows) within the LC. However, for evaluating LC costs and revenues of objects, specific mathematical LC models are necessary (Götze 2010).

These models can differ especially with respect to the underlying system of management accounting. They might base on: direct costing concepts, flexible standard costing based on margin, discounted cash flow methods of investment appraisal. Since technology decisions often implicate long-term investments and long-term effects, LCC based on discounted cash flow methods of investment appraisal, especially the net present value method (Götze et al. 2008), is recommended (see 2.3, L2). It allows for making short-term as well as long-term monetary effects of technology alternatives comparable by incorporating the time value of money. NPV usually represents the present value of all future cash inflows (CIF) diminished by the cash outflows (COF). These figures are used here as well—if the common terms costs and revenues are used, they have to be understood in the sense of cash flows.

Both CIF and COF incur at different points in time (t) within the LC, whose end is indicated by T, and have to be discounted to the beginning of the planning period (t = 0), when technology decisions are made. For discounting cash flows, the factor q^{-t} (q = 1 + i, i = discount rate) is used (Eq. 1).

$$NPV = \sum\nolimits_{t=0}^{T} (CIF_t - COF_t) \cdot q^{-t} \tag{1}$$

For a systematic modeling of all monetary effects occurring in the integrated product and component LC, the differentiation of LC costs and revenues by type and time and with reference to the product breakdown structure is suggested here. An applicable starting point for structuring monetary effects is provided by DIN EN 60300-3-3 (2005), which recommends estimating the relevant costs of different categories for each component in each LC phase (which should be refined to periods in LC phases). However, the allocation of cost categories to LC phases of a product seems to be quite rough and insufficient. Especially, the process dimension would be neglected. Costs of different categories can incur for various processes within one phase (e.g., material costs of different production processes in the manufacturing phase as well as of infrastructure maintenance processes). Thus, it is suggested to introduce a further level—the level of LC-related processes—resulting in a LC-related cost and even revenue breakdown structure enabling the

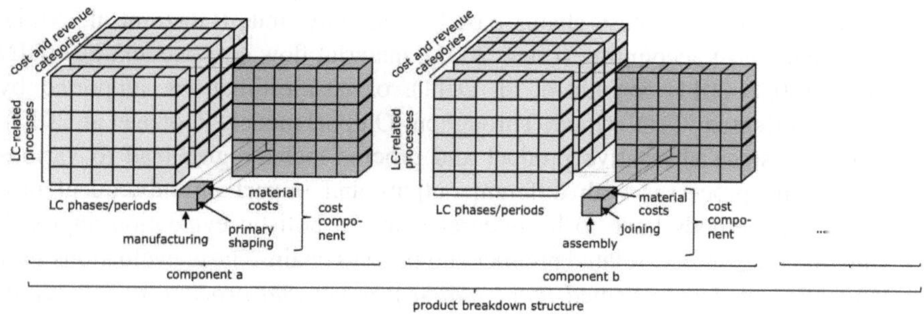

Fig. 8 Differentiation of LC costs and revenues (based on DIN EN 60300-3-3 2005)

identification and classification of all relevant cost and revenue categories according to product and process structure (Fig. 8).

For the forecast of the LC-related monetary effects, particular methods are needed. Since the applicability of methods of (standard) cost accounting is limited (to a large extent because they imply the assumption of given structures, products, processes etc.), instruments of (development-concurrent) *pre-calculation of costs* have to be considered. In literature, various pre-calculation methods such as engineering-based, analog or parametric cost methods are proposed (DIN EN 60300-3-3 2005; Ehrlenspiel et al. 2010). Most of these methods focus the analysis of historical costs and/or rely on costs estimated by experts on the basis of experiences with similar products or technologies. However, especially for highly innovative and new technologies there are usually little cost data available. Additionally, the estimation of costs incurring in later LC phases (such as service/ usage and after-use) is neglected. Furthermore, despite of the existence of methods and tools for *market research*, which enable the anticipation of customer needs and users behaviors (Meffert et al. 2008; Ehrlenspiel et al. 2010), there is a lack of methods for estimating revenues. Summarizing, a need of further methodological support is obvious.

Process-based evaluation methods might be approaches for reducing these problems. Because costs are driven by processes to a high degree, processes of the LC phases, their resource consumption (input), the way they are run (throughput) and their output should be examined in order to understand cost incurrence. In literature, different process-based evaluation methods such as activity-based costing or transaction costing are suggested (Miller and Vollmann 1985; Cooper and Kaplan 1988; Horváth and Mayer 1989; Kaplan and Anderson 2007). These methods have in common that cost driving processes as well as "cost drivers" (activity units which determine costs) are identified and relations between process quantity, resource consumption and costs are disclosed. Therefore, LC-oriented evaluation of product and process technologies can draw on these methods as well as on advanced approaches such as supply chain costing (LaLonde and Pohlen

1996), evaluation of process chains based on activity units (Götze et al. 2011), environmental cost accounting (Loew 2003), material flow cost accounting) (DIN EN ISO 14051 2011; Sygulla et al. 2011), or cost estimations supported by technically oriented "ITO" (Input-Throughput-Output) models (Götze et al. 2012).

However, since alternative product and process technologies lead to various manufacturing processes with different outputs and impact on subsequent processes, these methods have to be integrated into a holistic evaluation approach tackling the relevant LC-related product and process chain. Those evaluations may refer to *integrated LC-oriented input–output product and process models* representing the technology decision problem in an aggregate way by combining product and process models (see 2.2). According to a typical systems framework used for conceptualization of evaluation models (Cameron and Gani 2011), Fig. 9 shows an example of an integrated LC-oriented input–output product and process model.

Such an integrated LC-oriented input–output product and process model may help decision-makers to evaluate technologies considering targets and (monetary) consequences of the expected behavior of suppliers, subsequent customers, or other SC partners. Moreover, these models and the corresponding evaluations could be a basis for a cost-oriented optimization of the entire SC and the distribution of cost savings to participating partners. Based on this, they seem to facilitate the development of a supply chain life cycle costing (SC LCC) approach.

Finally, methods and tools for **LC assessment** have to be mentioned. These approaches are designed for analyzing and evaluating environmental impacts, e.g., carbon footprints caused by products and processes (DIN EN ISO 14044 2006; 14044 2006; Baumann and Tillman 2004). In this respect, they also contribute to the LC-oriented evaluation of and decision making about product and process technologies. Since environmental and economic analyses have common bases (some of the relevant inputs and outputs are identical), further research should strive for integrating these different domains (for an integrated methodology based on LCA and activity-based costing see da Silva and Amaral 2009).

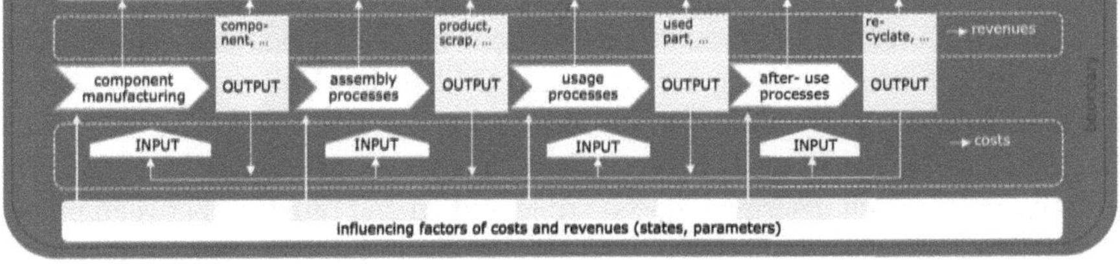

Fig. 9 Example of an integrated LC-oriented input–output product and process model

3 Case Study—Life Cycle-Oriented Evaluation of Alternative Technologies for Manufacturing Mountain Bike-Frame Components

In order to demonstrate the framework, its pillars, and its applicability, the case of a Full-Suspension-MTB of Ghost Bikes GmbH (a limited liability corporation) is used. The company faces the challenge to decide about the use of three technologically different alternatives of manufacturing a frame component. Thus, these alternatives have to be analyzed and evaluated with regard to the technical and economical LC performance (Fig. 10).

Currently, there are two versions available on the market, one made of aluminum (manufactured by drop forging) and one consisting of carbon fiber reinforced plastics (CFRP)) with thermosetting matrix (manufactured by manual wet lay-up) (Kroll et al. 2012). Furthermore, an innovative process for manufacturing three-dimensional contoured thermoplastic sandwich structures (t3S) is developed at Chemnitz University of Technology (CUT), combining two established mass manufacturing processes—thermoforming of semi-finished continuously fiber reinforced thermoplastic composites and injection molding (Müller et al. 2012).

Within the procedure model (see Sect. 2.3), initially, the relevant components have to be determined. In the case study, the component "linkage", transmitting the loads from the seat stays to the damping system, its LC phases (manufacturing, assembly, service/usage, and after-use) as well as its effects on the end customers are focused (L1). In the second step, target figures and evaluation methods are defined. On the one hand, they include technical performance criteria (among others, mechanical, chemical and thermal requirements, the weight, and the available mounting space). In order to identify mechanical requirements, loads acting on the frame component are determined under realistic usage conditions by field tests (jumps, pedaling) as well as laboratory tests. On the other hand, LC profit measured by NPV is established as main decision criterion (L2).

Within the steps L3-L5/S1-S5, the three technology alternatives are analyzed and evaluated with regard to their technical requirements and NPVs: The geometry of all components is modeled by using CAD. The stiffness as well as the strength properties of the three linkage versions are analyzed with the help of FEA, based on loading conditions resulting from typical usage scenarios throughout the LC of

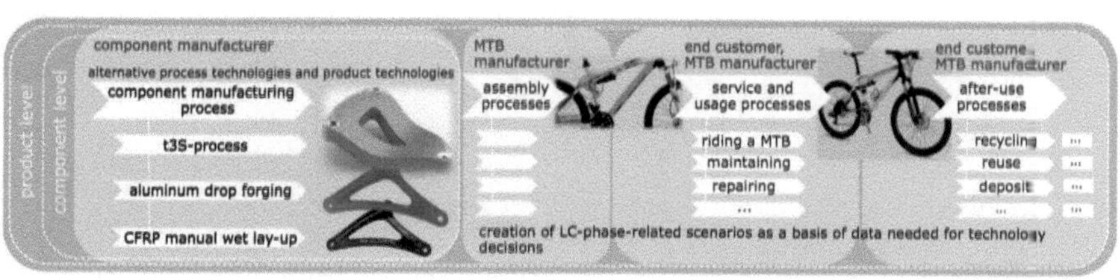

Fig. 10 Manufacturing alternatives of a MTB-frame component

the product. Within the FEA, specific technological sub-alternatives of the t3S-version) (e.g., different materials, varying geometry, laminate stack-ups) are analyzed as well (at the level of sub-models). Furthermore, the results are verified by mechanical testing within laboratory tests. The results from subordinate analysis and evaluation show that the currently used versions (aluminum and CFRP) as well as the innovative thermoplastic part (t3S) meet minimum requirements in terms of weight, stiffness, and strength. All other requirements such as esthetic properties or corrosion resistance are assumed to be met equally good/bad.

For calculating the NPV, it is necessary to identify the technology-dependent cash flows which are relevant from the perspective of the MTB manufacturer as decision-maker (Hertel et al. 2011). Ghost Bikes GmbH purchases the frame component from a supplier (component manufacturer). Moreover, the company has warranty obligations during the service/usage phase as well as recycling/disposal obligations after the use of the products. Other cost items such as manufacturing and assembly costs etc. are assumed to be identical. Since warranty and liability as well as recycling/disposal costs play a minor role and, thus, can be neglected for the sake of simplicity, the focus is on the purchase costs (or COF) and their influencing factors. Revenues (CIF) mainly result from the sale of the product (MTB) and depend on sales volumes and retail prices which mainly result from the benefit expected by the customers (Fig. 11). This expected benefit is influenced by the technical features of the MTB. Especially the weight seems to be a decisive buying criterion and the end customer's willingness to pay is supposed to increase in line with reducing weight of MTB. At this point, the component "linkage" affects revenues: The weight of the component directly influences the weight of the MTB and, thus, indirectly the revenues.

On account of this, it has to be estimated how the candidate manufacturing alternatives resulting in linkages (and bikes) with different weights influence the revenues. Since there was no complete information about weight-dependent retail prices available, an own data base had to be created. Therefore, the relationship between retail prices and weight of products of Ghost Bikes GmbH (derived from the Ghost Bikes product catalogue 2012 for national sales) was analyzed.

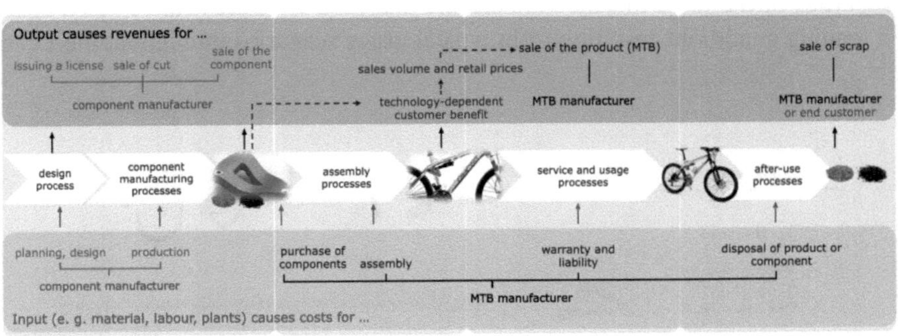

Fig. 11 LC-related cost and revenue effects of technology decisions in the case of MTBs

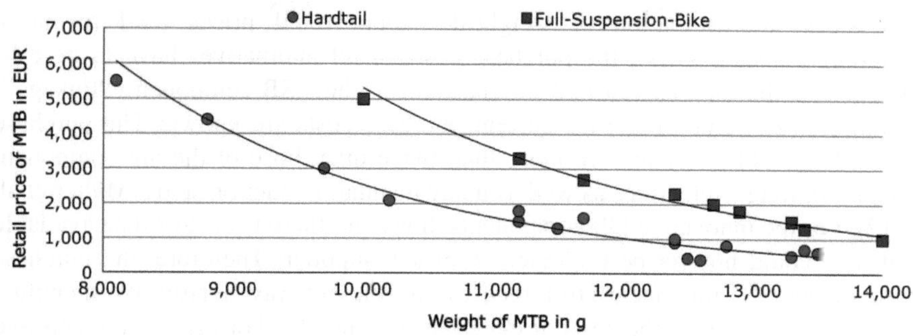

Fig. 12 Weight-dependent retail prices of bikes

Figure 12 shows the retail prices of several GHOST cross-country MTBs and their corresponding weights differentiated into Hardtail-Bikes (HTB) and Full-Suspension-Bikes (FSB).

FSB have to be considered separately because they show different characteristics and are made up of more components (damping system, mechanical joints, bearings) resulting in fundamentally higher weights. As indicated, light bikes are more expensive than comparatively heavy bikes, so the graph reflects the aforementioned relation between weight and price. The mathematical relations between retail prices and weight of MTB in gram (w) can be described by exponential functions. Similar relations between weight and retail prices were identified for products of a second company, indicating the significance of results.

By setting up the first derivative of the exponential functions, the decrease of the retail price per additional gram for each initial bike weight can be determined (shown in Fig. 13). By using these derivatives, the shift of retail price can be approx. forecast for each initial weight (and a given amount of weight increase or decrease).

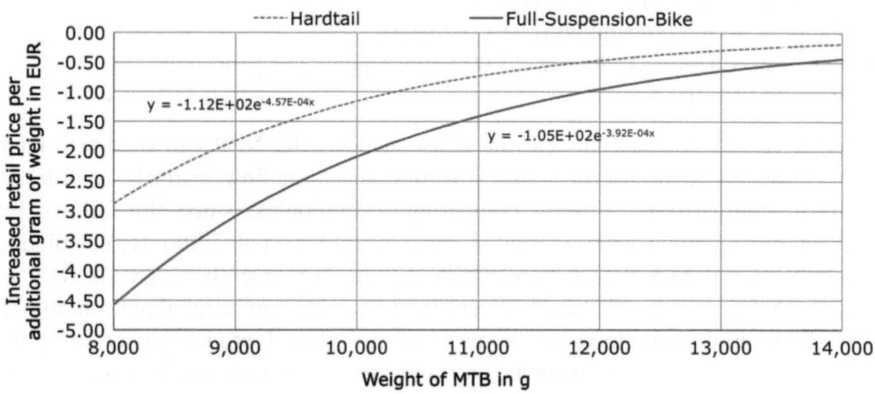

Fig. 13 Retail price differences in dependence on initial weight

Besides the estimation of weight-dependent retail prices (and revenues, assuming constant sales), the purchase costs of all alternatives have to be estimated. For the two established alternatives of the FSB-component "linkage", aluminum (alu) and CFRP, the current purchase costs are known: The purchase costs of the CFRP-version are more than twice than those of the alu component because of long cycle times as well as many manual production steps (Müller et al. 2012). For the innovative t3S components, however, there is no cost data available and they could not yet be requested from any supplier. Therefore, the purchase costs of the t3S component are forecast in a different way: Firstly, the manufacturing costs (including the costs of materials consumed, labor involved, plants and tools used) are estimated from a component manufacturer's perspective with the help of input–output process models (see Sect. 2.4). Secondly, a profit mark-up of the component manufacturer is assumed and added.

For the data forecast concerning t3S, various approaches are used (at this point, it should be noted that the case study is primarily intended to demonstrate the application and applicability of the procedure model—it is not claimed to represent reality with the accurateness necessary in a real-life decision situation):

- process times are measured during the manufacturing of prototypes at CUT,
- material prices are requested from suppliers, and
- influencing factors such as economic life time, costs of tools and machines, and labor cost rates are estimated by experts on the basis of experiences with similar technologies.

For the concrete estimation of data, assumptions have to be made: Amongst others, a gradual decrease of purchase costs of the highly innovative t3S component by 2 % p. a. within 10 years is expected (enabled by experience curve effects). Additionally, Ghost Bikes GmbH assumes increasing purchase costs of the CFRP component (about 7 % p. a.) due to past experiences regarding the rise of labor and shipping costs in low-wage countries. Furthermore, a series of other economic assumptions are made:

- profit mark-up of the component: 20 %
- planning period considered (t_0-T): 10 years
- annual production volume (x): 5,000 components per year, and
- discount rate (i): 5 %.

After having forecast all relevant consequences, the NPV can be determined. Here, it seems to be useful to use a differential cash flow analysis in order to simplify comparisons between competing technologies, since this allows for omitting the identical aspects of the alternatives (Götze et al. 2008). By calculating the NPV of differential cash flows (NPV_{diff}), the difference between the NPV of two technologies (e.g., t3S/alu) is determined as a measure of relative profitability. The differential cash flows are a result of the annual production volume, the purchase costs of the components (pc) and the relevant weight dependent retail price difference ($RPD^{t3S/alu} = RP^{t3S}$-RP^{alu}) (as shown in formula 2 for the example of t3S/alu):

$$NPV_{diff}^{t3S/alu} = \sum_{t=0}^{T} x((-pc_t^{t3S} + RPD^{t3S/alu}) - (-pc_t^{alu})) \cdot q^{-t} \tag{2}$$

Accordingly, the NPV_{diff} shows the relative profitability of the components in dependence on the weights obtained by the use of different technologies. If the NPV_{diff} is positive, then the t3S component has a higher NPV than the alu component and is, therefore, relatively more profitable (referring to the example in formula 2).

The NPV method (with differential cash flows) was used for a pairwise comparison of all manufacturing alternatives. As shown in Fig. 14, both, the t3S and the CFRP technology, are relatively profitable compared to the alu component (because for all initial weights (w) the NPV_{diff} are positive). This is primarily due to the fact, that the CFRP component has only 54 % and the t3S variant 66 % of the alu parts weight. The comparison of the t3S component with the CFRP technology reveals that the t3S is only relatively profitable for initial weights higher than approx. 11,800 g (as indicated in Fig. 14).

As these results show, the purchase costs and/or the weight of the t3S component have to be reduced, in order to obtain profitability also for bikes with a weight below approx. 11,800 g. Since the estimated purchase costs of the t3S are already lower (about approx. 40 %) than those of the CFRP alternative while its weight is higher, it is nearby that the weight of the t3S should be reduced. These findings motivated for an analysis of the components of a t3S linkage. As a result, the metallic inserts used in the original design (with a large proportion of component weight (12.5 %) and of material costs (20 %)) were identified as a starting point for weight reduction. Consequently, an alternative solution was developed replacing metallic inserts by featuring an in situ molded plastic load application element. Because this modified t3S component shows less weight, the competitiveness of the t3S process could be improved. Provided that this modified version is technically feasible (to be screened within L3-L5/S1-S5), the innovative t3S

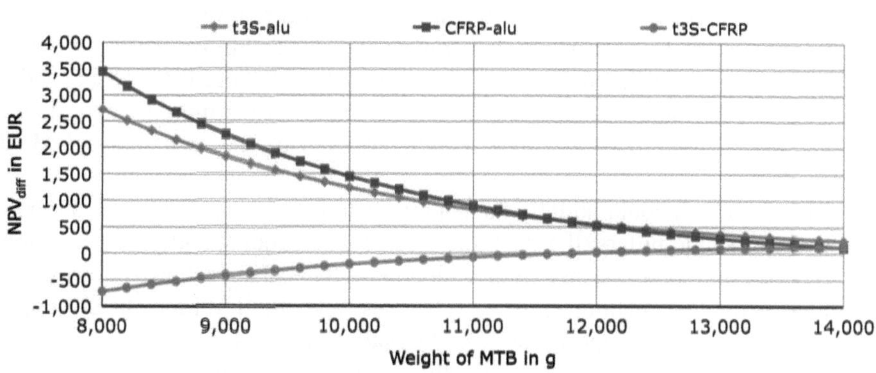

Fig. 14 Differential NPVs in dependence of initial weights (pairwise comparison of t3S, alu, and CFRP)

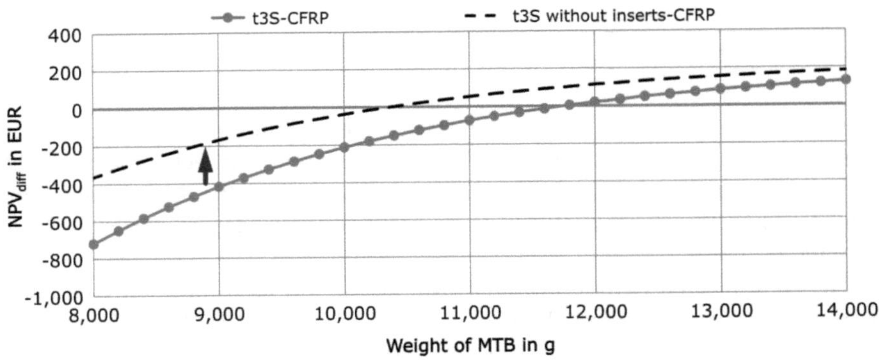

Fig. 15 Differential NPVs in dependence on initial weights (pairwise comparison of original t3S, CFRP, and modified t3S)

component with a molded plastic load application element (presented by the black broken graph in Fig. 15) is economically advantageous compared to the CFRP component beginning by a weight of approx. 10,400 g.

If the MTB manufacturer chooses the NPV as the main decision criterion, the t3S technology would be advantageous for Full-Suspension-MTBs with an initial weight higher than approx. 10,400 g (L6). Summarizing, the case study clearly shows the potential of the procedure model—the economic analysis allows for the comparison of technology alternatives and inspired a further development of an innovative technology leading to a higher competitiveness (presented by a higher NPV) of this technology.

4 Conclusion

For supporting technology decisions, a life cycle-oriented evaluation of product and process technologies has to be recommended. Contributing to the methodology of such evaluation, an integrated framework, consisting of generic product, life cycle and process models, a decision theory-based procedure model, and a method and tool set, is presented in this chapter. Furthermore, the principle applicability of the procedure model is demonstrated by the case study of MTB components.

Further research could address the sophistication of the framework as a whole as well as of single pillars: the refinement of generic product and process models and product and process modeling techniques, the enhancement of costs and revenues estimation techniques in early phases, and the development of supply chain life cycle costing approaches. Additionally, practical experiences with the framework should contribute to its validation and specification regarding particular fields of application.

Acknowledgments The authors thank the European Union and the Free State of Saxony for their funding (the presented results were mostly achieved within the research project *Economic process for serial production of three-dimensional contoured thermoplastic sandwich structures under flexural stress*). We also thank the Deutsche Forschungsgemeinschaft (DFG) for funding this work within the SFB 692. In particular, we thank our cooperation partner Ghost Bikes GmbH who provided data for the case study.

References

Almeida D, Peças P, Ribeiro I, Teixeira P, Henriques E (2011) Application of life cycle engineering for the comparison of biodegradable polymers injection moulding performance. In: Glocalized solutions for sustainability in manufacturing, proceedings of the 18th CIRP international conference on life cycle engineering. Berlin, Heidelberg

Baumann H, Tillman AM (2004) The Hitch hiker's guide to LCA: an orientation in life cycle assessment methodology and application. Studentlitteratur, Lund

Busch T, Beucker S (2004) Ressourceneffizienz als Managementstrategie für integrierte Produkt- und Prozessbewertung. UmweltWirtschaftsForum 12 2:30–34

Cameron I, Gani R (2011) Product and process modelling, a case study approach. Elsevier, Amsterdam

Cooper R, Kaplan RS (1988) Measure Costs Right: Make the Right Decisions. Harvard Business Review September–October:96-103

DIN 6789 Part 2 (1990) Dokumentationssystematik, Dokumentensätze technischer Produktdokumentationen. Deutsches Institut für Normung e. V. (ed). Berlin

DIN 8580 (2003) Fertigungsverfahren. Deutsches Institut für Normung e. V. (ed). Berlin

DIN EN 60300-3-3 (2005) Zuverlässigkeitsmanagement. Teil 3-3 Anwendungsleitfaden Lebenszykluskosten. Deutsches Institut für Normung e. V. (ed). Berlin

DIN EN ISO 14040 (2006) Environmental management—Life cycle assessment—principles and Framework. Deutsches Institut für Normung e. V. (ed). Berlin

DIN EN ISO 14044 (2006) Environmental management—Life cycle assessment—Requirements and guidelines. Deutsches Institut für Normung e. V. (ed). Berlin

DIN EN ISO 14051 (2011) Environmental management—Material flow cost accounting – General framework. Deutsches Institut für Normung e. V. (ed). Berlin

Ehrlenspiel K, Kiewert A, Lindemann U, Hundal MS (eds) (2010) Cost-efficient design. ASME Press and Springer, Germany

Eversheim W, Kloocke F, Schmitz JW, Winands N, Brandenburg F (1997) Potentials and procedure for the integrated planning of production technologies in the early phases of product development. Eur J Oper Res 100:367–378

Gadatsch A (2010) Grundkurs geschäftsprozess-management. Springer Vieweg, Wiesbaden

Gaitanides M (1996) Prozessorganisation. In: Kern W, Schröder HH, Weber J (eds) Handwörterbuch der Produktionswirtschaft, 2nd edn. Schäffer-Poeschel, Stuttgart, pp 1682–1696

Gielingh WF, Suhm AK (1993) IMPACT reference model, an approach to integrated product and process modelling for discrete parts manufacturing (Research Reports Esprit, Project 2). Berlin

Göpfert J (2009) Modulare Produktentwicklung, 2nd edn. Deutscher Universitätsverlag, Norderstedt

Götze U (2010) Kostenrechnung und Kostenmanagement, 5th edn. Springer, Heidelberg

Götze U, Hache B, Schmidt A, Weber T (2011) Methodik zur kostenorientierten Bewertung von Prozessketten der Werkstoffverarbeitung. Mat.wiss. u. Werkstofftechnik 42 7:647–657

Götze U, Northcott D, Schuster P (2008) Investment appraisal. Methods and models. Springer, Heidelberg

Götze U, Schmidt A, Weber T (2010) Vorgehensmodell zur Abbildung und Analyse des Lebenszykluserfolges von Werkstoffen—Konzeption und beispielhafte Veranschaulichung. Materialwiss. und Werkstofftechnik 41 6:464–475

Götze U, Schubert A, Bierer A, Goller S, Sygulla R (2012) Material- und Energieflussanalyse—Methodik zur Optimierung von Prozessen und Prozessketten. In: Neugebauer R (ed) Proceedings of the International Chemnitz Manufacturing Colloquium ICMC 2012/2nd International Colloquium of the Cluster of Excellence eniPROD, 17./18.04.2012 Chemnitz

Hertel A, Götze U, Schmidt A (2011) Life cycle costing für neue Technologien—das Beispiel innovativer Faser-Kunststoff-Verbunde (FKV). Bus Innovation 3:34–43

Hesselbach J, Herrmann C (eds) (2011) Glocalized solutions for sustainability in manufacturing. In: Proceedings of the 18th CIRP international conference on life cycle engineering, Technische universität braunschweig. Berlin, Heidelberg

Horváth P, Mayer R (1989) Prozeßkostenrechnung. Controlling 4:214–219

ISO/DIS 14955-1 (Draft International Standard) (2012) Machine tools—environmental evaluation of machine tools—Part 1: design methodology for energy-efficient machine tools. International Organization for Standardization. Geneva

Janz D, Sihn W, Warnecke HJ (2005) Product redesign using value-oriented life cycle costing. CIRP Annals—Manufact Technol 54(1):9–12

Kaplan RS, Anderson SR (2007) Time-driven activity-based costing. Boston, Harvard Business Review

Klein B (2010) Kostenoptimiertes Produkt- und Prozessdesign. Hanser, München

Kroll L, Schramm N, Kaufmann J, Kausch M (2012) Mehr als Carbon: Neue Materialansätze für die Entwicklung zukünftiger Hochleistungssportgeräte. 1. Kongress Materials for x-treme Sports. München

Kuo TC, Zhang HC (1995) Design for manufacturability and design for “X”: concepts, applications, and perspectives. In: IEEE/CPMT Int'l electronics manufacturing technology symposium, 446–459

LaLonde BJ, Pohlen TL (1996) Issues in supply chain costing. Int J Logistics Manage 7(1):1–12

Lambert DM, Cooper MC, Pagh JD (1998) Supply chain management. Implementation issues and research opportunities. Int J Logistics Manage 9(2):1–19

Loew T (2003) Environmental cost accounting: classifying and comparing selected approaches. In: Bennett M, Rikhardsson PM, Schaltegger S (eds) Environmental management accounting—purpose and progress. Springer, Dordrecht

Mazhar MI, Kara S, Kaebernick H (2005) Reusability assessment of components in consumer products—a statistical and condition monitoring data analysis strategy. In: 4th Australien LCA Confernence, Sidney:1–8

Meffert H, Burmann C, Kirchgeorg M (2008) Marketing, Grundlagen marktorientierter Unternehmensführung, 10th revis. edn. Gabler, Wiesbaden

Miller, JG, Vollmann, TE (1985) The hidden factory. Harvard business review September–October: 142–150

Müller M, Meyer M, Päßler E, Kaufmann J (2012) Numerische Simulation einer Fahrradrahmenkomponente in thermoplastischer Sandwichbauweise mit Wabenkern. In: Tagungsband zur 13. Chemnitzer Textiltechnik-Tagung Wertschöpfungspotenzial Textiltechnik, Chemnitz

Nahm Y, Ishikawa H (2004) Integrated product and process modeling for collaborative design environment. Concurrent Eng 12(1):5–23

Pahl G, Beitz W, Feldhusen J, Grote KH (2007) Engineering design—a systematic approach, 3rd edn. Springer, London

Peças P, Ribeiro I, Folgado R, Henriques E (2009) A life cycle engineering model for technology selection: a case study on plastic injection moulds for low production volumes. J Clean Prod 17:846–856

Pfaffmann E (2001) Kompetenzbasiertes Management in der Produktentwicklung: Make-or-Buy-Entscheidungen und Integration von Zulieferern. Deutscher Universitätsverlag, Wiesbaden

Reuter M (2007) Methodik der Werkstoffauswahl: Der systematische Weg zum richtigen Material. Hanser, München

Silva PRS, Amaral FG (2009) An integrated methodology for environmental impacts and costs evaluation in industrial processes. J Clean Prod 17:1339–1350

Supply Chain Council (2010) Supply-chain operations reference-model. Overview of SCOR Version 10.0. Pittsburgh

Sygulla R, Bierer A, Götze, U (2011) Material flow cost accounting—proposals for improving the evaluation of monetary effects of resource saving process designs, In: Beitrag zur 44th CIRP international conference on manufacturing systems

Ulrich K (1995) The role of product architecture in the manufacturing firm. Res Policy 24:419–440

VDI-Richtlinie 2221 (1993) Methodik zum Entwerfen und Konstruieren technischer Systeme und Produkte. Düsseldorf

VDI-Richtlinie 2860 (1990) Montage- und Handhabungstechnik. Düsseldorf

Venkatesan R (1992) Strategic sourcing: to make or not to make. Harvard Bus Rev 70(6):98–107

Yan XT (2003) A multiple perspective product modeling and simulation approach to engineering design support. Concurrent Eng Res Appl 11(3):221–234

Life Cycle Engineering Framework for Technology and Manufacturing Processes Evaluation

Inês Ribeiro, Paulo Peças and Elsa Henriques

Abstract Nowadays, the performance of products and processes along the life cycle is a competitive issue for the industrial development as well as a permanent challenge for researchers. This paper proposes a comprehensive life cycle framework to support the selection of sustainable technology and manufacturing processes. Considering both cost and environmental dimensions, process-based models are used to feed data and structure information to assess the life cycle cost and environmental performance of alternative technological processes. This approach is extremely useful when dealing with decision making processes inherent to products development and to the selection of materials and/or technologies in early design stages. In parallel, a technical evaluation of candidate alternatives is also proposed. Based on cost, environmental and technical dimensions, two integrating analyses are proposed to support informed decisions by mapping the best alternatives. This framework is applied to a case study regarding alternative mould designs to produce a part through injection moulding.

1 Introduction

Society awareness for environmental issues has fostered the shifting of the companies' strategies towards the integration of environmentally conscious technologies and products in their business chain. In fact, in the beginning of the 90s, Fiksel (1993) followed some progressive U.S. manufacturing companies that started implementing environmental quality programs. His observations revealed that design for the environment practices provided competitive advantage by reducing the costs of production and waste management, encouraging innovation

I. Ribeiro (✉) · P. Peças · E. Henriques
IDMEC, Instituto Superior Técnico, Universidade de Lisboa, Lisbon, Portugal
e-mail: ines.ribeiro@tecnico.ulisboa.pt

E. Henriques et al. (eds.), *Technology and Manufacturing Process Selection*,
Springer Series in Advanced Manufacturing, DOI: 10.1007/978-1-4471-5544-7_11,

in product simplification, and attracting new customers. This trend continued until nowadays, having been developed several life cycle design approaches to support sustainable design, both economic and environmentally.

The concept of life cycle design/engineering was first introduced by Alting and Legarth (1995) as "the art of designing the product life cycle through choices about product concept, structure, materials and processes". Going beyond the traditional decision process, based on technological and economic performances, this concept introduces the decision in a life cycle perspective also regarding the environmental performance of a design. In fact, Life Cycle Engineering (LCE) emerged in response to the need to develop life cycles causing the lowest possible environmental impacts, while still offering economic viability. LCE stands for the need to consider from the early product concept its complete projected life, including market research, design phases, manufacturing processes, qualification, reliability aspects and customer service-maintainability-supportability issues (Wanyama et al. 2003; Keys 1990; Alting and Legarth 1995). The development of increasingly sophisticated products (systems or facilities) in shorter timeframes is a tough challenge that can be better achieved by a holistic understanding of products and processes life cycle (Alting and Legarth 1995; Ishii 1995).

It is convenient to develop the LCE approach in the early design phase. However it can involve a large set of alternative materials and other design options which means an enormous effort in terms of life cycle analyses (e.g. Life Cycle Cost—LCC and Life Cycle Assessment—LCA). Providentially, there are several factors that limit the number of alternatives to study in the LCE approach. In fact, the company technological and strategic framework restricts the number of alternatives to the ones possible to handle, to have access and to process, among other constrains. Additionally, the product logic and specific requirements limit the alternatives domain to a few options (Peças et al. 2009; Pousa et al. 2009). Despite this focus on a range of specific alternatives, the spectrum of possible design variables is still large. So, to efficiently and effectively apply the LCE approach, the design problem must be well established based on the definition of a few set of design alternatives for further analysis and comparison. This allows focusing the LCE analysis on the most promising design alternatives.

The analysis of the technical, economic and environmental performance provides a solid foundation for designers to understand the trade-offs and implications of product design alternatives (Wanyama et al. 2003; Fava et al. 2000). The reduction to three performance dimensions can be considered as a drawback as compared with Design for X (DfX), which fosters the consideration of a wide spectrum of disciplines. The answer to this issue relies in the way the three dimensions are analyzed. For the economic performance the use of methodologies like LCC is recommended. The estimation of all the costs associated with a product throughout the product's life, from "cradle" to "grave" integrates the analysis of the impact of design for cost, design for maintainability, design for assembly, etc. Accordingly, the use of methodologies like LCA to estimate the environmental performance includes the disciplines of design for environment, design for recycling, design for standards, etc. Finally the designs for reliability, for service, for use, among others are surely integrated in the technical performance assessment (if not already integrated in cost with the available prediction models).

Several methods have been proposed for modelling and/or evaluating all or part of a system life cycle. Takata and Kimura (2003) proposed flexible means to represent technical information relevant to close loop manufacturing management, especially with respect to maintenance. A simulation system was developed for life cycle design and management, including functions of modelling and controlling the life cycle processes. Brissaud and Tichkiewitch (2001) proposed a comprehensive model that can be used to globally optimize the phases comprising the product life cycle. Their model deals with the problem of incidents happening along the life cycle of a product and is based on the capitalization of quality discrepancies in product life cycle. That is, product design is improved by organising continuous information feedback loops regarding quality problems from product usage to the designers and manufacturers. It also introduces learning procedures in a design system. Concurrently, Westkämper (2002) proposed a platform for integrating the management of the product assembly/disassembly with its life cycle. According to the author, business strategies are more and more aiming perfecting technical systems, optimizing product usage and maximizing added value over the entire lifetime. Therefore, a life cycle management platform is proposed to address the manufacturers need to manage products, adding value to the after sales processes.

Several authors proposed tools to support the conceptual design, the development, and the design assessment in terms of environmental cost and impact. Park et al. (2002) developed a life cycle predicting method using an Artificial Neural Network (ANN) model, especially focused on energy and maintenance costs in conceptual design stage. Kaebernick et al. (2003) proposed a simplified life cycle assessment for the early design stages of products based on the analysis of LCA case studies. They developed the product's Environmental Performance Indicator by using two sets of energy-based and material-based Impact Drivers. A similar study by Duflou et al. (2003) presented two strategies to support LCE application in the conceptual design phase; an Ecodesign Knowledge System and Eco-Cost Estimating Relationships. The first supplies problem-specific guidance based on classifiying both knowledge and user situations according to a common set of domain models. The second allows estimating the environmental impact of design concepts based on a limited number of functional parameters. A more recent study was carried by Pousa et al. (2009) who developed simplified LCC and LCA models to foster the design of sustainable plastic injection moulds. Focused on a specific area, they developed a life cycle predictive model, retrieving cost and environmental results with a low number of design inputs. Zhang et al (2004) proposed web-based applications for closed loop manufacturing and product environmental impact assessment, which takes end-of-life dispositions into consideration. All these methods and tools support the integration of life cycle design in products or tooling in early design phases minimizing the cost and time typically required for this type of analysis.

Finally, some authors have proposed the traditional approach of analysing the three performance dimensions (economic, environmental and technical) by attributing importance weights to each dimension (Betz et al. 1998, Saur et al. 2000). Considering the same dimensions, a slightly different approach is offered by Ribeiro et al. (2008, 2009). Their models, aiming the comparison of engineering alternatives, follow the LCE principles, using LCC, LCA and Multi Attributes Decision-Making (MADM) for the three dimensions of the analysis. The integrated analysis is performed through the use of ternary diagrams in which the best alternatives domains are mapped. A similar approach was already proposed by Betz et al. (1998) using 3D-graphics. However, ternary diagrams allow a clear visualisation and easier interpretation of the performance in each dimension as well as the identification of the "best domains" of each alternative. This innovative way of presenting the results promotes the communication and discussion, facilitating the decision making in the design process. In fact, the decision making in engineering design is also a negotiation process among numerous perspectives of the involved team. These models have been applied in the automotive sector (Ribeiro et al. 2008) and in the plastic injection moulding sector (Peças et al. 2009, 2010).

A recent paper (Peças et al. 2013) describes an evolution of this approach, proposing and adapting its application for the materials selection process. For the specific case of materials selection, the authors refer that the analysis of the functional/technical performance dimension is somehow redundant. In fact, any candidate material has to meet necessarily the technical specifications of the product being designed. Furthermore, for the sake of minimum cost, designers are used to look for materials that just meet the requirements, avoiding those whose technical properties exceed significantly the minimum requirements. Of course, well selected high performance materials might result in products that exceed their minimum requirements. But, if any material brings in a benefit indeed this should be reflected on the economic and/or environmental performances. The result is the development of a Materials Selection Engine (MSE) intending to contribute to a more informed decision-making process in materials selection (Peças et al. 2013). Starting from a large set of materials that fit the application, the selection procedure reduces the number of candidate materials based on technical performance requirements and follows through the analysis of their cost and environmental impact along the different life cycle phases. At the end, more than retrieving the best material, the results, as regards total life cycle environmental impacts and costs, are mapped in a 2D decision space. In the following section the LCE framework on the basis of the two mapping models of "best design alternative" domains is presented together with the illustration of the nature of results that can be obtained to support engineering design decisions.

2 LCE Framework

The proposed LCE framework (Fig. 1) structures all relevant information that needs to be available within the design phase in a single decision supporting tool that integrates the technical, economic and environmental performance

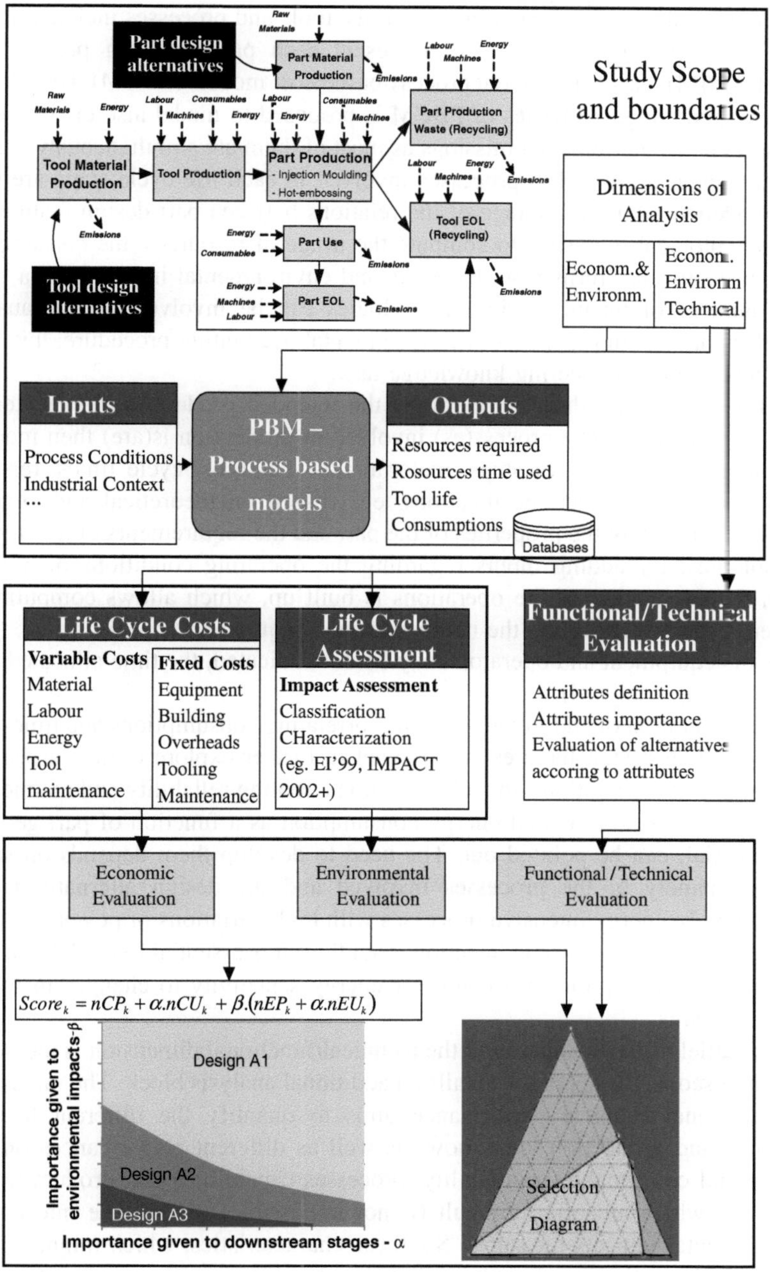

Fig. 1 LCE framework

dimensions. In several consecutive and interacting blocks of information and computing tools, the several life cycle stages of the product are considered. It begins with the boundaries definition of the life cycle in accordance with the definition of alternatives regarding materials, tools and processes included in each phase. The subsequent step is to represent each process using process-based models (PBM), derived from the process-based cost models (PBCM). Usually used for cost modelling, the basics of PBCM is proposed to model also environmental resources consumption and emissions as most of them are simultaneously cost and environmental drivers. The processes involved in each life cycle stage are therefore modelled taking advantage of the relations between part design features and the technological requisites to compute the usage of resources, the consumptions and emissions and, further on, the costs and environmental impacts. As a significant number of manufacturing technologies can be involved and because the modelling is not simply a cost or environmental accounting procedure, this is can be a demanding engineering knowledge task.

PBM begins with the description of the intended product (part(s) material(s) and geometry(ies)). The process(es) involved in production is(are) then modelled regarding the sequence of steps, material flow balances, cycle times, resources (equipment and labour) specifications, etc., based on theoretical and empirical relations correlating the properties of the part and the requirements of the involved technologies. By adding inputs regarding the operating conditions of a certain plant, the description of the operations is built up, which allows computing the needed resources regarding the number of tools, equipment, operators, etc. (or, as far as the equipment and operators may be not dedicated, the time allocated to the product being analysed).

The integration of additional relations correlating consumptions and time-use of resources with design features are relevant to further explore critical aspects. As examples of these relations, models to estimate tooling reliability and maintenance performance, cycle time and energy consumption as a function of part geometry and material, can be pointed out. The need to develop them depends on several aspects, namely of the processes involved and the design alternatives under evaluation. In energy intensive processes with high variations of power use over a production cycle (e.g. plastic injection moulding) it is essential to model efficiently the energy consumption to account for energy sensitivity to changes in the part design features.

In parallel, if the evaluation of the technical/functional dimension is included in the initial scope, the approach entails an additional analysis block. The inclusion of the functional/technical performance aims to quantify the different levels of technical and production know-how, as well as different performances on technology and consumables availability, processes capabilities and product time-to-customer, which are often difficult (if not impossible) to translate into costs or environmental impacts measures. So, these characteristics, which might differentiate each alternative, must not be included in the other two dimensions of analysis.

Having defined the scope and modelled all the processes it is possible to compute the required resources to produce the part(s) in each design alternative. The deep level of process parameterization of the PBM based tool permits a myriad of sensitivity analysis forming a block of information that will be used as input for the LCC and LCA analysis—the third step. So, in the third step, price factors are attributed to each cost driver to assess the economic performance of an alternative throughout the life cycle. The environmental impact assessment is also performed to the data inventory retrieved from the PBM, achieving a global quantitative evaluation of the environmental dimension of each alternative, similarly to the cost dimension. Finally, with the outputs regarding the environmental and economic evaluation of the alternatives, two integrating graphic-based comparisons are proposed. The first (Peças et al. 2013) is proposed to address two aspects subject to controversy when dealing with the dual nature of the outputs. On one hand it intends to deal with the question of adding costs (although considering an interest rate) supported by the different stakeholders involved in the whole life cycle in different timeframes, namely the upstream stakeholders (part designer and producer) and the downstream stakeholders involved in the use and end-of-life (EoL) phases. On the other hand, it assists the separation between costs and environmental impacts that forces the decision maker to choose a design alternative with two dimension of analysis, often with opposite performances evolution. The second possible graphic output regards the integration of the results of the analysis in a ternary diagram of three dimensions: cost, environmental and technical evaluation of the alternatives. The ternary diagrams to map the best design alternatives were first proposed by Ribeiro et al. (2008) applied to a material selection problem.

The proposed LCE framework relies on the integrated analysis of these three dimensions. For supporting focused decisions, like the selection of materials and technologies, the functional/technical dimension is generally exploited in a first preliminary stage for screening the candidate alternatives that meet the technical thresholds, and the economic and environmental impacts dimensions are the ones present on the final decision-making process. When the functional/technical performance of alternatives is itself a selection criteria and it surmounts any cost or environmental measure the three dimensions are integrated in the same stage (stage 3) of the decision-making process.

2.1 Study Scope and Boundaries Definition

The product life cycle phases must be identified and all the processes involved characterized in a detail that depends on the study aim. It means that the inputs/outputs of each life cycle phase and their dependence on the design alternatives under analysis need to be identified. If dedicated tooling is involved in the product manufacturing (e.g. injection moulds, stamping dies, etc.) their related manufacturing process should also be considered. Additionally, there are some life cycle stages that may occur outside an industrial context, as for example the product use or some EoL scenarios such as landfill or recycling. As these stages involve stakeholders outside the design and manufacturing context and often occur in a later timeframe, they should be evaluated individually. Finally, the scope regarding the dimensions of the analysis is defined regarding the type of comparison aimed. If three dimensions are required, because technical/functional performance effectively differentiate the alternatives in a way that cannot be purely reflected in costs and environmental impacts, an additional technical/functional analysis must be developed.

2.2 Functional/Technical Dimension

Several decision making based methods can be applied to perform the comparative analysis of functional/technical performance of the product (system, tooling or technology) design alternatives, such as graphic theory and matrix approach and fuzzy multiple attribute decision-making methods (MADM). In common, all of them rely on the know-how and expertise of professionals and users to determine the relevant functional attributes for the application, and, on a comparison basis, assess the performance of the alternatives within this set of attributes. Figure 2 outlines the basic methodology used by Peças et al. (2010) based on MADM methods and pairwise comparison for the relative evaluation of the functional/ technical performance of the candidate alternatives.

2.3 Economic Dimension

The economic performance assessment is developed according to LCC methodology. LCC is essentially an evaluation tool in the sense that it gets on important metrics for choosing the most cost-effective solution from a series of alternatives. The general proposed LCC model with its general structure and inputs is presented in Fig. 3. It uses as input the data derived from the PBM, providing the cost of each process for each cost category, the cost of each life cycle phase and the costs incurred throughout the life cycle.

2.4 Environmental Impact Dimension

The environmental performance analysis is performed using LCA, as a structured method to quantify potential environmental impacts over the entire life cycle of products (systems or services), that can also be integrated with PBM inputs and

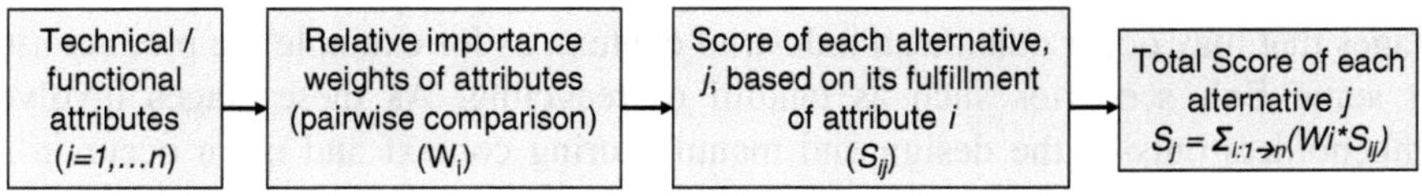

Fig. 2 Basic methodology used for the functional assessment. More complex methods can be used

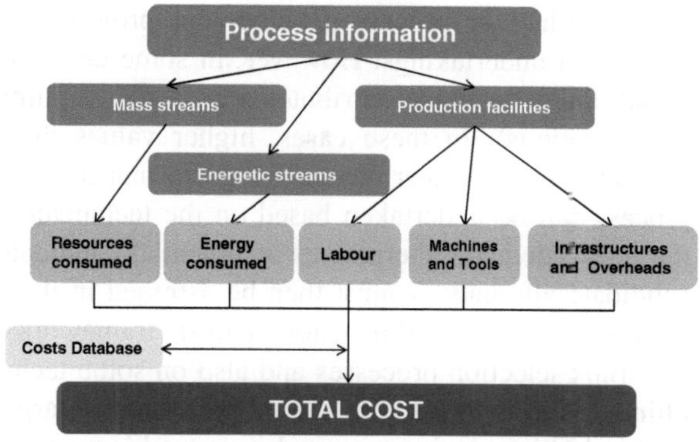

Fig. 3 LCC model structure

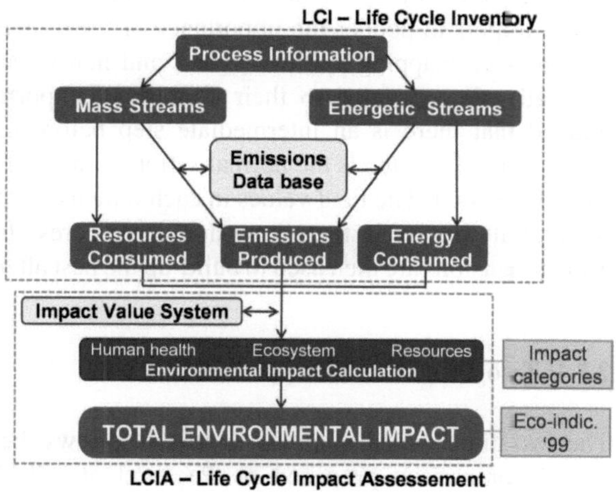

Fig. 4 LCA model structure

outputs. In fact, the mass, energy and emissions determined for the cost computing are used as input on the LCA model, representing the Life Cycle Impact phase. For the Life Cycle Impact Assessment phase, 11 environmental impact categories are considered, combined in three major impact areas: Human Health, Ecosystem Quality and Resources. After aggregating all the emissions and resources consumption from the life cycle into these impact categories, the three achieved scores are pondered into a single value, called the eco-indicator 99 (Goedkoop 2000). The general proposed model is presented in Fig. 4.

2.5 Life Cycle Integrated Performance

The LCE final analysis depends on the decision-making context. In the cases in which the functional/technical performance varies among the alternatives, the three dimensions must be analysed simultaneously. This is the current case in some product or process development projects and also on some technology evaluation undertakings. However, in some decision-making processes the functional and/or technical attributes are mostly requirements that must be achieved (specifications). In these cases, higher values than the ones specified are not translated into a higher recognized performance. It means a preliminary screening process can be undertaken based on the technical/functional dimension to select the most adequate alternatives as potential candidates among a larger set. These candidate alternatives must then be assessed in the economic and environmental dimension. This two dimension analysis framework is proper to apply in most of materials selection processes and also on some technology evaluations (when the ultimate aim is to achieve a fixed performance target).

The LCE analysis framework proposed is based on the mapping the "best performance alternatives" rather than in common, fixed or pre-recommended importance weights to ponder the importance of the dimensions of analysis. The use of performance mapping permits a clear and non-forced view of the possible "best alternatives" correlated to their domain of importance (weights). It should be noticed that there is an intermediate step before mapping the three or the two dimensions. This step is the normalisation of the performance of the alternatives in each dimension. The final values in each dimension are normalised for the value of the best alternative in that dimension. The scores obtained for each alternative in each dimensions are then used to build-up the best alternative performance mapping.

2.5.1 Two Dimensional Approach: CLUBE

The two dimensional mapping approach follows the CLUBE graphic-based integrated comparison proposed by Peças et al. (2013) to deal with cost parcels supported by different stakeholders along the life cycle and with the different nature of costs and environmental impacts measures, often with opposite evolutions.

The first step of the integrated comparison is the normalisation of the costs and environmental impacts of each life cycle phase. The normalised values are then aggregated in Production related (upstream phases) and User related (downstream phases) values. For each design alternative ($A_1,\ldots, A_k,\ldots$), the normalized production related costs (*nCP*), the normalized user related costs (*nCU*), the normalized production related environmental impacts (*nEP*) and normalized user related environmental impacts (*nEU*) are computed ($Score_k$) according to Eq. 1 for each alternative:

$$Score_k = nCP_k + \alpha.nCU_k + \beta.(nEP_k + \alpha.nEU_k) \quad (1)$$

in which α is the importance given to the downstream phases and β the importance given to the environmental impacts of the integrated life cycle. Both are relative to the production related costs, which being entirely supported and paid by the producer has a full importance (100 %) for any stakeholder involved.

So, in the (α, β) space the domains of the "best alternatives" (the ones with the highest Score for each pair of α and β values) can be mapped. The proposed characterization helps the understanding of the influence of candidate alternatives on the performance dimensions of the product life cycle. Some alternatives can be visible for large ranges of α and β values meaning that their best performance is not significantly affected by changing the relative importance the decision maker attributes to upstream and downstream phases and environmental impacts. Others will not be "visible" in the performing mapping, meaning that its performance is lower than at least one of the other alternatives. Further information regarding this method can be found elsewhere (Peças et al. 2013).

2.5.2 Three Dimensional Approach: Ternary Diagrams

When a technical/functional evaluation is also included in the integrated analysis, the performance evaluation can be implemented through ternary diagrams, where each axis represents one dimension of analysis. For each alternative a single indicator is obtained for each dimension of evaluation allowing the direct incorporation of the technical, economic and environmental performances into a multi-criteria decision problem.

According to Eqs. 2 and 3, the normalized Life Cycle Costs (*nLCC*), Life Cycle Impact Assessment (*nLCIA*) and Technical/Functional Evaluation (*nTF*) are computed for each design alternative (A_1,..., A_k,...):

$$Score_k = w_1.nLCC + w_2.nLCIA + w_3.nTF \quad (2)$$

$$\text{and } w_1, + w_2 + w_3 = 100\,\% \quad (3)$$

were w_1, w_2 and w_3 are the importance given to the economic, environmental and technical/functional dimensions, correspondingly.

The result is then a global evaluation, presented in a ternary diagram, clearly showing the ''best application domains'' of the alternatives as a function of the importance of the three dimensions of analysis. Within this approach, the difficult task of materializing the relative importance of the three dimensions into a set of weights is overcome. One should note that the performance of an alternative is a relative quantity and it depends on the set of alternatives being considered. Therefore there is no universally best alternative for a given application, which reinforces the need for tools to support decision making.

3 Application Case

The case study was developed in a Portuguese company, Fapil, where all the data regarding exogenous industrial parameters was gathered. Values for material and energy unit costs were obtained for Portugal during 2012. Economic data

Table 1 Designation of the design alternatives and main process parameters

Ref.	Mould design type	Injection cycle time (s)	Mass per injection cycle (g/cycle)
16H	16 cavities, hot runners	16.3	47.9
16C	16 cavities, cold runners	20.9	56.4
32H	32 cavities, hot runners	17.1	95.9
32C	32 cavities, cold runners	22.8	114.4
96H	96 cavities, hot runners	19.5	287.6
96C	96 cavities, cold runners	25.4	414.0

regarding materials were also collected in Portuguese companies. The cost of normalized mould components were computed according to suppliers catalogues.

In accordance to the purpose of the application case the boundaries of the analysis were defined. The analysis is focused on the mould life cycle, meaning that the cloth peg use and EoL phases were ignored, as they do not change with the mould design options. However, the cloth peg production, indeed the mould use phase, and the required volume of plastic material, indeed the plastic material consumption in the mould use phase, were both taken into account. The alternatives of different types of runners change the amount of plastic material consumed and recycled, affecting the materials acquisition phase and the EoL phase of process scrap (Table 1).

3.1 Economic Dimension

The processes within the boundaries of the defined life cycle scope were modelled and assessed through process-based cost models in order to allow posterior sensitivity analysis related with the impacts of changes in design, materials, manufacturing processes, production volume, etc. The material cost is considered from a manufacturing point of view. So, it takes in all the material mass required for the injection process multiplied by its commercial specific price. The plastic material required for the injection process but not included in the final product (plastic waste in the mould feeding system) can be reused in the injection process. Nevertheless, for quality reasons a limit of 10 % of material recycling was allowed for each injection. The remaining material was considered to be sold or incorporated in other products as recycled plastic. Regarding the EoL, the mould was considered to be sent to recycling and so sold as scrap alloy steel.

Finally all costs (expressed in monetary units—MU) were gathered and the total LCC of each alternative was computed (Fig. 5) for a production volume of 4 million cloths pegs (4 Mpegs) per year, involving the injection of 8 million body parts.

For this level of production volume the importance of the plastic material costs is quite significant for all the alternatives. The effect of the production rate, deeply

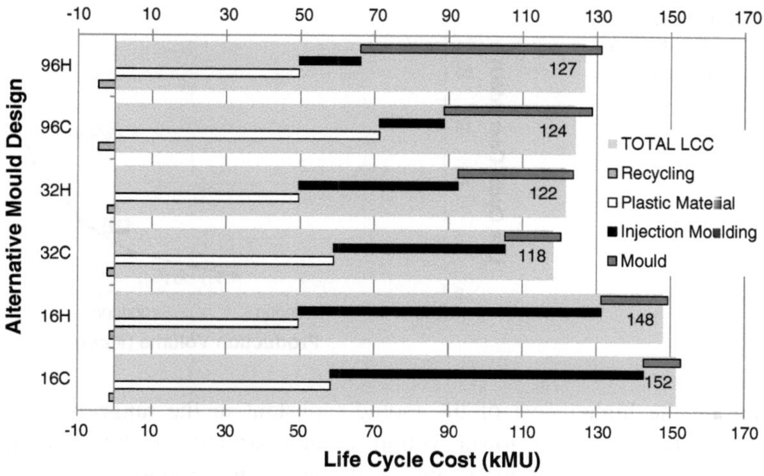

Fig. 5 LCC of the mould design alternatives for 4Mpeg (mould includes mould production and mould material cost)

dependent on the number of pegs produced per injection cycle (number of moulding cavities) is evident. For alternatives with 96 moulding cavities the injection moulding costs are really low. Nevertheless the very high costs of these moulds increase its overall LCC costs. The alternatives with 32 moulding cavities have a balanced performance achieving the best economic life cycle performance: the mould is slightly more expensive than the ones with 16 cavities but efficient enough to avoid high injection moulding costs. The EoL costs are not significant in any alternative.

The effect on the type of injection runners is visible in the amount of material required in the case cold runners are used and in the increasing of the mould cost in the case of hot runners. For the moulds with 96 and 32 cavities the cold runners allow for a lower LCC but the opposite happen for the mould with 16 cavities. Nevertheless the number of moulding cavities has a higher impact on the economic performance than the type of runners.

Taking advantage of the potential of process-based cost models the sensitivity analysis of costs to changes in the production context is possible. The best mould design alternatives with the evolution of the production volume are presented in Fig. 6. The LCC model takes into account the technological modifications required in the production system for each mould alternative, namely it selects the most adequate injection machine depending on the mould size and clamping force, meaning that the machine to deal with the 96 cavities mould is significantly larger and more expensive than the one required to deal with the 16 cavities mould.

For high production volumes (more than 4.81 Mpegs) the lowest LCC is achieved for the pegs bodies injected in 96 cavities moulds with hot-runners. It means that the lowest cycle time per part of these moulds, allows lower injection moulding costs that together with lower material consumption (hot-runners)

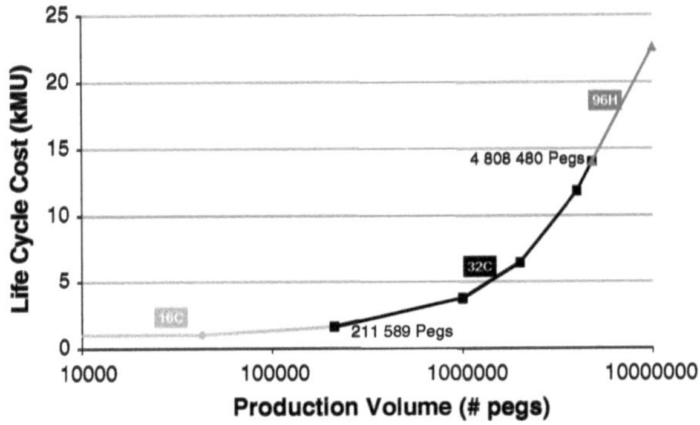

Fig. 6 Best (lowest LCC) alternatives for different production volumes

compensate the higher cost of the mould (constant in the range of production volumes considered). For lower volumes, until around 211.6 kpegs, the 32C alternative is the best one, due mainly to the increasing importance of the mould cost for the cloths peg life cycle cost. For lower production volumes the mould productivity largely reduces its impact and the LCC value becomes essentially driven by the mould cost. Therefore for low production volumes the 16C alternative is the best one since it has the lowest mould cost.

3.2 Environmental Impact Dimension

In order to evaluate the environmental performance of the different alternatives, the LCA model was applied (Fig. 7). From the results achieved it is possible to observe the extremely reduced environmental impact of the mould. The plastic

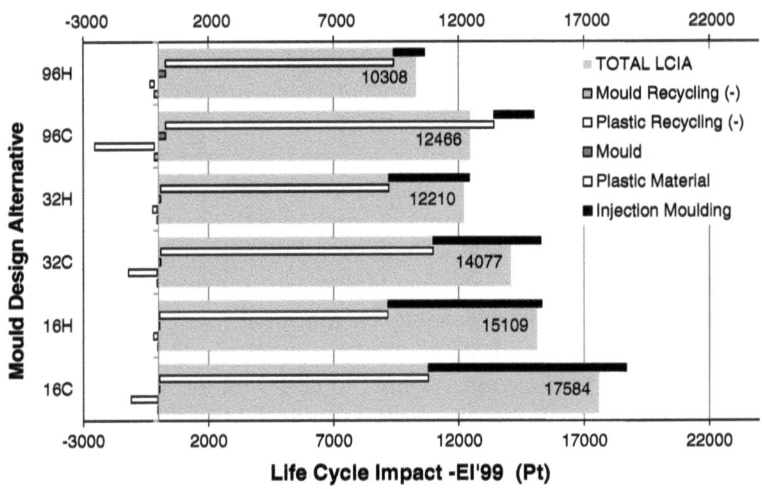

Fig. 7 LCIA of the mould design alternatives for 4 Mpegs

material has the major slice of impact mainly in alternatives with 96 and 32 cavities. In fact, as observed for the costs, the environmental impact of plastic material varies in a relative strict range.

The alternatives with lower environmental performance are the ones that use cold runners, due to the extra material required for the feeding channels and the limit of 10 % of recycled material in the injection process. It must be noticed that since this wasted material was considered for recycling, there is a recovering of impacts for the alternatives using cold runners (negative impact figures). This fact motivates a relative balance of the overall impact of the consumed materials, which results in a higher importance of the impact of the injection moulding process (energy consumption) in the alternatives ranking. Therefore, alternatives with lower cycle time per part have a lower environmental impact. The alternatives with cold runner moulds present always higher impact than the similar ones with hot runners, due to higher impact of injection moulding (more material has to be melted and cooled) and material consumption.

3.3 Functional Analysis Dimension

For the analysis of the functional dimension, the requirements selected, not fully reflected on costs or environmental impacts, are related with the moulds production and its performance in injection moulding (Table 2). The number of cavities and the use of hot runners increase the mould complexity. The effect of this complexity is only partially reflected in the production cost accounting of the LCC model. In this dimension are essentially reflected the non-tangible effects of increasing the number of production steps and the components to integrate in the mould and the subsequent increasing of the lead time of mould production and of the potential of mistakes and rework.

The same approach was followed for the performance of the moulds in the injection moulding. The use of higher number of cavities will reduce the number of cycles, so the need for maintenance is lower for the same number of injected parts. The use of hot runners improves the injection process capability since it's easier to control the process. Also the decreasing of the number of cycles with the

Table 2 Functional dimension assessment (S_{ij}: 1-lowest; 10-highest performance based on moulds cavities characteristics)

Requirements	Weigth (W_i)	Score of each alternative (S_{ij})					
		16H	16C	32H	32C	96H	96C
Mould complexity	35 %	10	9	6	5	2	1
Mould reliability	25 %	1	3	5	7	8	10
Mould capability	40 %	2	1	4	5	10	9
Total		4.55	4.3	4.95	5.5	6.7	6.45

number of cavities affects the performance in this requirement. Both mould reliability and capability were not considered in the LCC model used.

The 96H mould is the one with better performance in the functional dimensions followed closely by the 96C mould. The number of cavities and the use of cold runner decrease the overall performance of the functional dimension.

3.4 Integrated Analysis

This section presents the type of results obtained with the integrated analysis. The first integrated approach, CLUBE, addresses two decision problems: the question of adding costs allocated to the different stakeholders involved in the whole life cycle and the separation between costs and environmental impacts. It was proposed to map the performance of the design alternatives in the two performance dimensions: cost and environmental impacts. Results of this graphic proposal show different tool design alternatives when different sets of weights regarding environmental impacts and downstream costs are chosen. The second integrated analysis includes a third dimension, the technical/functional analysis, presenting the results in a ternary diagram. Besides the addition of the third dimension, this approach also differs from the first regarding the equal importance given to the different stakeholders. The overall results from LCC, LCA and technical/functional performance analyses are normalized to serve as inputs of the ternary diagram.

The two integrated analyses can lead to different best alternatives, being the adequate analysis dependent on the critical aspects of the problem. The first deals with situations where all alternatives are technically or functionally equivalent, or when technical aspects are already included in the cost or environmental dimensions. Furthermore, it allows the separation between upstream and downstream economic/environmental impacts. The second deals with decisions between alternatives with different technical/functional performances and when the importance given to different stakeholders is not an important issue.

3.4.1 Two Dimensional Approach: CLUBE

So, to map the best alternatives according to the importance given by the involved stakeholders to the environmental impacts and costs, the integrated CLUBE comparison was performed. Table 3 presents the normalized values for the production costs (nCPk), user costs (nCUk), production environmental impacts (nEPk) and user environmental impacts (nEUk) of each alternative. The upstream phase regards the mould production and the downstream phases the part production (mould use phase) and EoL. Results of the integrated comparison are presented in Fig. 8 and consider the expected production volume of 4,000,000 pegs per year.

Table 3 Normalized values of each alternative (4 Mpegs produced)

	16H	16C	32H	32C	96H	96C
nCP_k	5.5	10.0	3.2	6.7	1.5	2.5
nCU_k	5.0	4.6	7.1	6.3	10.0	7.5
nEP_k	10.0	10.0	2.2	2.2	1.1	1.1
nEU_k	9.2	7.7	9.6	8.0	10.0	7.0

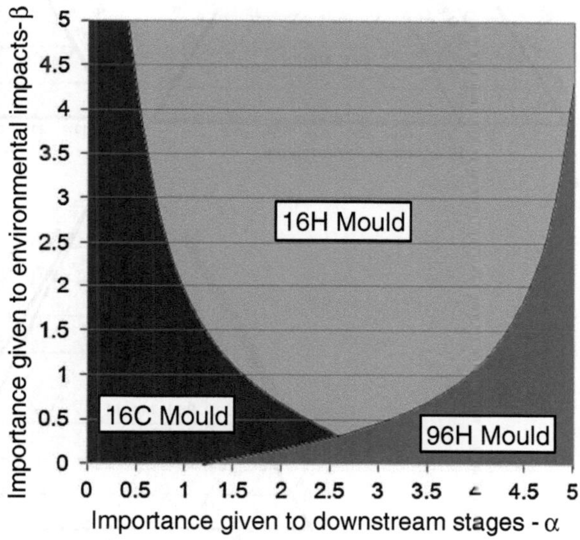

Fig. 8 Best alternative mapping using CLUBE approach

Three alternatives appear as best ones. The mould with higher production cost (96H) is the best alternative when high importance is given to the cost of the downstream use phase (higher α values). The mould with 16 cavities and hot runners (16H) becomes the best one with the increasing of the environmental impact importance (higher β values).This is due to the energy and material savings during part production with this alternative. The mould alternative with 16 cavities and cold runner (16C) is the best alternative when low importance is given to the downstream phase and environmental impacts (low values of α and β). This is in accordance with the fact this mould is the best choice if only mould material and production are considered.

3.4.2 Three Dimensional Approach: Ternary Diagrams

Adding the third dimension to the analysis (technical/functional dimension) and assuming full importance of all life cycle stages in the cost and environmental dimensions, a ternary diagram is performed to map the best alternatives (Fig. 9). The normalization of the performance obtained in each dimension is present in Table 4 for 4 Mpegs.

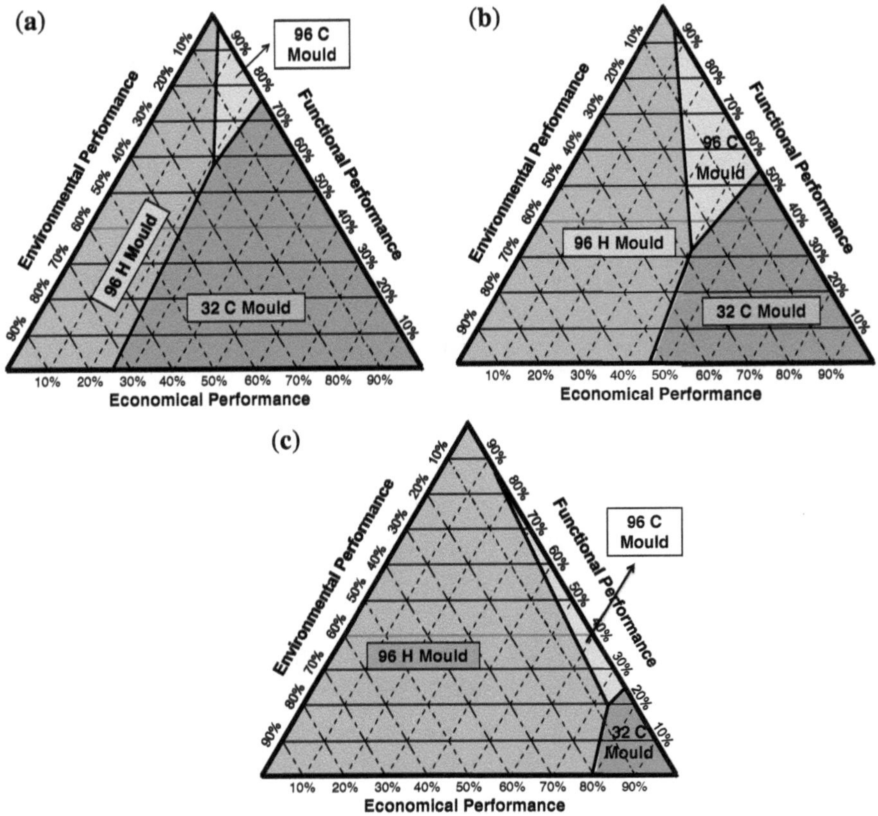

Fig. 9 Ternary selection diagrams representing the "best" solutions map for different production volumes: **a** 0.5 Mpegs; **b** 2 Mpegs and **c** 4 Mpegs

Table 4 Absolute and normalized values for each performance dimension assessment (4 Mpegs Produced)

		16H	16C	32H	32C	96H	96C
LCC	Value (MU)	147995	151500	121652	118326	126733	124229
	Normal	8.0	7.8	9.7	10.0	9.3	9.5
LCA	Value (EI'99)	15109	17584	12210	14077	10308	12466
	Normal	6.8	5.9	8.4	7.3	10.0	8.3
FA	Value	4.55	4.3	4.95	5.5	6.7	6.45
	Normal	6.8	6.4	7.4	8.2	10.0	9.6

As it can be seen in Fig. 9, with the increasing of the production volume the efficiency of this mould in the injection moulding enlarges the domain in which its performance is better than the others. The mould with 32 cavities with cold-runners has the best performance in the economic dimension due to a compromise of average injection moulding efficiency and mould cost. With the increasing of the production volume the importance of the mould cost tends to reduce so the selection domain for this alternative narrows.

The mould with 96 cavities and cold-runners has higher performance than the 96 H mould, in the economic dimension, and the 32 C mould, in the functional and environmental dimensions. Also its functional performance is lower but very close to the 96H mould. The consequence is its appearance in the solutions space when low importance is given to the environmental dimension and average importance is given to the other dimensions.

4 Conclusions

The presented LCE framework regards a comprehensive analysis of design alternatives in terms of the life cycle environmental and economic impacts, and also in terms of their functional performance. These analyses are gathered in a single solutions space, in which the mapping of the best alternatives becomes possible. The comparison of the alternatives is visually represented, providing a common communication tool to support the discussion and the decision among the design team members.

Two different approaches are proposed to systematize to comparison of design alternatives. The first evaluates only the economic and environmental dimensions, separating the cost and environmental impacts of design alternatives into upstream and downstream phases. It deals with situations where all alternatives are technically or functionally equivalent, or when technical aspects are already included in the cost or environmental dimensions. The second aggregates also a technical/functional dimension, evaluated together with the life cycle economic and environmental impacts. It deals with decisions between alternatives with different technical/functional performances. Each point in the mapping diagrams is representative of a set of importance weights given to the different dimensions of analysis. So, depending on the companies' strategies the best alternative selection can be identified in an informed way. Finally a case study regarding alternative mould designs to produce a part through injection moulding is presented to illustrate the proposed LCE framework and both alternative mapping approaches.

Most of manufacturing sectors are very intense in introducing continuously new technologies and new ways of products and parts manufacturing. So, the framework presented is a valuable tool to assess, for several production and life cycle scenarios, the performance of those innovations even in stages where the existing information is limited.

References

Alting L, Legarth JB (1995) Life cycle engineering and design. CIRP Ann-Manuf. Technol 44(2):569–580

Asiedu Y, Gu P (1998) Product life cycle cost analysis: State of the art review. Int J Prod Res 36(4):883–908

Betz M, Schuckert M, Herrmann C (1998). Life cycle engineering as decision making support in the electronics industry. In: Proceedings of the custom integrated circuits con-ference. Ed: IEEE

Brissaud D, Tichkiewitch S (2001) Product models for life-cycle. CIRP Ann-Manuf. Technol 50(1):105–108

Duflou J, Dewulf W, Sas P, Vanherck P (2003) Pro-active life cycle engineering support tools. CIRP Ann-Manuf. Technol 52(1):29–32

Fava JA, Brady K, Young SB, Saur K (2000) Sustainable strategies using life cycle approaches. Environ Pro 19(2):61–64

Fiksel J (1993) Design for the environment: an integrated systems approach. In: Proceedings of IEEE international symposium on electronics and the environment, pp 126–131 Ed: IEEE

Goedkoop M, Effting S, Collignon M (2000) The Eco-indicator 99: a damage oriented method for life cycle impact assessment manual for designers. PRé Consultants. http://www.pre.nl/

Ishii K (1995) Life-cycle engineering design. J Vib Acoust 117(B):42–47

Kaebernick H, Sun M, Kara S (2003) Simplified lifecycle assessment for the early design stages of industrial products. CIRP Ann-Manuf Technol 52(1):25–28

Keys LK (1990) System life cycle engineering and DF“X”. IEEE Trans Compon, Hybrids, Manuf Technol 13:83–93

Park JH, Seo KK, Wallace D, Lee KI (2002) Approximate product life cycle costing method for the conceptual product design. CIRP Ann-Manuf Technol 51(1):421–424

Peças P, Ribeiro I, Folgado R, Henriques E (2009) A life cycle engineering model for technology selection: a case study on plastic injection moulds for low production volumes. J Clean Prod 17:846–856

Peças P, Henriques E, Ribeiro I (2010) Integrated approach to product and process design based on life cycle engineering. In: Silva A, Simões R, Global IGI (eds) Handbook of research on trends in product design and development: technological and organizational perspectives. doi:10.4018/978-1-61520-617-9.ch021

Peças P, Ribeiro I, Silva A, Henriques E (2013) Comprehensive approach for informed life cycle-based materials selection. Mater Des 43:220–232

Pousa C, Ribeiro I, Folgado R, Peças P, Henriques E (2009) LCC and LCA simplified models to foster the design of sustainable plastic injection moulds. In: ElMaraghy W (ed) 16th CIRP international conference on life cycle engineering—LCE 2009. University of Windsor, Cairo, pp 99–104

Ribeiro I, Peças P, Henriques E (2009) Life cycle engineering applied to design decisions, a case study. In: Proceedings of the 16th CIRP international conference on life cycle engineering, Cairo

Ribeiro I, Peças P, Silva A, Henriques E (2008) Life cycle engineering methodology applied to material selection, a fender case study. J Clean Prod 16:1887–1899

Saur K, Fava JA, Spatari S (2000) Life cycle engineering case study automobile fender designs. Environ Pro 19(2):72–82

Takata S, Kimura T (2003) Life cycle simulation system for life cycle process planning. CIRP Ann-Manuf Technol 52(1):30–37

Wanyama W, Ertas A, Zhang HC, Ekwaro-Osire S (2003) Life-cycle engineering: issues, tools and research. Int J Comp Integ Manuf 16(4):307–316

Westkämper E (2002) Platform for the integration of assembly, disassembly and life cycle management. CIRP Ann-Manuf Technol 51(1):33–36

Zhang HC, Li J, Shrivastava P, Whitley A, Merchant ME (2004) Web-based system for reverse manufacturing and product environmental impact assessment considering end-of-life dispositions. CIRP Ann-Manuf Technol 53(1):5–8

Proposal for an Architectural Solution for Economic and Environmental Global Eco-Cost Assessment: Model Combination Analysis

Nicolas Perry, Alain Bernard, Magali Bosch-Mauchand, Julien Le Duigou and Yang Xu

Abstract This chapter highlights the complementarities of cost and environmental evaluation in a sustainable approach. Starting with the needs and limits for whole product lifecycle evaluation, this chapter begins with the modeling, data capture and performance indicator aspects. Next, the information issue, regarding the whole lifecycle of the product, is addressed. In order to go further than economical evaluation/assessment, the value concept (for a product or a service) is discussed. Value can combine functional requirements, cost objectives and environmental impact. Finally, knowledge issues are discussed, addressing the complexity of integrating multi-disciplinary expertise into the whole lifecycle of a product.

N. Perry (✉)
Arts et Métiers ParisTech, I2M-UMR 5295 33400 Talence, France
e-mail: n.perry@i2m.u-bordeaux1.fr

A. Bernard
Ecole Centrale de Nantes, IRCCyN-UMR CNRS 6597, Nantes, France
e-mail: alain.bernard@irccyn.ec-nantes.fr

M. Bosch-Mauchand · J. L. Duigou
Université de technologie de Compiègne, ROBERVAL-UMR 7337, Compiègne, France
e-mail: magali.bosch@utc.fr

J. L. Duigou
e-mail: julien.le-duigou@utc.fr

Y. Xu
Department of Information Management, Peking University, Beijing, China
e-mail: yang.xu@pku.edu.cn

E. Henriques et al. (eds.), *Technology and Manufacturing Process Selection*, Springer Series in Advanced Manufacturing, DOI: 10.1007/978-1-4471-5544-7_12,

1 Introduction

Sustainable concerns are increasing in the industrial sector. Different aspects, environmental, economic and social aspects, have to be considered (see Fig. 1). Most industries have turned "green due to regulatory constraints or marketing targets". As for quality management, industries have often adopted these changes as non-pro-active actors. There has been a shift from ISO 9,000 to 14,000. However, few industries have linked clear strategic policies to their priorities and their project's return on investment potential. Product definitions, manufacturing possibilities, logistics strategies and end of life alternatives offer many ways to work toward sustainability.

The social side of the sustainable approach is hard to deal with and is outside the scope of this chapter. However, this aspect should be taken into account very quickly in order to develop new service opportunities that meet consumer demand and optimize the product use ratio (real used time versus overall life time) and their environmental effect (Brissaud and Lelah 2010). There is a huge challenge to consider here, namely consumer and engineer tutoring. People have to learn to reduce consumption and pollution in order to adapt to the world's limited natural resources. Solutions have been found in green manufacturing and green alternatives, i.e. products that create less pollution at all stages of the product life cycle whilst ensuring minimal consumption of non-renewable resources. In addition, consumer tutoring has to focus on the way people use products and resources in their daily lives (e.g. water, light, etc.).

Cost and environmentally oriented industry decisions are therefore linked. Indeed, when engineers have to work in an environmentally-friendly way, their natural reflex is to aim to reduce the quantity of materials used and energy consumption. In this way, they not only reduce the product's impact on natural resources but they consequently also reduce material and energy costs within the global cost. Section 2 of this chapter will discuss the latter.

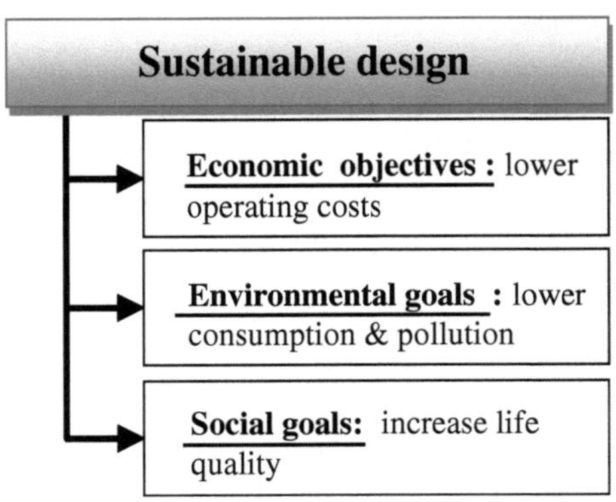

Fig. 1 Sustainable design goals

ISO 14062 defines eco design as the integration of environmental constrains in design and product development. Johansson (Johansson 2001) emphasizes that eco design encompass several concepts and definitions. He identifies the following terms: DfE stands for Design for Environment eco design, green design, life cycle design, eco-effective product design, to which can be added ecological design (Ventère 1995) or environmentally conscious design (Ritzén and Beskow 2001). Eco design can be seen as ambiguous, as it expresses, depending on the definition, a strategy, a process, an activity or even a product. Nevertheless Johansson notices that it is tacit or explicit in all definitions that the objective is to minimize the overall environmental impact of the product throughout its life cycle by adopting preventive actions during the design phase of the product.

In most cases, use and maintenance stage of the product involves the greatest impacts or costs. Then, overall cost of ownership is now the target of the designer and the marketing departments, and the same is true for environmental design and the use of Life Cycle Assessment (LCA, also called eco-balance or cradle-to-grave analysis) (EPA 2010). Figure 2 presents the different design interests for sustainable product/service development. A life cycle cost is one of the facet of sustainable design concerns and evaluations, in other worlds sustainable value. Section 2 will discuss the needs of an integrated Product Lifecycle Management system in order to evaluate efficiently all the stages impacted. Product information is unclear or unknown in the early phases when decisions are made and 80 % of the final costs have been determined (Blanchard 1978; Fabrycky and Blanchard 1991). The same problem arises for environmental consequences.

To define Product Lifecycle Management (PLM), product and process modeling are required. These models provide the basis for analysis and optimization of different solutions. The third section will present a value-based analysis approach such that not only cost, on the one hand, or environmental concerns on the other hand, can be taken into account, but value evaluation and value definition can also be proposed. This section will also introduce the links between value analysis and a PLM information system for sustainable analysis.

In order to ensure reliable evaluations, the data must reflect the reality. In addition, the aggregation rules must be adapted to the product portfolio, to organizational behavior and the evaluation criteria. In order to take advantage of previous or similar

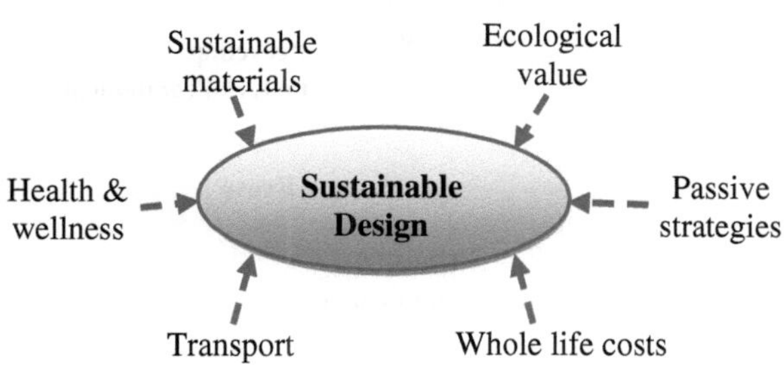

Fig. 2 Sustainable design interests

projects, it is necessary to look for the best practices for project guidelines and to locate the most important knowledge used. The last section will illustrate the use of roadmap methodologies and knowledge value evaluation to enhance and ensure the success of eco-design approaches in parallel to product cost assessment.

2 Cost and Environment Similarity and Complementarities

As for ISO 9000, ISO 14000 standards for environmental management systems are being developed to formalize the LCA method components (Viadu et al. 2006). Figure 3 presents a classic Product Lifecycle process. Each stage of the loop includes cost, and environmental impacts (consumption and pollutions). The aim of product life cycle costing and LCA is to evaluate performances on an overall cycle and sometimes on multi-cycles.

In the case of environmental impact, there are no data available to compare fixed impact due to design and global life cycle impact (such as for cost). But, excluding the usage phase, we assume that the ratio is quite similar. For a whole lifecycle evaluation, cost or environmental indicator definition and estimation are equally difficult. This section emphasizes the need for integrated information models and expert viewpoints to tackle the evaluation of the cost and environmental impacts over the whole life-cycle of a product or a service.

2.1 Full Lifecycle Model

Total lifecycle modeling is unachievable. Indeed, specific lifecycle phases have not yet a complete definition due to the possible detail of the basic activities (that

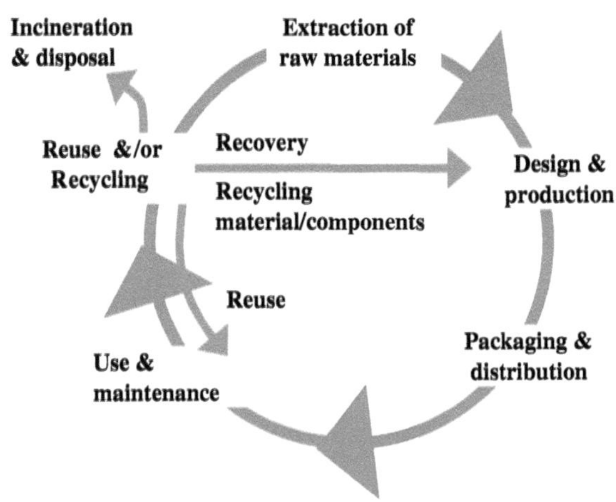

Fig. 3 Product lifecycle process

consume costs or affect the environment). Moreover, costs become shared results for a network of stakeholders (Mevellec and Perry 2006). They shift from a productive industry (mainly direct costs linked with manufacturing costs) to a cognitive and worldwide networked industry (with major allocations related to indirect costs linked with research and development stages) (Bouin and Simon 2000). As a result, the product lifecycle phases are already partially formalized. These phases can be more easily populated and monitored and indeed, the process definitions (required by ISO 9000 standards) provide a good basis for extracting and aggregating manufacturing costs. However, in a world where innovation and R&D projects maintain competitivity, the associated overheads costs are not easy to assess with real data. At the end of the product lifecycle, there is no rule that guides designers concerning the impact of whole costs on the final estimate. Depending on the alternatives, some financial advantages can be introduced into the loop. For example, reuse as second life sub-systems or material recycling can generate positive financial flow and reduce the global bill.

The same problems arise from environmental indicators. They have to take consumption of resources into account (mainly raw materials and energy), different types of pollution and emissions (solid, liquid, gaseous) and their impacts (human, eco-system, ground, water, atmosphere, etc.). As for cost analysis, some life phases or resource consumption can be monitored easily, such as power supply factories, distribution in a known supply chain, etc. However, in a continuously moving network of enterprise, many measurements depend on the network's dependencies. Consequently, evaluations may be inaccurate during the product development stage. The criteria that are in fact used to choose suppliers are far from the scope of environmental issues. Moreover, the end of life may have a great impact. Depending on the existing recycling paths, or developed technology, this impact could be positive and enhance the global environmental dependence. Burning or landfill solutions will no longer have a future. Industry and designers have to consider this impact in their future designs and developments. Automotive regulations for 2015 will limit the percentage of CO_2 emission but also impose a high ratio of recycling for end of life vehicles.

The use phase of a product can be hard to evaluate. In a Business-to-Business relationship, this phase is quite well defined and could lead to good evaluations, whereas Business-to-Consumer products could lead to unusual uses which in turn lead to unexpected costs or environmental consequences. In the case of a LCA, the use and maintenance phase may be the most noxious and designers and industry can have little impact upon it. This is where the designer's options start to be limited. Efficient information and customer tutoring can lead to achieving real sustainable products.

Even if it seems impossible to completely define the whole lifecycle, similarities and complementarities arise from the two modeling points of view: cost and environment. In each case, product evolutions have to be modeled and evaluated. Energy and material consumption are required data for both. Product transformation models are also sources of common rating. Thus, process and product models are used to perform cost analysis and product LCA through different stages of manufacturing, use, and end-of-life options. The system can be analyzed using process flow diagrams which show that the inventory of environmental impacts and resources used is comparable. This provides joint cost and environmental analysis (Hendrickson et al. 1998; Satish 1999).

2.2 Full Lifecycle Information

Most of the time, the expected lifecycle information is defined only partially or not at all in the early phases when decisions are made (Guinee 2002). As a result, it is hard to develop cost or environmental design strategies which could guide designers efficiently, due to these unreliable values. Specific risk analysis evaluation should be done at the key stage of the product-process development. A contingency analysis would allow the variability of the results to be measured and highlight the main incident factors (Wimmer et al. 2004). These methods are still under validation from an environmental point of view.

It seems possible to have detailed information made available for some stages like manufacturing, packaging and transport or recycling processes. However, even in these cases, the real data are not so easy to capture (Perry et al. 2007). Nowadays, the supply chain is worldwide, and the reality of modeled processes and data collection are hard to guarantee (Degos 1998). This is the case for cost evaluation and the environmental aspect despite the standard framework imposed on the suppliers.

Consequently, calculations must be made using unknown data and in most cases have to be interpreted as relative values. Thus ranking a new product or product process alternative might be risky.

2.3 Multi-Data Aggregation

Another common issue remains regarding the need to calculate with multiple kinds of data. In the case of LCA, the environmental impacts included are: global warming, acidification, energy use, non-renewable consumption, water eutrophication, gas and toxic emissions to the environment, etc. This combination of multiple and non-homogeneous data highlights the issue of indicator design and equivalence definition. Some research proposals have started working on unified metric units. For instance, decibels have been proposed as a possibility (Coatanea et al. 2007). This solution has no unity dependence and indicates the contribution or losses of the value (the decibel is calculated as a ratio compared to a nominal value) (Seager and Theis 2004). The energy equivalent calculation is another possibility. This thermodynamic concept measures material and energy resource consumption for each impact (Szargut et al. 1988).

In the same way as having a unique cost indicator, Perrin promoted the single value added unit methodology (Perrin 1963 and 1996). This proposal tries to find an independent cost unit that could facilitate true representativeness and hence the final aggregation. In fact, Perrin realized that the analytical accounting system is not adapted to industrial reality. Using the same philosophy of cost independence, target costing or activity-based costing approaches were developed and adapted for use and integration in design methodologies (Gosselin and Mevellec 2003; Innes et al. 2000).

Based on these studies, the concept of value promoted by Porter emerges as a global and transitional concept applied to both costs and environmental analysis (Norman and Ramirez 1993; Porter 1998). Indeed, traditionally, value includes different factors such as cost, quality, delay, and enables value chain evaluation and optimization to be carried out (Kaplinsky 2004; Mauchand et al. 2010). This notion of value is easily extended to environmental aspects.

3 Lifecycle Engineering and Product Lifecycle Management Based on Value Evaluation

As mentioned in the previous section, whole lifecycle evaluation means formalization and information at all stages of the product development. Nevertheless, the product itself cannot be the only focus. The processes that support product development, manufacturing, using step and end of life dismantling also have to be taken into account. As a result, the information system that supports such approaches must consider both product/process as well as different stakeholder viewpoints (Mevellec and Lebas 1998).

3.1 The Value Framework for Cost and Environment Combined Analysis

To ensure an efficient twin-eco evaluation (economic and ecological), it is necessary to quantify the alternatives for product and processes. This quantification will be functional, economical and environmental. In order to take stakeholders' viewpoints into account, each aspect has to be weighted. The final choice will be made according to the strategy or the enterprise objectives.

Value is a concept that enables different factors to be analyzed independently or in combination. According to the European standard (EN-12973), the value analysis is effective in improving the performance and taking into account other factors than cost. The evaluation of the performance of a company is not limited only to its financial performance. The value is the consumer's overall assessment of the utility of a product based on perceptions of what is received and what is given (Zeithaml 1988). Butz and Goodstein (Butz and Goodstein 1997) define customer value as the emotional bond established between a customer and a producer after the former has used one (or more) product (s) or services produced by the latter and found it of added value.

According to Porter (Porter 1998) a value chain is formed of a set of value generating activities and is used to diagnose a competitive advantage as well as to find ways to strengthen it. The value chain according to the AFNOR FD X50-158 is defined as all activities of the organization divided into cost elements and contributing to the final value of the product or service. Extending this concept, Elhamdi defines a value network as a group of partners collaborating in order to create value (Elhamdi 2005). He argues that value is linked to a beneficiary party. Hence, the generated value of the network does not concern only the company itself, but all other parties who are beneficiaries thus who derive interest or satisfaction. Performance and value indicators, presented in Fig. 4, come from a reflection on the benefits of product manufacture for each benefiting entity (Mauchand et al. 2010).

Mauchand proposes a product-process data model focusing on value chain modeling and evaluation (Mauchand et al. 2010) (Fig. 5). This model needs to integrate lifecycle concepts in order to enrich the value concepts with environmental concerns. For example, the process can be extended to product stages, and will represent all the steps illustrated in Fig. 3. Labrousse links the Product Process Resource model to the Functional Behavior Structure view (Labrousse et al. 2004). This solution gives the opportunity of managing both value and value chain evaluation (using the model in Fig. 5) and the dynamic aspect of the life cycle evaluation.

For any product (set of N functions), there are different technical solutions to meet the needs. In addition, for each solution, the process alternatives (composed of a set of activities) can lead to the product development and use. For each path, a value chain can be defined, as illustrated in Fig. 6.

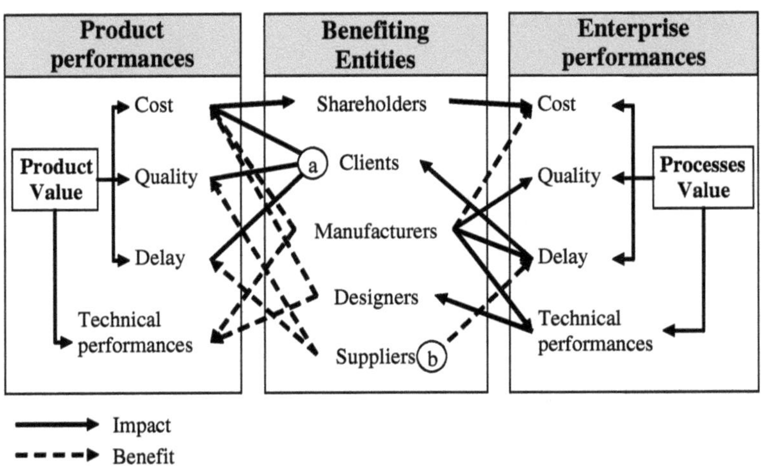

Fig. 4 Performances that affect value and their interactions with benefiting entities

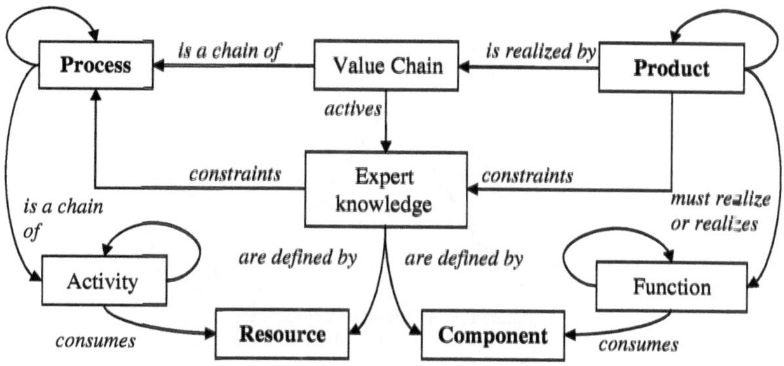

Fig. 5 Structure of the concepts for industrial system modeling (Mauchand et al. 2010)

Using this method, Mauchand proposes a Value Chain Simulator (VCS) that can compare solutions (Mauchand et al. 2010). Depending on the weights applied related to the benefiting entity interest, the solution will balance high technical performances oriented possibilities, low cost (or adapted market) solutions and environmentally friendly proposals. The structure and basic elements of the VCS are illustrated in Fig. 7.

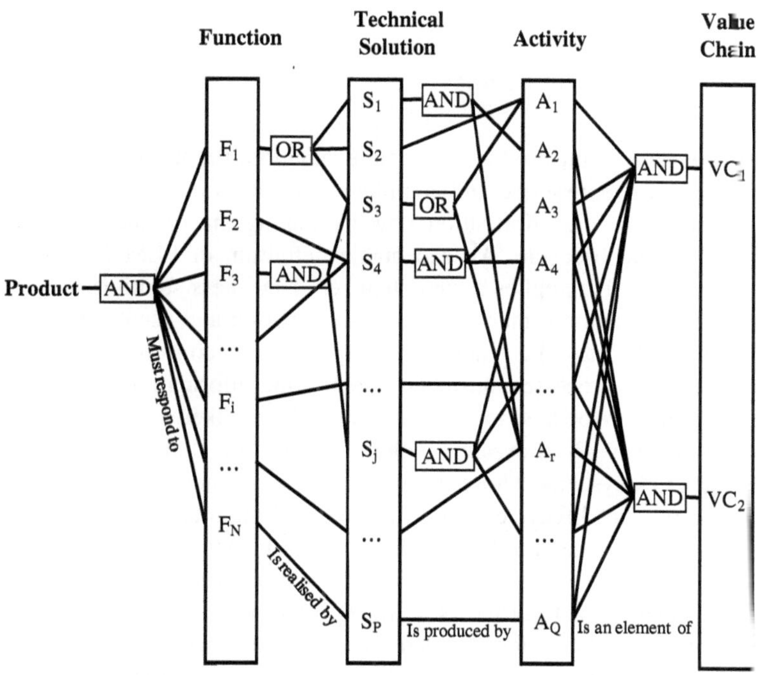

Fig. 6 Choice process of value chain alternatives

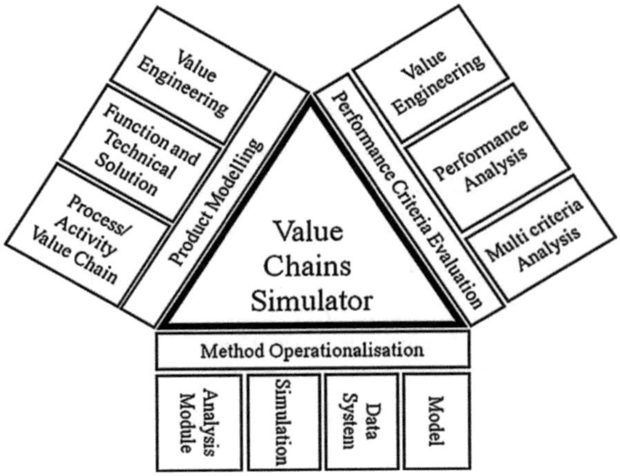

Fig. 7 Value chains simulator architecture (Mauchand et al. 2010)

Despite all the excellent qualities of this proposal, there is still something missing in terms of lifecycle simulation with such tools. Indeed, the model and data system required for the simulation are scarcely complete. This tool has been dedicated mainly to the manufacturing phase and must be adapted to the other product lifecycle stages.

3.2 PLM System Definition

In order to ensure a full product lifecycle evaluation, the life model and life use phases have to be represented and completed with relevant data.

PLM can ensure the information flow by linking the product definition files through the whole lifecycle. An acceptable definition of PLM is: "A strategic business approach that applies a consistent set of business solutions in support of the collaborative creation, management, dissemination and use of product definition information across the extended enterprise from concept to end of life—integrating people, processes, business systems, and information" (Amann 2002). PLM is generally associated to a set of applications linked to the product development. In this way, PLM systems include or tend to include PDM, MPM, ERP, CMMS and LCA systems. Those information systems store information for each stage of the product lifecycle except the use and maintenance phase. Concepts such as closed-loop PLM tried to overload this issue, extracting information from the product during the use step using wireless technologies. The concept of closed-loop PLM can be defined as follows: a strategic business approach for the effective management of product lifecycle activities by using product data/information/knowledge which can compensate PLM to realize product lifecycle optimization dynamically in closed loops with the support of product embedded information devices (PEID), product data and knowledge management (PDKM) system (Jun et al. 2007).

PLM systems rely on a data model composed of business objects that intervene in business processes and in product portfolios. Several modeling methods and languages have been developed to model these objects (Bernard and Perry 2003). Some of the many languages used to represent these objects and related activities are SADT or IDEF3, Business Process Modeling Notation (BPMN) (White 2004) or Functional Behavior Structure (FBS) coupled with Product Process Resources and External effects (PPRE) (Bernard et al. 2005). The establishment of patterns, based on this language, describes an approach to represent the processes. CIMOSA (Kosanke and Zelm 1999), ARIS (Scheer 1998), GRAI (Doumeingts et al. 2006) and PERA (Williams 1994) are modeling languages and modeling methodologies that must be adapted for PLM implementation.

Le Duigou proposed a PLM structure adapted to SME's. Supported by the French Technical Institute of Mechanical Industries (CETIM), the aim of this proposal was to provide a PLM solution for SME's. With this PLM information system, they can get into an extended enterprise structure with measured investments and time (Le Duigou et al. 2009). Based on a Product—Activity—Resource—Organization meta-data structure (see Figs. 8 and 9), this proposal has to be aligned with the previous value-based proposal, in order to be used to assess the product lifecycle model.

This PLM proposal is based on SME needs and requirements analysis (Le Duigou et al. 2012). Consequently, it is not completely adapted to the cost and environmental evaluation. Indeed, the different indicator measures can be implemented at all the levels: product, activity, resources and organization. It appears that if these data are available, the activity and the resource views could quickly give pertinent ratings. In the case of the product, the different lifecycle steps are represented by the different activities linked to the product (design, manufacture, use, disassembly, etc.). In the case of the use and maintenance phase, alternative uses (i.e. non-nominal) are represented by alternative activities in use and maintenance process. This makes it possible to evaluate the product and the impact of customers (depending on their behaviors). This example gives an idea of what a PLM system with evaluation facilities could be.

4 Knowledge Management for Virtual Engineering-Based Evaluation Discussions

In order to ensure high quality and efficient evaluations, the model should not only be adapted to the whole lifecycle, but the calculated rank should also be proposed with contextual information and data that reflect reality. Calculation and aggregation rules, data source reliability and model representations must be available for

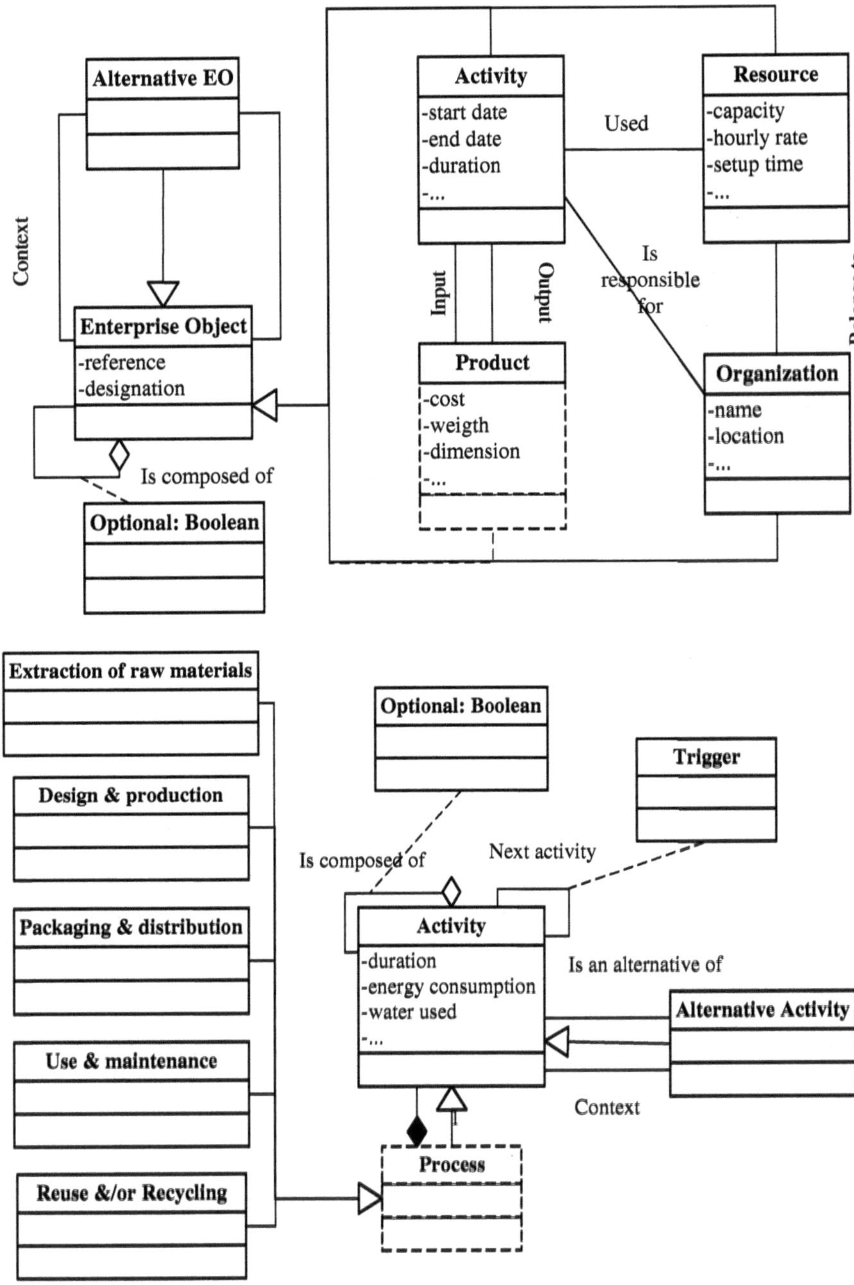

Fig. 8 Product activity resource organization meta-model (Le Duigou et al. 2011)

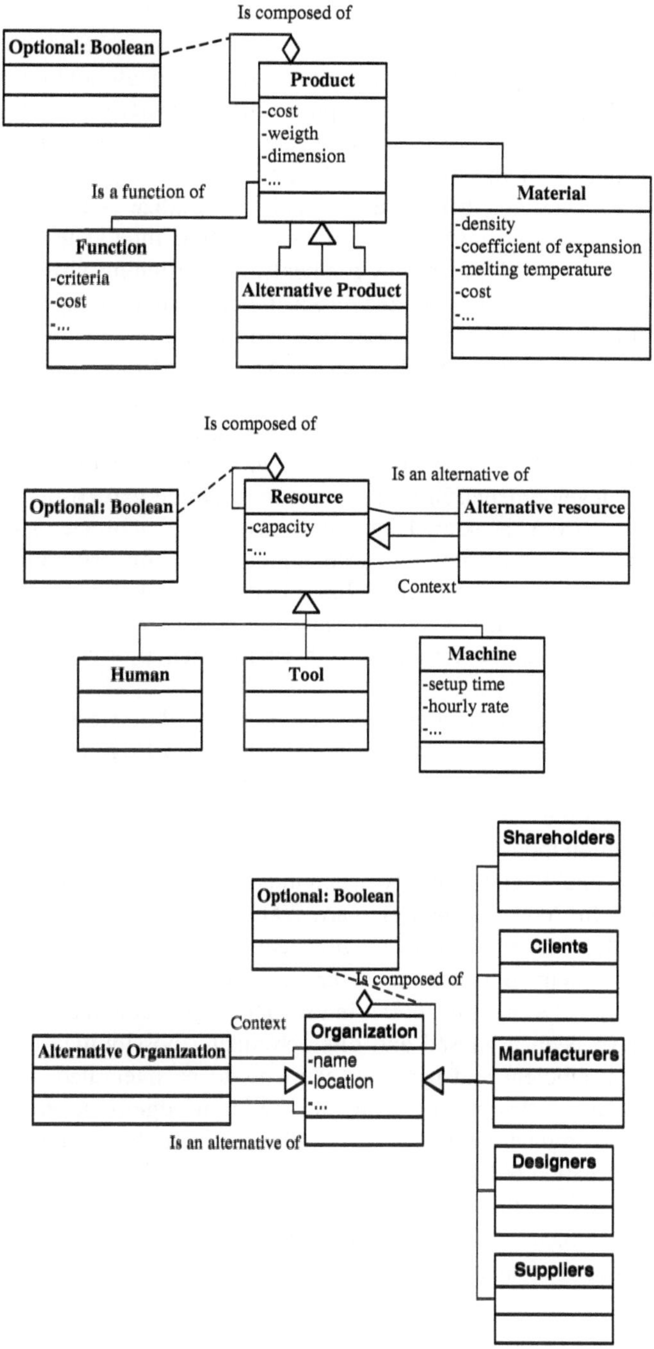

Fig. 9 Product, activity, resource and organization models (Le Duigou et al. 2011)

the contextualization of results. Consequently, knowledge from different experts must be integrated into knowledge-based systems. These systems must be interoperable with all the specific tools from the modeling phase and data capture to evaluation and results comparison or optimization. Virtual engineering environments allow the integration of all the lifecycle models (Bernard 2005). Engineers have new media to interact with the different numerical representation and simulation models. They use them to define and industrialize complex systems that must integrate more and more perspectives in a short time. The challenge is in improving product development environments and designing virtual engineering platform software that takes all the product and system lifecycle phases into account, and integrate knowledge (Bernard et al. 2007).

Consequently, knowledge tracking, identification and formalization, from different forms of expertise, at different levels of detail, must be carried out and integrated into knowledge-based engineering platforms (Ammar-Khodja et al. 2008). Specific methods ensure the coherence and consistency of these knowledge-based system developments. In order to ensure multiple expertise coherence and interoperability (from the knowledge and software point of view) various integration models exist, and ontology-based approaches seem very promising for the future 2.0 technologies (Bigand et al. 2007; Bachimond et al. 2002). For instance, a specific ontology definition of concepts like cost has already been proposed (H'Mida 2002) and can be combined with environmental or sustainability ontology (Missikoff et al. 2002).

Exchanged documents and previous projects are the information repository areas that can be exploited to enrich the expected knowledge (of costs and environmental evaluation) (DuPreez et al. 2005). From these documents, key knowledge can be identified. Xu proposes a knowledge value rating system to optimize the best evaluation models, representative methodologies or efficient software that can be used to quickly and precisely match the product or system cross evaluations (Bernard and Xu 2009; Xu and Bernard 2009). With this proposal it becomes possible to select the relevant evaluation techniques, according to the level of product development, information maturity, perspectives and target constraints. Such an operational system is not yet in use. The basic components of knowledge evaluation have been proposed and offer promising possibilities to browse and select the most efficient and pertinent elements to be integrated into the global knowledge database. The wish to integrate the knowledge of several experts into all phases of the product life cycle leads to a huge system that is unmanageable and unusable. Information reduction coupled with intelligent information technologies (i.e. 2.0) can reduce these risks. This is why many actors of the worldwide community have focused on methods and tools for effective knowledge life-cycle management (Bernard and Tichkiewitch 2008).

5 Conclusion

This chapter highlights the complementarities of cost and environmental estimates. The same needs and limitations for whole lifecycle evaluation appear for cost or environmental applications. Some representations of the lifecycle phase are missing from the modeling level due to absent data or unknown solutions for these phases. The data capture level for simulation lacks accuracy or sensitivity analysis for evaluating the quality of the results in terms of confidence or main factor impact. The performance indicators, the cost or environmental impact, can be analyzed separately or combined under a common such as the value concept. PLM possibilities, dedicated to data management and product information management relating to lifecycle, can therefore be adapted to support the different eco-calculations (from an economic and/or ecological point of view). In addition, to ensure a good level of result contextualization and best practice integration, expert knowledge must be included in a knowledge database. These knowledge databases are structured to support the definition and the development of agile virtual engineering platforms. Modeling tools may differ from one phase to another and the kind and quality of information will be at different levels. In order to maintain coherence and ensure agility with future software integration in the engineering method, ontology-based systems can offer solutions for service-oriented architecture for platform development.

This type of global approach cannot be addressed in a single project or test case, but results from development strategies for the different components identified in a system and their integration into a coherent global proposal.

References

Amman K (2002) Product lifecycle management 'empowering the future of business. Cimdata report, available at www.cimdata.com

Ammar-Khodja S, Perry N, Bernard A (2008) Processing knowledge to support knowledge-based engineering systems specification. Concurrent Eng Res Appl(CERA) 16(1):89–101. doi:101177/1063293X07084642

Bachimont B, Isaac A, Troncy R (2002) Semantic commitment for designing ontologies: a tool proposal. Lect Notes Comput Sci 2473:211–258. doi:10.1007/3-540-45810-7_14

Bernard A, Perry N (2003) Fundamental concepts of product/technology/process informational integration for process modeling and process planning. Int Journal of Comput Integr Manuf 16(7–8):557–565. doi:10.1080/0951192031000115723

Bernard A (2005) Virtual engineering: methods and tools. Proc Inst Mech Eng Part B-J Eng Manuf 219(5):413–421. ISSN 0954–4054

Bernard A, Labrousse M, Perry N (2005) LC universal model for the enterprise information system structure. Innovation in life cycle engineering and sustainable development. Springer, New York, p 429–448 doi:10.1007/1-4020-4617-0_30

Bernard A, Ammar Khodja S, Perry N, Laroche F (2007) Virtual engineering based on knowledge integration. Virtual Phys Prototyping 2(3):137–154 doi:10.1080/17452750701677509

Bernard A, Xu Y (2009) An integrated knowledge reference system for product development. CIRP Ann—Manuf Technol 58(1):119–122. doi:10.1016/j.cirp.2009.03.013

Bernard A, Tichkiewitch S (eds) (2008) Methods and tools for effective knowledge life-cycle-management. Springer, Berlin, p 586. ISBN: 978-3-540-78430-2

Bigand M, Bourey JP, Perry N, Mauchand M (2007) Case studies in model integration. Int J Comput Integr Manuf 20(7):619–626. doi:10.1080/09511920701566434

Blanchard BS (1978) Design and manage to life cycle cost. M/A Press, Oregon, p 255. ISBN 10-0930206002

Bouin X, Simon FX (2000) Les nouveaux visages du contrôle de gestion, Approches Techniques et Comportementales. Dunod, Paris, p 292. ISBN 2-10004887-2

Brissaud D, Lelah A (2010) A framework for the design of sustainable product-service systems. In: Conference on Advances in Production Management Systems, APMS 2010 Como, Italy. ISBN: 978886493007-7

Butz HJ, Goodstein L (1997) Measuring customer value: gaining the strategic advantage. Organ Dyn 24(3):63–77. ISBN 0090-2616

Coatanea E, Kuuva M, Makkonen P, Saarelainen T (2007) Early design evaluation of products artifacts': an approach based on dimensional analysis for combined analysis of environmental, technical and cost requirements. In: Advances in life cycle engineering for sustainable manufacturing businesses–14th life cycle engineering CIRP Conference proceedings Part 4-C2:365–370. doi:10.1007/978-1-84628-935-4_63

Degos JG (1998) La comptabilité, Collection Dominos In: Serres M, Farouki N (eds) Flammarion, Paris. ISBN 2-08-035538-4 (in French)

Doumeingts G, Vallespir B, Chen D (2006) GRAI Grid decisional modelling. In: Handbook on architectures of information system, part one Springer, Berlin, pp 321–346. doi:10.1007/3-540-26661-5_14

Du Preez N, Perry N, Candlot A, Bernard A, Uys W, Louw L (2005) Customised high-value document generation. CIRP Ann 54(1):123–126. doi:10.1016/S0007-8506(07)60064-X

Elhamdi G (2005) Modelisation et simulation de chaines de valeurs en entreprise : Une approche dynamique des systèmes et aide à la décision : SimulValor PHD thesis in Ecole Centrale Paris (in French)

EPA (2010) Defining Life Cycle Assessment (LCA), wwwgdrcorg/uem/lca/lca-definehtml (Nov 2010)

Fabrycky WJ, Blanchard BS (1991) Life-cycle cost and economic analysis. Prentice Hall, New Jersey, p 352. ISBN 0135383234

Viadiu FM, Fa MC, Saizarbitoria IH (2006) ISO 9000 and ISO 14000 standards: an international diffusion model. Int J Oper Prod Manage 26(2):141–165. doi:10.1108/01443570610641648

Gosselin M, Mevellec P (2003) Development of a cladogram of cost management systems, 6th manufacturing accounting research conference. Twente, The Netherlands

Guinee JB (2002) Handbook on life cycle assessment–Operational guide to the ISO standards. Kluwer Academic Publishers, Boston, p 692. ISBN 1-4020-0228-9

Hendrickson C, Horvath A, Joshi S, Lave L (1998) Economic input-output models for environmental life-cycle assessment. Policy Anal 32(7):184–191. doi:10.1021/es983471i

H'Mida F (2002) Contribution à l'estimation des coûts en production mécanique : l'approche entité-coût appliquée dans un contexte d'ingénierie intégrée, PhD Thesis of Metz University, mars 2002 (in French)

Innes J, Mitchell F, Sinclair D (2000) Activity-based costing in the UK's largest companies: a comparison of 1994 and 1999 survey results. Manage Acc Res 11:349–362. doi:10.1006/mare.2000.0135

Johansson G (2001) Environmental performance requirements in product development: an exploratory study of two development projects. Ph.D Thesis of Linköping University

Jun HB, Kiritsis D, Xirouchakis P (2007) Closed-loop PLM, advanced manufacturing—an ICT and systems perspective. In: M.Taisch (ed) Taylor and Francis, London pp 79–87. ISBN 978-0-415-42612-2

Kaplinsky R (2004) Spreading the gains from globalization: what can be learned from value-chain analysis? Problems of Economic Transition, M.E.Sharpe 47(2):74–115

Kosanke K, Zelm M (1999) CIMOSA modelling processes. Comput Ind 40:141–153. doi:10.1016/S0166-3615(99)00020-2

Labrousse M, Bernard A, Véron P (2004) In: Horwvath I, Xirouchakis P (eds) Generic FBS concept for process/product/resource integration, tools and methods of competitive engineering, vol 1. Millpress, Rotterdam, pp 384–394. ISBN 90-5966-018-8

Le Duigou J, Bernard A, Perry N, Delplace JC (2009) Global approach for technical data management, application to ship equipment part families. CIRP J Manuf Sci Technol 1(3):185–190. doi:10.1016/j.cirpj.2008.10.005

Le Duigou J, Bernard A, Perry N (2011) Framework for product lifecycle management integration in small and medium enterprises networks. Compu-Aided Des Appl 8(4):531–544

Le Duigou J, Bernard A, Perry N, Delplace JC (2012) Generic PLM system for SMEs: application to an equipment manufacturer. Int J Product Lifecycle Manage 6(1):51–64. doi:10.1504/IJPLM.2012.046429

Mauchand M, Siadat A, Perry N, Bernard A (2010) VCS: value chains simulator, a tool for value analysis of manufacturing enterprise processes (a value-based decision support tool). J Intell Manuf . doi:10.1007/s10845-010-0452

Mevellec P, Lebas M (1998) Managing simultaneously cost and value: the challenge. The role of management accounting in creating value. IFAC Publication, New York, p 13–25

Mevellec P, Perry N (2006) Whole life cycle cost: a new approach. Int J Prod Lifecycle Manag Inderscience 1(4):400–414

Missikoff M, Navigli R, Velardi P (2002) Integrated approach to web ontology learning and engineering. Computer 35(11):60–63. doi:10.1109/MC.2002.1046976

Norman R, Ramirez R (1993) From value chain to value constellation: designing interactive strategy. Harvard Bus Rev 71(4):65–77

Perrin G (1963) Prix de revient et contrôle de gestion par la méthode GP (in French). Dunod, Paris

Perrin J (1996) Cohérence, pertinence et évaluation économique des activités de conception, in Cohérence Pertinence et Evaluation. Ecosip Economica, Paris. ISBN 2-71-783023-1

Perry N, Bernard A, Delplace JC (2007) Concurrent cost engineering for decisional and operational process enhancement in a foundry. Int J Prod Econ 109(1–2):2–11. doi.org/10.1016/j.ijpe.2006.11.001

Porter M (1998) Competitive advantage: creating and sustaining superior performance, Harvard Business Review. Free Press, New York. ISBN 0-684-84146-0

Ritzén S, Beskow C (2001) Actions for integrating environmental aspects into product development. J Sustain Prod Des 1:91–102. doi:10.1023/A:1014446421205

Satish J (1999) Product environmental life-cycle assessment using input-output techniques. J Ind Ecol 3(2–3):95–120

Scheer AW (1998) ARIS handbook on architectures of information systems. Springer, New York, pp 541–566

Seager TP, Theis TL (2004) A taxonomy of metrics for testing the industrial ecology hypotheses and application to design of freezer insulation. J Cleaner Prod 12(8–10):865–875. doi:10.1016/j.jclepro.2004.02.032

Szargut J, Morris DR, Steward FR (1988) Exergy analysis of thermal, chemical and metallurgical processes. Hemisphere Publishing, NY, p 322

Ventère J (1995) La qualité écologique des produits – Des écobilans aux écolabels, 166p (in French). Éditions Sang de la Terre, AFNOR, Paris

White SA (2004) Introduction to BPMN, http://wwwbpmnorg/

Williams TJ (1994) The purdue enterprise reference architecture. Comput Ind 24:141–158. doi:10.1016/0166-3615(94)90017-5

Wimmer W, Zust R, Lee KM (2004) Ecodesign Implementation: a systematic guidance on integrating environmental considerations into product development. Springer, Dordrecht, p 145. ISBN 1-4020-3070-3

Xu Y, Bernard A (2009) Knowledge organization through statistical computation: a new approach. Knowl Organ 36(4):227–239

Zeithaml VA (1988) Consumer perceptions of price quality, and value: a means-end model and synthesis of evidence. J Mark 52:2–22

The Ecodesign of Complex Electromechanical Systems: Prioritizing and Balancing Performance Fields, Contributors and Solutions

S. Esteves, M. Oliveira, F. Almeida, A. Reis and J. Pereira

Abstract In the Product Development (PD) of complex mechanical and electromechanical systems, such as machine-tools, mapping the relationships between technical behavior, environmental and cost impacts brings new challenges. Having technological advances, cost drivers and environmental performance under surveillance, manufacturers and designers are expected to provide eco-efficient systems keeping a competitive price. This introduces a new set of design functions with increased complexity due to the new interdependent variables, requiring complementary technical, environmental and cost assessments. The redesign study of a sheet metal forming machine-tool, a press brake, is here used by the authors to present the first assessment of a proper methodology to support the main decision processes and to illustrate the technical and technological trade-offs faced by a PD team in order to achieve the global design objectives. The design process aimed to reduce the environmental footprint of the machine-tool and simultaneously to improve the bending process accuracy. A Voice of Customer (VOC) study was carried out in order to assess the receptiveness of press-brake users to the targeted product specifications and price changes sensitivity. The research effort was mainly focused in defining and testing the applicability of different assessment and measurement tools for the Ecodesign of a press-brake.

S. Esteves (✉) · M. Oliveira · J. Pereira
Instituto de Engenharia Mecânica e Gestão Industrial (INEGI), Campus FEUP,
Rua Dr. Roberto Frias, 400, 4200-465 Porto, Portugal
e-mail: sesteves@inegi.up.pt

F. Almeida · A. Reis
DEMec, Faculdade de Engenharia, Universidade do Porto Campus FEUP,
Rua Dr. Roberto Frias, s/n 4200-465 Porto, Portugal

E. Henriques et al. (eds.), *Technology and Manufacturing Process Selection*,
Springer Series in Advanced Manufacturing, DOI: 10.1007/978-1-4471-5544-7_13,

1 Introduction

Till recently, technical performance and cost were the main drivers for products and systems design. In the last decade, attention has been extended to the assessment of their environmental profile. The continuous issuing of green standards and directives are increasing the pressure exerted not only by the users but also pushing for greater responsibility of manufacturers and other market stakeholders, forcing the improvement of the environmental performance of their products accordingly.

At the technical level, the new trends in complex mechanical and electromechanical systems, and machine-tools in particular, are higher automation, less human intervention, as well as higher efficiency and/or accuracy. This is often reached throughout the introduction of new energy and controlling sources and shall be achieved with no additional cost effort. In order to include the environmental profile function, the following factors must be carefully analyzed:

1. The full set of resources used to obtain the machine-tool on itself, accounted as input–output substances associated to the components production and their assembly (materials- and manufacturing processes-related);
2. The main functionality of the machine-tool, which determines the resources required during operation, namely electricity, accounted as the specific process energy (SPE) and other process- or operation-related resources, accounted as input–output substances associated to the use of the machine-tool (consumed directly in the process, by auxiliary systems during operation or in maintenance operations);
3. The disposal of the resources referred in the above points, at their end-of-life.

While for the technical and cost dimensions, internal procedures and tools are typically well established, the account of environmental impact function during the design phase implies the use of specific assessment methods. Life-Cycle Assessment (LCA) tools appear as the most broadly accepted for this purpose, being particularly suited to compare products and systems that perform the same function (Azevedo et al. 2011). Product designers are then requested to be capable of working with a set of tools which enable to account the environmental profile as a decision criterion at different design stages, since the main environmental contributors (reported to be responsible by about 80 % of the impact (reported in European Commission Energy Efficiency) are decided in these phases. Thus, a growing number of industrial sectors, including machine-tools manufactures, seek the optimization of the global performance of their products and services. Again considering the complexity of this type of systems, the need of robust global decision criteria is crucial to achieve a successful design or redesign.

1.1 Technical Performance Assessment Tools

Increasing performance has been a market drive in machine-tool industry. Higher dynamic structural behavior, dimensional stability, reduced manufacturing complexity, good damping properties and high mass to avoid rigid body movements are the main current technical improvements. To support the objective assessment of the fulfillment degree in such factors, key performance indicators (KPI) and the respective target values must be established, specifying the valuable technical improvements expected. Such KPI parameters are defined according to the type of operation and expected accuracy level, determining the type of control parameters and characterization or measurement methods.

Process, power and control subsystems are the main technical units integrated in a machine-tool. In an Ecodesign perspective, when defining their level of integration, an appropriate modular structure should be considered, determining the maintainability, upgradability, reusability, and recyclability of the different modules, without compromising the machine lifetime as a whole (Chung et al. 2011; Umeda et al. 2008). Thus, modularity is an important approach in machine-tool design in order to reduce environmental impacts and costs during entire life cycle of a product.

Figure 1 presents an approach to the modular design, as proposed by Shamsuzzoha (2011).

The first step in designing a modular structure of a complex system is the product decomposition in single components or independent sub-systems. This is a task to be followed in all technical, environmental and cost analysis, and the level

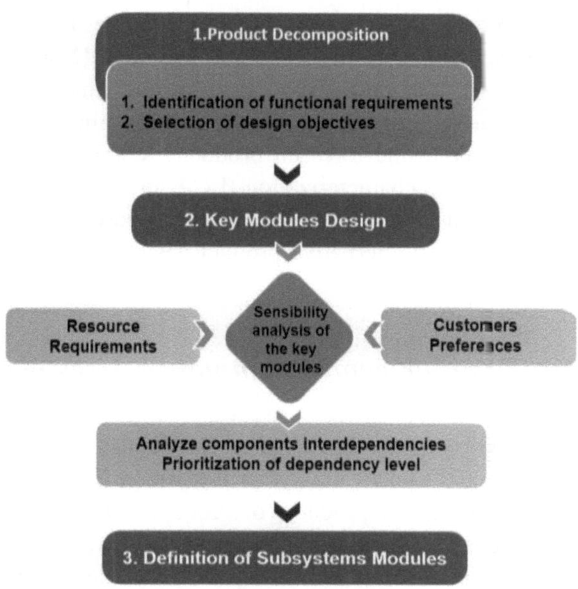

Fig. 1 Modular design structure, according to the proposal of Shamsuzzoha (2011)

of detail is determinant for the results reliability. The product resources inventory process is an important step for mapping of relationships between subsystems functionalities and physical components. Those components/subsystems revealing significant interdependencies, such as application, access, assembly, development effort or reliability, might then be recognized as key modules candidates. The number and size of the modules is mostly dependent on the functional requirements and design objectives. Assessment tools such as a Design Structure Matrix (DSM) are well positioned to support these tasks (Shamsuzzoha 2011).

The evaluation of the key modules performance is the second and central step. The most appropriate modules present only functional interdependency between internal module components. Production, assembly and integration ability potential and the related time and cost efforts, as well as the product variants able to be assembled from one same group of modules, are assessed in detail. The criteria are either defined by the component requirements and/or customer preferences. The most promising modules are then defined in detail and their interdependencies are prioritized.

1.2 Cost Performance Assessment Tools

Life-Cycle Cost (LCC) tools evaluates the costs of an asset throughout its life-cycle and are used to demonstrate the resource and cost efficiency of the alternative system/sub-system/components solutions, indirectly increasing the relevancy of the LCA studies for decision making and previewing economic consequences of the alternative solutions (economic risk modeling). The methodology shall be based on standard procedures for accounting the direct monetary flows related to the input–output flow of resources considered in the respective component/sub-system resources inventory (Norris 2001). More than in LCA, timing is critical in LCC, and a specific time horizon scope shall be adopted, corresponding either to the 'technical' life-time of the machine-tool, its use-phase duration, or to the foreseen product lifetime. Proper parallel environmental and cost life-cycle characterization (LCA and LCC, respectively) enables the capture of important relationships and trade-offs between the environmental and economic performance of alternative product design solutions.

1.3 Environmental Performance Assessment Tools

The environmental impact assessment also requires the application of specific methods and tools. Life-Cycle Assessment is the reference tool to quantify environmental data of products or processes, thereby obtaining their environmental profile over the entire life cycle, from the extraction of raw material to its end-of-life. Together with LCC, this joint assessment helps to identify opportunities to improve environmental performance of products at various points in its life cycle, providing essential information to support decision-making strategic planning and prioritization in the design/redesign of products and/or processes. Such valuable info is also being used in the selection of environmental performance indicators, and is supporting product/system/processes marketing throughout the implementation of eco-labeling and the preparation of environmental statements.

Every LCA methods use qualitative, quantitative or semi-quantitative analysis, although the quantitative form is considered more suitable for detailed LCA studies (Todd and Curran 1999; Hochschorner and Finnveden 2003). However, LCA tools can be time and work consuming and thus have significant costs. In recent years, there has been a trend for the development of simplified methods for LCA. These are quantitative or semi-quantitative methodologies aiming to give quick answers and suggestions (Azevedo et al. 2011). Although these methods tend to be very universal and wide-ranging, given the broad applicability of these methodologies and the strong emergence of its use, they have a strong customization potential. In fact, these simplification techniques can be adapted to provide 'customized' or 'tailor-made' perspectives in studies of specific systems or sectors, enabling to include system-specific principles and practices more relevant and appropriate to the interested LCA end-user, while still producing valid and robust results, and keeping the LCA basic conditions regarding scope and methodology (Bala 2010). In line with this, Hochschorner and Finnveden (2003) highlighted the importance of the method applicability to the field of application as the most important selection criteria of the proper LCA method to adopt, in order to deliver the required information.

To ensure optimal reproducibility and applicability, documentation guidelines for data and metadata are included in this approach. Guidance on the definition of a functional unit and a reference flow as well as on the determination of system boundaries meets the generic LCA goal and scope definition requirements of ISO 14040 and ISO 14044. This was developed with the purpose of provide high-quality life-cycle inventory (LCI) data for all the stages of the life cycle of the product.

1.4 'All-in-One' Ecodesign Tooling

In such a Product Development (PD) scenario of new equipments, an Ecodesign approach is being used. Though there is no unique Ecodesign definition, this aims to minimize the environmental impact of a product/system life-cycle, at no expenses of product cost and quality (Vezzoli and Manzini 2008; Peças et al. 2009). In this sense, Ecodesign is indeed based on the identification, analysis and reduction of environmental impacts likely to occur over the lifetime of products. Thus during product design, factors affecting the environmental profile and general performance of a product/system at each life-cycle stage are considered, towards its global improvement. Thus, not only the product is to be analyzed but all input-output flow of resources and efforts related to its production, use and end-of-life stages, in order to achieve an effective impact reduction and not only their transfer between stages (Giudice 2006). For each type of resource and flow, specific measuring and other support tools shall be made available to provide for reliable and high-quality data, supporting grounded design decisions and the most appropriate selection of materials, processes or concepts according to the product specifications. The press-brake case study described in the following sections has been developed with this approach.

2 Methodologies

2.1 The Ecodesign Approach to the Press-Brake Case-Study

This case-study consist in design and develop a new press-brake, with full integration of Ecodesign practices. Such machine-tools have recently been classified on the basis of Article 15 (2) of Energy-using Products (EuP) Directive (2005/32/EC) and its amendment (2008/28/EC) in the list of the 25 EuP categories considered priority to be in compliance with the Ecodesign directive (European Commission 2007). In this ranking, this type of equipment appears in third place. The strategy behind this case-study is based on the differentiation of the final product, based on a methodology of Ecodesign, in order to face competition by anticipating new solutions that comply with regulatory measures.

The press-brake is clearly classified as an intensive-use product, due to the high energy consumption in service. Since this is a product of considerable size/weight, the consumption of raw materials is also relevant to its environmental impact as well as end-of-life. From this perspective the Ecodesign approach must focus in areas such as optimization of materials use, the development of more energy efficient systems and minimizing environmental impacts associated with the treatment and its final destination.

In order to handle the necessary trade-offs, a structured approach was developed in order to support design decisions, selection of solution alternatives, monitor performance evolution and maintain LCA and LCC comparability between current and new design scenarios. The referred structured approach comprises the following four main stages, structured as shown in Fig. 2:

(1) VOC (Voice of Customer) analysis, aimed to identify a complete set of customer needs and wishes, prioritized by in terms of importance;
(2) Specifications, building a basic hierarchy tree of design requirements, including the environmental ones;
(3) Product Design concepts and Multi-criteria and Balancing Assessment, generation of design alternatives to fulfill the different objectives, and their evaluation using LCA, LCC, performance KPI and balancing models/decision matrix;
(4) Final Design Solution.

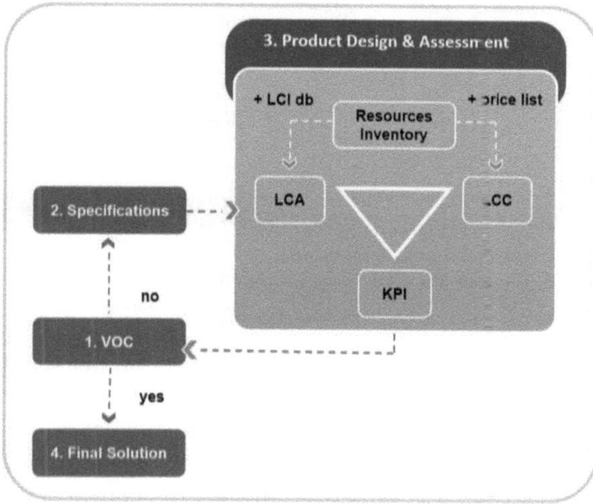

Fig. 2 Structure of the product development of an electromechanical system framed in ecodesign

The VOC analysis provides essential information for the PD process, introducing the customer main needs. Technically this consists in the organization of focus sessions with different stakeholders groups, including end-users but also engineering and sales teams on the manufacturer side, able to identify the different features to be improved or introduced in the new equipment. This information is then used as design inputs and converted into targets to assess the customer's satisfaction potential regarding the alternative specifications to be proposed for the new equipment. Such requirements' gathering sessions might also be completed with workshops, joining with experts on existing systems to pull together the state-of-the-art solutions available. Additional overview meetings might later be followed with the end-users to present the changes the new design concept plans to provide, anticipating their feedback and letting them know their needs were considered and understood.

In the Ecodesign process, product alternatives and life-cycle scenarios should be assessed and compared considering a directly comparable functional unit. The functional unit shall help to normalize and compare the data, and is to be used as input–output reference of the different processes integrating the life-cycle. In the case of the machine-tool development, the functional unit considered was the Assembly and Use of a hydraulic press-brake within lifetime. The flow diagram in Fig. 3 presents the respective system frontiers defined for this press-brake.

As soon as the critical specifications and system frontiers are set, the full set of resources related to all product life-cycle stages is to be compiled and assessed, in order to enable the identification of the main cost and environmental detractors by type and life-cycle stage and proceed with their prioritization in terms of potential technical and economic changeability.

All technology-driven changes presenting the most promising KPI values must also be assessed in terms of cost and environmental impacts. As the related technical

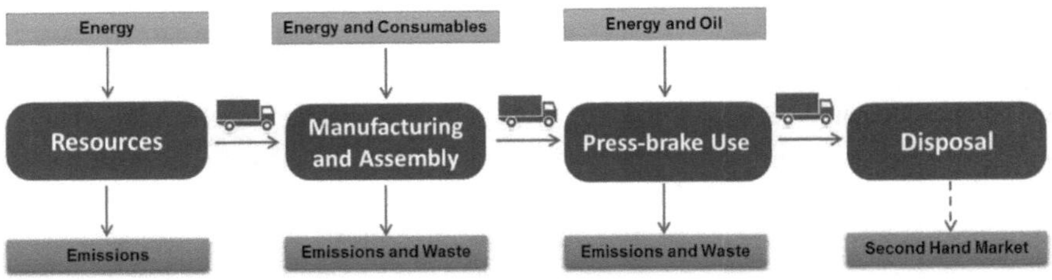

Fig. 3 System frontiers to be considered in the LCA of a hydraulic press-brake

benefits will mainly be visible during the use phase of the electromechanical system, the main resources flow involved, namely electricity and consumables consumption, must be carefully analyzed. As made evident in the following sections, these are mainly affected by the particular system system/sub-systems technology being used for the operation. Moreover, in what refers to energy consumption, and as reported by the authors (Santos et al. 2011), the utilization mode during production also plays an important role, though this is usually neglected on the common life-cycle inventory (LCI) databases. Considering the urgent need of low energy consuming systems, such factors must be analyzed in detail.

Attention must be given to each machine-tool assembled sub-system, particularly to the materials incorporated, in which steel has traditionally been dominating. The change in steel pricing policy and increasing steel cost are pressing overheads and margins at the machine-tools manufacturers and their components suppliers. As the need for alternative materials, less subjected to such market variations, becomes more evident, technical targets, process quality and environmental profile might be compromised. Given the typical lack of data available in the design early stages, a product resources inventory with significant detail must be accounted to support the technical, cost and environmental assessments. In fact, though all environmental detractors should be theoretically addressed, it may not be possible to address some particular points, given the need to comply with the technical specifications or cost limits introduced.

Table 1 presents some factors with significant environmental impact that should be considered in any design study of electromechanical systems, based on the experience built on several I&D projects in this field.

In what concerns the LCA and LCC of highly complex systems, such as a press-brake, it is essential to establish a methodology that simultaneously rules the data collection process and the execution of necessary simplifications to adopt. In this sense the following techniques have been adopted for the different assessment stages:

- Sub-systems structure approach: In order to facilitate data allocation and intuitive identification of the resources flow positioning per life-cycle stage, a proper product life-cycle tree must be defined (Table 2). The upper level, the main systems level, should be defined according to the internal assembly structure of the manufacturer and the life-cycle stage sequence of the product. This will

Table 1 List of main environmental factors to be considered in the design of electromechanical systems, related typical detractors and improvement potential actions

Environmental impact factor	Related detractors	Improvement potential
Consumption of resources	High diversity of resources and raw materials; Over-sizing	Minimize resources consumption in type and quantity; Maximize manufacturing processes efficiency
Selection of resources	Different raw material have different environmental impacts	Select raw materials with lower environmental impact among the likely technically and economically feasible alternatives
Manufacturing, assembly and disassembly processes	Processes are determined by the type of resources to be used Complexity and environmental impact (mainly regarding energy and consumables use during operation) are significantly distinct	Preview a 'lean' manufacturing approach towards reduction of cycle time, raw materials and consumables and higher global process efficiency, at each process/life-cycle stage
System usage	Utilization is not optimized, in functionality, availability and maintenance, with strong impact in energy consumption efficiency, product lifetime and related costs	Global mass reduction, sub-systems synchronization and combined dimensioning; Facilitate and minimize maintenance interventions/tool replacement; Improve usability and interface comfort
Disposal scenario	No disposal scenario is foreseen, at a full-system or sub-system level	In general, the easier the disposal, the easier the reuse/recycling. Prefer solutions with a mature and accessible EOL scenario

facilitate both the data transfer to the computing LCA and LCC as well as the results analysis in terms of the identification of the main contributors to the respective environmental and cost impacts.

- Product resources detailed inventory: In what concerns raw materials consumption, the component mass is the basic factor. Table 3 presents the data fields collected for each of the system components. For complex components, such as motors, valves, pumps or electrical cabinets, this task might be inhibited by the high complexity of the raw materials integrated and by the lack of data made available by the suppliers, thus seen as black boxes. In some cases, parallel LCA assessments might even be followed to assess the individual profile of such a complex component, supporting potential simplifications.
- Resources flow prioritization: In what concerns the functional unit and the system frontiers some resource flows might be selectively not considered, due to their negligible contribution to the global impacts. In the press-brake case, the transport of raw materials to the production site, due to the relatively short

Table 2 Subsystems pre-defined for data allocation of the press-brake under study

Life-cycle stage	Sub-system
Press-brake materials, manufacturing and assembly	Hydraulic system
	Main structure
	Guards
	Table
	Support systems
	Electric cabinet and CNC
	Standard components
Press-brake use	Energy consumption
	Oil consumption
	Transport—customer site
Press-brake disposal	Transport—end of life

Table 3 Data to be collected per each individual product component, building the product resources inventory required for LCA

Field	Field description
Level	Component hierarchic level on product tree
Code	Component identification (CAD 3D)
Designation	Component designation
Quantity	Component quantity
Volume	Component volume
Material	Raw material designation
Final mass	Component mass
Initial mass	Component original block mass
Processes	Component process flow

distance between the suppliers' sites and the assembler, and other oil-based consumables and lubricants, besides hydraulic oil have been neglected.

- Definition of a standard use scenario and measurement technique for electricity accounting: Anticipating the comparative evaluation of the performance between current and new machines, a reliable application scenario must be defined to assess the machine performance in the use phase. The energy consumption of the machine is one of the most important resource flows to assess in this phase. The energy consumption can either be estimated throughout theoretical calculations or directly measured with a standard data logger placed at the main power input on the electrical cabinet, operating the machine at the pre-defined use scenario.
- Definition of the environmental impact indicator to adopt: All data to be accounted will then be transferred in proper formats to be processed in proper computational tools enabling the conversion of the resource flow data inventoried into standardized environmental impact indicators. The normalized indicator Ecoindicator99 (EI'99) (Pré-Consultants 2011) is the most widely used. Different normalizing and weighing reference values can be used in its

Table 4 Normalization and weights used in the EcoIndicator 99 (EI'99) calculations (Pré-Consultants 2011)

	Hierachist (EI'99 H/A)		Egalitarian (EI'99 E/E)		Individualist (EI'99 I/I)	
	Normalisation	Weights (%)	Normalisation	Weights (%)	Normalisation	Weights (%)
Human health	0.0154 DALYs(0,0)	40	0.0155 DALYs(0,0)	30	0.00825 DALYs(0,1)	55
Ecosystem quality	5130 PDF*m2*a	40	5130 PDF*m2*a	50	4510 PDF*m2*a	25
Resources	8410 MJ	20	5940 MJ	20	150 MJ	20

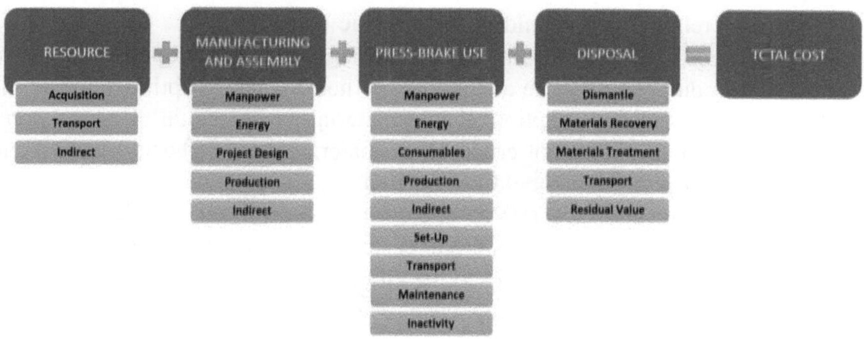

Fig. 4 Cost items considered in the LCC of the press-brake case-study

calculation, as shown in Table 4, and the Hierachist (H) approach is the most common. In the present case-study, the SimaPro 7.1 software (Pré-Consultants 2011) and the Ecoinvent 2.0 (Ecoinvent 2010) LCI databases were used for this calculation.

Finally, the LCA was complemented with the LCC analysis. The cost items considered in this study are presented in Fig. 4, per life-cycle stage.

3 Results and Discussion

3.1 *Voice of Customer (VOC)*

As advanced, the press-brake redesign process started with the VOC step. The following inputs have been gathered from several meetings with end-users joining significantly different products, processes and interests, as well as from internal meetings with engineering, maintenance/customer-support and sales teams:

1. Technology-related consumables and residues: Current product is based on a hydraulic driving system. The system driving concept is to be reviewed, towards lower energy consumption and the oil dependency should be carefully questioned. Whatever the solution to be adopted, the adoption of organic solutions (biodegradable) should be favored when selecting any oil and lubricant;
2. Modularity: As new technological solutions and functionalities become available and the production targets are pushed to much higher values in the strongly competitive market scenarios, the current machine-tool solutions lifetime is shorter and they faster becomes obsolete. Though some subsystems are considered to be quite operational and could even be integrated in other equipment, the current product is not upgradable. The lifetime of the new product should be extended and the quantity of materials, components and subsystems able to be replaced or reintegrated should be of strategic concern;
3. Raw/materials type and consumption: The design solutions adopted for some of the current components to be assembled are not efficiently optimized in terms of complexity and consumption of raw materials, from metal alloys to composites. Moreover, the use of crude-based materials has not been selective and justified. Their reduction should be targeted;
4. Human–machine interface (HMI) and on-line control service: Intuitive and user-friendly applications, with high customization potential to explore all enabled functionalities according to the end-user needs are highly appreciated. Efficient fault diagnosis and repairing is essential to maximize equipment availability. An open computer numeric control and proper instrumentation are essential to insure these requirements;
5. Eco-image: The Ecodesign framework should be made evident as a positive differentiation factor. The issuing of an Environmental Product Declaration (EPD) and an Eco-Label should be foreseen in the new product marketing plan;
6. Cost: The price of the new product for the same functionality level should not exceed that of the current product.

3.2 Specifications

The bending equipment under redesign is a commercial hydraulic press-brake of standard construction from 2006, made in Portugal. This equipment is specified to operate at a maximum bending capacity of 110 t and a maximum bending length of 3 m, with a main motor rated at 7.5 kW. It weighs about 7 t, and the structural parts are made of different types of steel, mostly structural steel and carbon steel, with some aluminum, polymer and composites present in small parts.

The technology of the main driving system leaded the specifications process. Though the data logger simulation is considered reliable for the energy consumption estimation, a parallel comparative study involving press-brakes of different technologies was followed. In this framework, several time studies were conducted to analyze actual machine usage and end-user needs. From these, the performance of commercial press-brakes and bending operation regarding energy consumption were discussed, and a model for estimation of the energy consumption needs for this machining operation was presented. The scans in Fig. 5 expose the drastic influence of machine-tool technology in the evolution of power and energy consumption with time, per operation cycle, as directly acquired from the conventional hydraulic press-brake and the targeted electric models.

In contrast to the conventional chipping operations, for discrete loading operations such as bending, the process rate must be described as a function of the frequency of production cycles, instead of the amount of material removed or being processed (Gutowski et al. 2006, Kuhrke et al. 2010, Kellens et al. 2012). Moreover, as demonstrated by the authors (Santos et al. 2011), the actual energy and usage of each machine are essential to estimate the specific process energy (SPE) required for bending operation or any other discrete loading operations. In this perspective, SPE should be accounted according to Eq. (1):

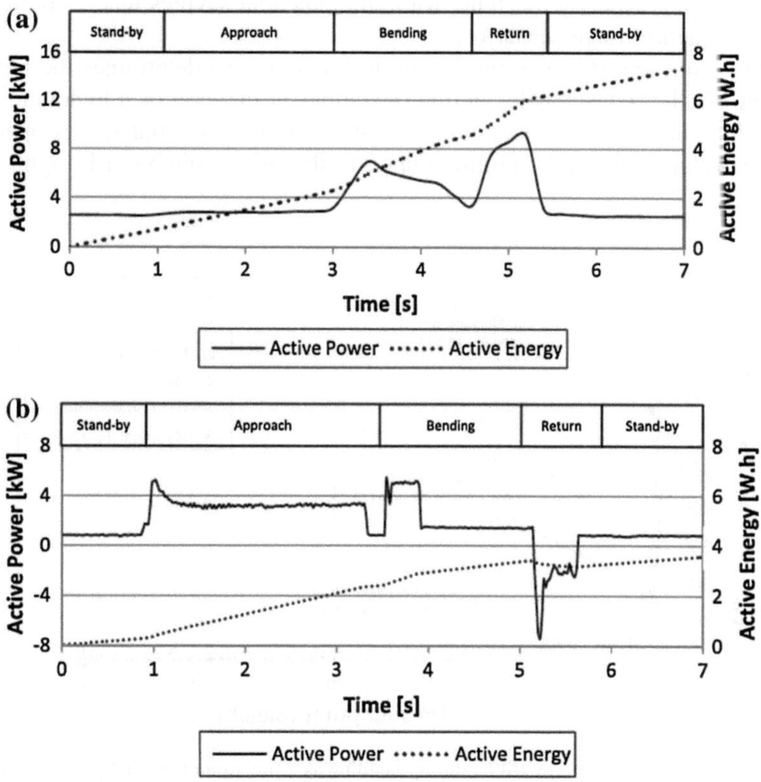

Fig. 5 Examples of active power and active energy per bending cycle recorded for: **a** a hydraulic press-brake and **b** an all-electric press-brake

$$B_{elect} = \frac{P_{idle}}{n} + q \qquad (1)$$

where P_{idle} is the active power consumed during stand-by mode, n corresponds to the throughput in cycles/h, constant q represents the cycle peak energy obtained at a pre-defined process load and loading time, excluding the fixed contribution, while the variable contribution $\frac{P_{idle}}{n}$ considers the total cycle time, which is dependent on the production throughput, opposed to the loading time in the chipping processes.

From this, the SPE as a function of the throughput was estimated for a set of hydraulic and all-electric press-brakes based on real consumption data measured directly on the machines. Figure 6 shows the estimation models obtained for all machines, working at the highest used loading capacity during the study and with a maximum throughput value of 720 cycles/h, as this is the theoretical limit for a machine working continuously at the smallest cycle time observed (5 s).

As depicted from Fig. 6, particularly for irregular and/or low usage scenarios, the electric-based drive technology is to be recommended, as this might lead to energy savings of about 90 % when compared to an all-hydraulic system, for a similar loading capacity machine, while the potential savings tend to be reduced for more intense usage scenarios.

Apart from the SPE, system driving technology also determines the type and amount of other consumables during operation. In the case of a hydraulic press-brake, hydraulic oil is a technology-specific resource essential for its operation. The environmental impact profile related to the oil consumption is significantly

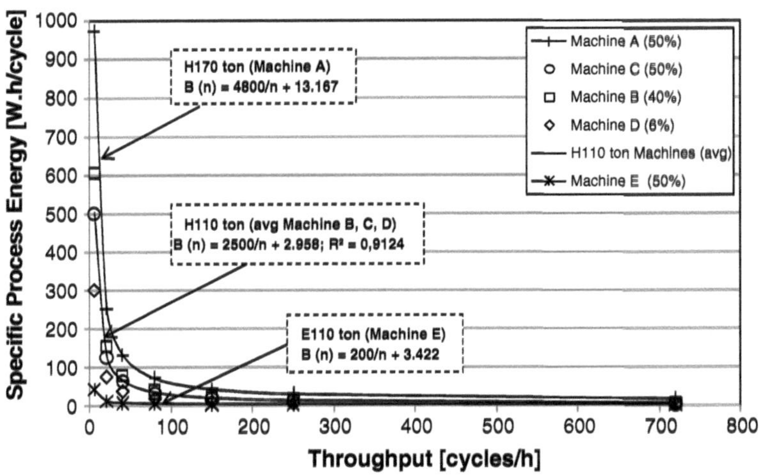

Fig. 6 Specific process energy (SPE) during bending as a function of throughput, obtained from energy consumption data measured directly on a set of bending machines (n: throughput [cycles/h]; Hxxx: hydraulic technology and Exxx: all-electric technology, indicating the respective maximum bending capacity; xx %: capacity loading tested in a specific machine)

affected by its no-renewable character, as this is a standard crude oil by-product, typically incinerated at the end-of-life.

From this a cradle-to-grave LCA of the equipment was followed to identify the main detractors to its environmental performance. The following settings were considered:

- Assembly: includes materials and manufacturing processes associated to all structural parts. Parallel individual LCA studies have been developed for main motor, hydraulic pump, valves and electric cabinet (black boxes concept), and proper simplifications were adopted. Secondary metals were used whenever applicable;
- Transport: considers the worst-case scenario, corresponding to the sales share for non-European market (30 % total), composed of about 260 km by road transportation and about 9000 km water transportation, in average;
- Use: a useful product lifetime of 15 years was assumed, based on technology relevance. Hydraulic oil and electricity have been accounted with a yearly consumption of 120 L and 5475 kWh, according to the respective model in Fig. 6. Hydraulic oil consumption is according to the preventive maintenance plan foreseen, while electricity input value was accounted assuming a yearly machine usage of 2000 h in a manual-intensive mode. The application scenario used included the following sequence steps: (1) Motor ON; initialization sequence; blind bending at 110 t; Motor OFF. An utilization rate of 65 % (stand-by 35 %) was used, as proposed by other for this kind of machines (Devoldere, Dewulf et al. 2007);
- The main inputs in the use phase, electricity and oil, were distinguished as different use-phases, to assist the analysis of their individual impact.
- Final disposal: a reuse scenario was considered, as this is the most probable scenario. This includes 400 km of road transportation and 9700 km of water transportation, assuming a transfer from Portugal to a potential second-hand user located in far countries. The contribution of an end-of-life scenario has here not been accessed. In practice, this would represent an extension of the lifetime, reflected by an increase on the use-phase inputs, i.e. SPE and hydraulic oil contributions.

Figure 7 shows the environmental profile generated and the relative contributions of each life cycle phase to the global environmental impact of the machine, based on the Eco-Indicator 99 (H,A) method, using SimaPro 7.1 with Ecoinvent 2.0 unit processes as LCI database.

The major contribution of the use-phase inputs is evident, as they represent more than 55 % of the total environmental impact, to which electricity consumption itself contributes with about 46 %. As these are fully associated to the drive system of the machine, the close analysis of this system, and the search for alternatives, became a top priority, confirming the indications of the research works previously referred. Other considerations, such as the replacement of the petroleum-based hydraulic oil by biodegradable alternatives for hydraulic-based

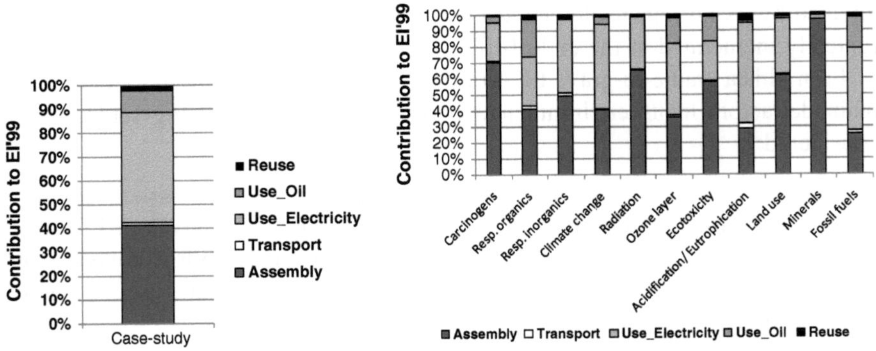

Fig. 7 Environmental profile and relative contributions of life-cycle phases to the global environmental impact of the press-brake under redesign, based on the Eco-Indicator 99 H/A method

systems were still open. The relative contribution of the current oil consumption to the global environmental impact of the machine is higher than 20 %, in at least 2 of the 11 middle-point impact categories analyzed. The main impact of the oil consumption, in absolute value of the indicator, is on the depletion of fossil fuels, and, in this category, the impact is similar to that of the total of assembly resources incorporated in the machine, which is considered quite significant.

3.3 System Redesign and Performance Assessment

Although energy efficiency improvements adopted along the last 20 years were seen to reduce energy requirements of machine-tools in approximately 50 %, the basic guidelines for energy savings during process, such as the specification of most energy efficient components and guidelines for effective energy management during machine-tool processing are still not established (Devoldere et al. 2008; Dietmair and Verl 2010; Oliveira et al. 2011). The examples given in the previous section support the strategies proposed to improve energy and global efficiency:

(1) The conversion of hydraulic to all electric systems;
(2) The maximization of the rate at which the physical mechanism can perform the desired operation, i.e., the optimization of machine usage;
(3) The enhancement of the awareness of the end-user regarding the importance of energy management. Independently on the many possible solutions targeting the automatic control of the machine-tool, enabling the user to obtain detailed and real-time data about the energy consumption is essential to accomplish the optimization of the machine-tool environmental profile during the use stage, as the user must be actively involved in this process.

(4) To match the power demand profile of the main energy-consuming sub-systems integrated, in what concerns the power consumption of the sub-systems, as realized from the a parallel study followed (Santos 2011).
(5) Sub-systems modularity should be the approach, increasing the product evolving potential towards the extension of the lifetime of the different energy-consuming functional subsystems and to maximize the recovery of the structural components, typically those integrating higher amounts of raw materials favoring more efficient raw-material recycling.

Some examples of high potential actions enhancing benign metal forming currently being developed and adopted are here pointed out:

- Detailed analysis of assembled sub-systems components
  Regarding the assembled sub-systems of the machine-tool, and considering the trend for all electric or electromagnetic versions of these, particular attention should be given to the use of advanced functional materials, particularly composites, and the increased use of additional electronic components, as these typically includes higher amounts of hazardous materials or raw materials which are hard to recover. Also on this analysis, the sub-system approach for improvement is recommended in order to favor a more detailed analysis of all components. In fact, while the significant impact of a housing material can be more evident from the volumetric contribution of the component, only a detailed sub-system analysis can insure that the determinant impact of a small volume component based on a hazardous material would not be missed. Although some mandatory related legislation is established for electronic components, the amount and combination of substances in a multi-component electromechanical sub-system is still relying on the environmentally conscious of the sub-system manufacturer.
- Mass reduction of moving parts
  In moving sub-systems/parts of machine-tools, the current replacement of standard materials by lightweight alternative materials simultaneously reflects the trend to optimized material consumption, general material reduction and the introduction of high-performance materials, such as reinforced polymer-based composites or low-density metals. Although this trend is often pointed out as a positive factor pushing for new dynamics to the sector, the issue of the environmental cost of the introduction of these alternative materials should be carefully analyzed. The lifetime and end-of-life disposition of such components/materials should be particularly regarded, although a lot of work is on-going regarding innovative end-of-life strategies for these materials.
- Mass reduction of structural parts
  In what concerns the assembly resources used for the machine-tool construction, the materials and process inputs associated with the base structure of the equipment tends to determine its environmental impact, due to its dominant volume and weight. When looking for high performance materials for high-accuracy processes/systems, innovative polymer concrete solutions, also

referred as mineral casting, are being introduced to replace the typical steel welded main structure towards a performance upgrade even in the most conventional machine-tools. Technically, this solution is indicated to overcome the static and dynamic stiffness and vibration damping requirements (Erbe et al. 2008), reflected on Fig. 5, while, indirectly, this has significant environmental, technical and cost benefits.

- Selection of sub-systems technology and power matching
  Although the sub-systems dimensioning and energy-consumption could be individually optimized for a specific range according to the application conditions, the former are quite technology dependent (Pusavec et al. 2010a,b). Besides, considering that one same machine-tool model typically operates in very distinct operating modes, it is important to insure that all sub-systems are properly synchronized in each operating condition, and the respective power-consumption profiles should be matched. This is expected to contribute significantly to the reduction on the power demand and to improve efficiency of auxiliary main-systems against the main energy-source sub-system.
  The impact of the motor power to the SPE values of the bending case also reflects some needs for improvement. Although the energy models presented are able to be tuned for different motor rated power levels integrated in press-brakes of different maximum bending capacity, this is indirectly associated to the maximum loading capacity of the machine. Obviously, in order to minimize power consumption, the motor rated power installed should be as low as possible. Moreover, in these hydraulic systems, where the stand-by consumption contributes significantly to the SPE, the motor power related energy savings are maximized with the increase on system usage.
  The properly addressing of the above considerations requires the establishment of effective strategies to combine improved technical and environmental performance. The schema on Fig. 8 presents the decision processed followed in the press-brake redesign study, to accomplish this purpose.
  The broad applicability of this decision process, in all the improvement perspectives to be integrated, provided for valuable inputs for: (1) the final product definition, (2) the identification of main improvement opportunities and (3) the development of an action plan for monitoring the new machine performance, for validation and continuous improvement purposes.

3.4 Final Design Solution

In-between the design alternatives for the different main sub-systems, the following focus areas haven been prioritized in this press-brake Ecodesign approach:

1. Alternative main driving system, insuring high accuracy. Tight synchronism and ON/OFF and stand-by control, towards significant reduction of energy consumption and higher efficiency during use;

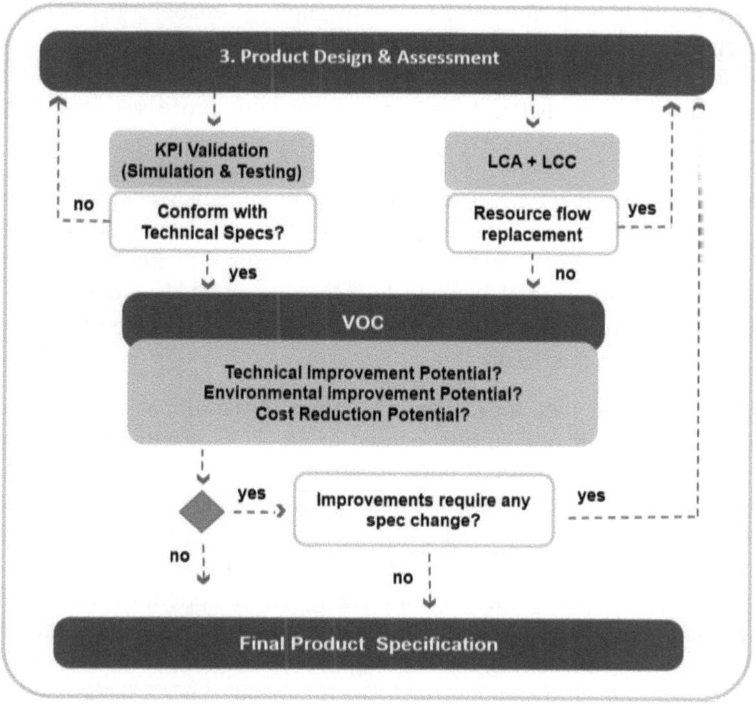

Fig. 8 Decision process flow applied in the ecodesign of electromechanical systems

2. Savings in raw materials and residues, throughout a more selective identification of raw materials, and the improvement of the on-line equipment control and monitoring, (integration of specific state-of-the-art instrumentation), favoring:

  - 'First time right' production, reducing setup waste;
  - Optimized HMI systems, built to facilitate use and remote control, targeting easier and shorter repairing and longer product lifetime.

3. General subsystems redesign, focused in easier repairing and dismantling, for reuse, upgrading or even recycling.

Following the strategy here described, the Product Development team was able to provide for a competitive equipment solution, with the following improved specifications:

1. Excellent accuracy standings, in line with the increasing geometrical complexity of the bending parts to be produced;
2. Significant increase on bending productivity;
3. Auto-correction of the structural elements deformation;
4. Intelligent control, minimizing the need of dedicated operator training;

5. Integration of hybrid 'Greendynamics' technology, enabling the direct control of each press-brake main hydraulic actuator by an electric servomotor through a valveless hydrostatic transmission;
6. Design concepts and materials and systems selection processes according to the Ecodesign directive (2005/32/CE).

Figure 9 resumes the general reduction, higher efficiency achieved in the critical resources flow, in terms of the main environmental detractors on the current solution, namely assembled materials, energy consumption and hydraulic oil.

In terms of environmental impact, the new equipment developed presented a significant lower impact, in about 31 %. The following factors are identified as the main contributors to this result:

- *Reduction of input materials*: The integration of additional components associated to the new functionalities focused in the VOC resulted in no significant reduction on the total mass of input materials;
- *Reduction on the number and diversity of components*: the modular concept developed for the new system and the resources optimization approach resulted in a decrease of about 15 % in the quantity of components;
- *Reduction of waste*: Together with the energy consumption, the most important achievement related to the technology conversion was the reduction in the amount of oils and lubricants during the use stage of the machine-tool. The technology conversion resulted in a particular decrease of more than 67 % in the consumption of hydraulic oil (from 160 to 50 l/year), with a significant direct contribution to the decrease of the environmental impact of this machine-tool, as shown in Fig. 7;
- *Reduction in energy consumption*: The technology conversion described for the driving system of the press-brake ('Greendynamics technology'), resulted in about 63 % energy savings.

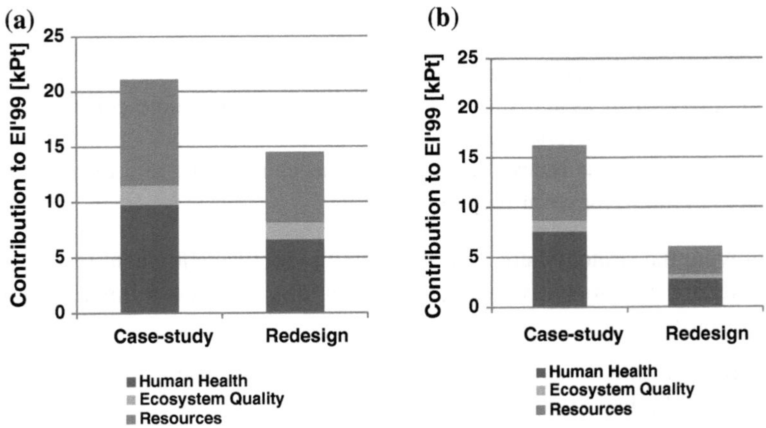

Fig. 9 Comparison of the environmental impact of the main detractors in old and new press-brakes, namely: **a** global impact; **b** energy consumption in use phase, based on the Eco-Indicator 99 H/A method

- In order to increase product performance, new functionalities and components has been integrated, that provide excellent accuracy standings, important increase on bending productivity and auto-correction of the structural elements deformation. In the other hand the modular concept developed allow a significant reduction on the number and diversity of components and a similar cost.

These results make evident the applicability and robustness of the methodologies here described to the fulfillment of the project objectives defined for the conversion of a hydraulic press-brake to a more environmentally-friendly technology.

4 Conclusions

The processes established in Product Development teams working in the design of electromechanical systems have been centered in the technical performance of the product. In the last decade, these processes have been evolving to integrate a multi-criteria decision process, including technical, cost and environmental systems profile. Highly complex systems, such as a machine-tool, integrate a high number and variety of unit sub-systems with significant different contributions to each performance dimension. The Ecodesign case-study here described contributed to explore new design tools and the identification of the critical steps to consider in a combined technical-cost-environmental design optimization, from which the following are to be highlighted:

- The main intervention should be oriented according to the customer view, identifying technical and cost targets;
- Technical and technological priorities shall lead the environmental assessment process. Typically, the environmental profile is determined by the technology of the electromechanical systems integrated, responsible for the energy-consumption and main consumables during the use phase;
- State-of-the-art alternative technologies offer compact, efficient and friendly solutions providing for higher customization of system units, enhancing the modular and synchronism approach of sub-systems. Strong benefits on the environmental impact of the system are evident: lower energy and other consumables utilization; easier repairing/disassembly/replacing, extension of the product lifetime.
- Traditional heavy main structure enveloping an intricate net of subsystems is to be avoided. Assembly materials should be properly selected according to the component type (moving or structural part), cost and environmental impact of the raw material type. Advanced composite materials are now more competitive to replace the conventional welded steel structures, but attention should be paid to the environmental impact of the raw materials involved, even when used in small amounts.

References

Azevedo M, Oliveira MI, Pereira JP Reis A (2011) Comparison of two LCA methodologies in the machine-tools environmental performance improvement process. In: Proceedings of 18th CIRP international conference on life cycle engineering, Braunschweig, 2–4 May 2011

Bala A, Raugei M, Benveniste G, Gazulla C, Fullana-i-Palmer P (2010) Simplified tools for global warming potential evaluation: when 'good enough' is best. Int J Life Cycle Assess 15:489–498

Chung W-H, Okudan G, Wysk RA (2011) Modular design to optimize product life cycle metrics in a closed-looped supply chain. In: Proceedings of the 2011 industrial engineering research conference, Reno 21–25 May 2011

Devoldere T, Dewulf W, Deprez W, Duflou JR (2008) Energy related environmental impact reduction opportunities in machine design: case study of a laser cutting machine,In: Proceedings of the CIRP international conference on life cycle engineering, Sydney, pp 412–419

Dietmair A, Verl A (2010) Energy consumption assessment and optimization in the design and use phase of machine tools, In: Proceedings of the 17th CIRP international conference on life cycle engineering, Anhui, 19–21 May 2010

Ecoinvent Centre, Ecoinvent 2.0 database. Accessed Oct 2010, www.ecoinvent.ch

Erbe T, Król J, Theska R (2008). Mineral casting as material for machine base-frames of precision machines. In: Proceedings of 23rd annual meeting of the american society for precision engineering and the twelfth ICPE, Portland, 19–24 Oct 2008

Giudice F, La Rosa G, Risitano A (2006) Product design for the environment: a life cycle approach. Taylor and Francis, New York

Gutowski T, Dahmus J Thiriez A (2006) Electrical energy requirements for manufacturing processes, In: Proceedings of 13th CIRP international conference on life cycle engineering, Leuven, May 31–Jun 2, pp 623–628

Hochschorner E, Finnveden G (2003) Evaluation of two simplified life cycle assessment methods. Int J Life Cycle Assess 8(3):119–128

Kellens K, Dewulf W, Overcash M, Hauschild M, Duflou JR (2012) Methodology for systematic analysis and improvement of manufacturing unit process life cycle inventory (UPLCI)—CO2PE! initiative (cooperative effort on process emissions in manufacturing) Part 1: methodology description. Int J Life Cycle Assess 17:69–78. doi:10.1007/s11367-011-0340-4

Kuhrke B, Schrems S, Eisele C, Abele E (2010) Methodology to assess the energy consumption of cutting machine tools, In: Proceedings of the 17th CIRP international conference on life cycle engineering, Anhui, May 19-21

Norris GA (2001) Integrating life cycle cost analysis and LCA. Int J LCA 6(2):118–121. doi:org/10.1007/BF02977849

Oliveira MI, Santos JP, Almeida FG, Reis A, Pereira JP, Rocha AB (2011) Impact of laser-based technologies in the energy-consumption of metal cutters: comparison between commercially available system, In: Proceedings of 14th international conference on sheet metal, Leuven, 18–20 Apr 2011. Published on key engineering materials, vol 473, pp 809–815, Trans Tech Publications, Switzerland. doi:10.4028/www.scientific.net/KEM.473.809

Peças P, Ribeiro I, Folgado R, Henriques E (2009) A life cycle engineering model for technology selection: a case study on plastic injection moulds for low production volumes. J Clean Prod 17(9):846–856. ISSN 0959-6526. doi:10.1016/j.jclepro.2009.01.001

Pré-consultants, Report: 'The eco-indicator 99—a damage oriented method for life cycle impact assessment'. http://www.pre.nl. Accessed Oct 2011

Pré-consultants, Simapro 7.0 software. http://www.pre.nl. Accessed Oct 2011

Pusavec F, Krajnik P, Kopac J (2010a) Transitioning to sustainable production—Part I: application on machining technologies. J Cleaner Prod 18:174–184

Pusavec F, Krajnik P, Kopac J (2010b) Transitioning to sustainable production—Part II: evaluation of sustainable machining technologies. J Cleaner Prod 18:1211–1221

Santos JP, Oliveira MI, Almeida FG, Reis A, Pereira JP Rocha AB (2011) Improving the environmental performance of machine-tools: influence of technology and throughput on the electrical energy consumption of a press-brake. J Cleaner Prod 19(4):356–364 doi:10.1016/j.jclepro.2010.10.009

Shamsuzzoha AHM (2011) Modular product architecture for productivity enhancement. Bus Process Manag J 7:21–47

Todd JA, Curran MA (1999) Streamlined life-cycle assessment: a final report from the SETAC North America streamlined LCA workgroup: society of environmental toxicology and chemistry (SETAC) and SETAC foundation for environmental education

European Commission (2007), Study for preparing the first working plan for the ecodesign directive; report: ENTR/06/026; Dec 2007

Umeda Y, Fukushige S, Tonoike K, Kondoh S (2008) Product modularity for life cycle design. CIRP Ann Manuf Technol 57:13–16. doi:10.1016/j.cirp.2008.03.115

Vezzoli C, Manzini E (2008) Design for environmental sustainability, Springer, ISBN-13:978-1849967419

Composite Fiber Recovery: Integration into a Design for Recycling Approach

Nicolas Perry, Stéphane Pompidou, Olivier Mantaux and Arnaud Gillet

Abstract In industry, the use of composites, and more specially carbon fiber/thermoset matrix ones, is ever increasing. However, end-of-life solutions for these materials are still under development. In this chapter, a solution linking design strategies with a recycling process based on the solvolysis of the matrix by water under supercritical conditions is proposed. The needs and multi-disciplinary skills required for (i) taking recycling possibilities into account from the early stages of the product design, and (ii) the necessity to standardize its recycling capabilities with design requirements, will both be discussed. The present chapter highlights the need for designers to take a functional approach into consideration, including material characterization, limits of the recycling process, constraints and opportunities. The first lessons learned from experiments using this technique will be shown.

1 Introduction

Today, reducing greenhouse gases and pollution is one of our society's main challenges as it strives for sustainable development. In this way, the key focus for transport industry is to make lighter vehicles in order to decrease both energy consumption and CO_2 emissions. Thus, composites provide good opportunities for combining high material properties with an increased freedom for defining parts' geometry; as a consequence, their use is ever increasing. Aerospace and aeronautics industries have integrated composites for long, and at different levels

N. Perry (✉)
Arts et Métiers ParisTech, I2M, UMR 5295, Talence 33400, France
e-mail: n.perry@i2m.u-bordeaux1.fr

S. Pompidou · O. Mantaux · A. Gillet
Université de Bordeaux, I2M, UMR 5295, Talence 33400, France

E. Henriques et al. (eds.), *Technology and Manufacturing Process Selection*,
Springer Series in Advanced Manufacturing, DOI: 10.1007/978-1-4471-5544-7_14,

(e.g. organic matrix based composites for cold applications, and metallic or ceramic matrix based ones for high temperature parts). However today, composites' low potential of recyclability limits their use in the automotive industry. Indeed, regulations in this sector impose a 95 % recycling ratio for an out-of-use vehicle. Moreover in a global and eco-friendly approach, one must analyze and take into account end-of-life solutions for systems, at an early stage of their development process.

The term *de-manufacture* has become more and more common, especially in the electronics industry. This is a recycling process for materials and products that includes end-of-life strategies and logistics in product development (Berry 1996; Gaustad et al. 2010): design engineers have to balance safety, energy efficiency and cost. Unfortunately, they rarely get to the point of thinking about what will happen to the product at the end of its useful life (Kriwet et al. 1995; Vallet et al. 2010). However, as new materials and technologies are developed, the challenge that recyclers face in, safely and economically, by recycling those products, grows ever more difficult (Calcott and Walls 2005). Recycling processes have to balance the technical, economic and environmental aspects of the end-of-life proposal. Recycling a product means:

- to have a recycling technology available,
- to get a dismantling solution and an access to the product,
- to have a disposal for life-ending composites (i.e. used parts, eventually polluted), clean parts (e.g. offcuts from machining) or even raw materials (e.g. unused carbon fabric, prepregs, etc.).

Composite applications have opened up a new field of research and development in the domain of recycling processes. For example, water under supercritical conditions gives the opportunity to recover thermoset matrix based composites' reinforcement, like carbon fibers (certainly one of the most interesting reinforcement to recycle, both environmentally and economically; Duflou et al. 2009). This process will also open up new opportunities for second-life composites. In this way, we are focusing on a research area that aims at integrating recycling constraints in the design stage of composite parts. At the same time, we hope to promote discussion between designers and recyclers for innovating in the definition of new recycled composite products. This means that information and skills from both sectors will be shared. However, it also implies that materials and mechanical knowledge have to be developed, for both designers and recyclers. Therefore, it is necessary to include a third party in the discussion: experts in material and mechanical characterization.

The first section of this chapter presents an overview of composite recycling possibilities and the technical and economic reasons for their development. The second part will focus on the design for recovery issue and the specificities related to composite design. The third section will explain our understanding in terms of skills, needs and know-how required for addressing this issue. Before concluding, the last section will illustrate some feedback and the first lessons learned regarding eco-design for composites.

2 Composites Recycling: Motivations and Solutions

2.1 Motivations and Interests

Future regulatory constraints are driving industries to develop efficient end-of-life alternatives, based on technical and economic constraints. For carbon and aramid based products, high prices (e.g. carbon prepreg: approx. € 180/kg; Kevlar©: approx. € 150/kg) and the world shortage in raw materials production, are the leitmotiv for finding technical and cost-effective recycling solutions. In such cases, second use of composite fibers will be dedicated to the manufacture of medium or low loaded parts (non-structural in many cases). Indeed, the recycled fibers and reprocessed semi-products have to achieve full acceptance and gain the trust of users (i.e. designers), regarding their material qualities and performances. Actually upon seeing the "recycled" label, many stakeholders still tend to think of low quality.

However, as far as recycled carbon fibers are concerned, this is far from the reality. Methods exist today for recycling carbon fibers and pre-impregnated (prepregs), and the resulting recyclate retains up to 90 % of a new fiber's mechanical properties. In some cases, this method even enhances the electrical properties of the carbon recyclate which can deliver a performance superior to the initial material's (Perry et al. 2010a, b). So it is necessary to create demand for recycled reinforcement, by packaging it in a useful or attractive form to end-users. Besides, by retaining good material properties, this constituent can be reused as raw, in a so called *second generation (2G) composite*.

2G-composites (i.e. materials based on a 2G-reinforcement) are obviously more environmentally friendly. Indeed, they mainly ensure to decrease the use of petroleum for their production, and they also keep the potential of a next recycling loop. Lastly, 2G-composites are cheaper, which can lead to broaden the composites materials' applications field.

2.2 Trails and Solutions

To make such a recycled product cycle viable, the key factors are the amount of waste deposit and its availability. In order to ensure the efficiency of the recycling path, that is to say, to guarantee and improve the flux regularity, infrastructures for both collection and identification must be established. Lastly, end-users must have confidence in the quality of the product in terms of robustness and value. Thus, mechanical, physical and chemical studies must be led in order to enrich the recycled material data, and comfort the properties of recycled fibers, semi-products and structures. This has been done successfully in the plastics industry; for example, identification labels were added to plastic parts to facilitate collecting and sorting. Most composite manufacturers already carry out waste management procedures, with impetus from the REACH regulation; as a result, they recycle waste materials in the very workshop, when collection and processing solutions are proposed.

Otherwise, thermoset matrix-based composites' designers and manufacturers are currently taking two different directions. On the one hand, they are trying to increase the use of green or bio-composites; on the other hand, they are developing or improving recycling technologies. In the first place, the advantage of using bio-composites lies in their small environmental impact. They are made of:

- a bio-polymer matrix reinforced by synthetic fibers (e.g. glass, carbon, Kevlar©);
- a petroleum-derived matrix reinforced by natural fibers (e.g. kenaf, flax, hemp, bamboo, coconut stems);
- biopolymers (e.g. PLA or PHA) reinforced with natural fibers.

Natural fibers are currently tested to estimate their reinforcement properties. They promise to be the future solution for organic matrix composite parts (Feng 2010; Mohamad 2010). For example, regarding mechanical properties, some can easily compete with glass fibers, as their specific density ranges from 20 to 45, and their tensile stress from 400 to 1,500 MPa. Furthermore, solutions already exist for high-performances composites, e.g. by improving weaving processes (Weager 2010). Besides, biopolymers or bio-compounds (a combination of bio-polymer and petroleum derived polymer) have also been studied (Bourmaud and Baley 2009). For example, they come from PLA derived from cornstarch, or PA11 from castor seeds and its ricin protein. On the one hand, their carbon footprint is reduced by using a bio-renewable material, and on the other hand their recyclability potential is increased.

In addition, end-of-life impact awareness helps develop recycling technology. Thermoset matrix can be removed either by burning or grinding techniques. This is cheap but very aggressive for the carbon fibers (Mantaux et al. 2004, 2009). Complex thermal, chemical and mechanical processes are needed to obtain high quality recycled carbon fibers. Pyrolysis and solvolysis are two of these very promising solutions, as summarized in Table 1 (Pimenta and Pinho 2011). However, there are some limitations; for example, it is impossible to recycle different categories of matrix simultaneously. At the same time, specific coatings (e.g. metallic cladding for electric behavior) are not compatible with some processes. Thus, specific requirements must come from the recycling stages if they are to be efficient. The most obvious (but absolutely necessary) thing is to extract and free all the metallic inserts, even before grinding. Moreover before recycling, products have to be dismantled and adapted to the recycling process reactor. These reactors are mainly cylindrical; cutting operations are therefore compulsory.

In southwest France, composite recycling will increase soon in terms of quantity due to the creation of two dismantling platforms:

- TARMAC platform (acronym for *Tarbes advanced recycling and maintenance aircraft Company*) is dedicated to civil aircraft applications in collaboration

Table 1 Summary analysis of different recycling processes (adapted from Pimenta and Pinho 2011)

Process	Mechanical	Thermal		Chemical
		Pyrolisis	Fluidized bed	
Advantages	Recovery of both fibers and resin	High retention of mechanical properties	High tolerance to contamination	Very high retention of mechanical properties and fiber length
	No use or production of hazardous materials	Potential to recover chemical feedstock from the resin	No presence of residual char on fiber surface	High potential for material-recovery from resin
			Well established and documented process	
Drawbacks	Significant degradation of mechanical properties	Possible deposition of char on fiber surface	Strength degradation between 25 and 50 %	Common reduced adhesion to polymeric resins
	Unstructured, coarse and non-consistent fiber architecture	Sensitivity of properties of recycled fibers to processing parameters	Fiber length degradation	Low contamination tolerance
	Limited possibilities for re-manufacturing	Environmentally hazardous off-gases	Unstructured (*fluffy*) fiber architecture	Reduced scalability of most methods
			Impossibility of material-recovery from resin	Possible environmental impact if hazardous solvents are used

with Airbus, EADS Sogerma. TARMAC first focuses on the re-use and certification of replacement parts in aircraft maintenance;
- P2P platform, close to Bordeaux, deals with the disassembly of ballistic weapons.

Otherwise, in order to manage the end-of-life of structures, a consortium of aerospace manufacturers (EADS Astrium Space Transportation, Snecma Propulsion Solide, etc.) has been working on the RECCO project (acronym for *Recycling carbon fiber reinforced composites*). The solvolysis process has been chosen for removing the thermoset matrix. In this technological and industrial background, all the stakeholders involved are analyzing an early integration of the recycling constraints and possibilities, in the design process of carbon composite parts. The next sections will explain the integration levels we face into develop a *design for recovery* approach.

3 Composite Design for Recovery

In order to take the end-of-life information into consideration from the product design phase, we naturally opted for *design for X* approaches. In our case, we worked on design for *recovery* (rather than *recycling*); taking disassembly into consideration, it should lead designers to propose recovery solutions for products or composite parts. In this approach, all the recycling requirements are considered as input (data) that must be taken into account in the product's functional specifications. Consequently from a semantic point of view, we also encourage the use of the term *recovery* instead of *recycling*, to emphasize the second life or second use of the product or of its constituents, after the recycling phase.

Lastly, we shift from the *cradle to grave cycle* to the spirit of *cradle to cradle.* Thus, even if we are using a *design for recycling* methodology or using the term *recycling*, we consider it to be a dynamic state, like a *rebirth for future use,* and not as a static and final goal. This means to consider future product design or future second life material from the recycling level. In other words, we are in two dual areas of research:

- firstly, design for end-of-life can be summarized as *design for recovery*;
- secondly, from recycling to design, the research deals with *robust material recycled for design*, that is to say constituents for which mechanical properties remain intrinsically invariant.

The challenges in *design for recovery* are to protect the environment, and create sustainable means for preserving our resources and reducing energy loss and pollution. It has two very basic goals. The first is to eliminate or reduce the use of hazardous or toxic materials which may present a serious threat to the environment, or put a recycler's workforce in jeopardy. The second is to discourage the use of materials that are simply not recyclable, or manufacturing techniques that make a product non-recyclable with current processes. The best time to address these issues is during the design stage (Ferro and Amaral 2005). Indeed, addressing a product's end-of-life is essential at the very beginning. Adopting this premise helps to ensure an efficient recycling chain, that goes far beyond the scrap processor in a mill, a smelter or an extruder, that would take the recycled materials to make new ones. Design for recycling is a mindset that all design engineers must embrace if they hope to have their products considered as environmentally friendly. As mentioned before, design for recycling is driven by governmental mandates like the European Union's waste electrical and electronic equipment directive (WEEE) or end-of-life vehicle (ELV) directive; for example in Europe, the rate of re-use and recovery should reach 95 % by 2015, and 85 % for re-use and recycling, in average weight per vehicle and per year.

There is more than environmental compliance at stake here. As new materials are developed (such as carbon fiber-based composites), they bring about a new threat in terms of recycling. And as new constituents are introduced into products and are replacing some that have been recycled for generations, they negatively affect the products' recyclability (both practically and financially), and can have a devastating impact; like this, even recyclable materials can pose a problem when used in combination. Take for instance a product that is made from many different types of plastics; today's recycling technology can only sort two or three different types of polymer materials using a mechanical solution, at best. Composites effectively become non-recyclable, or at least the plastics fraction of that product will be non-recyclable (Seager and Theis 2004; Perry et al. 2010a, b).

Therefore, in order to address the global problem, we are working on different levels: (i) design teams, (ii) design methods, and (iii) design tools. In this way, we hope to reduce the gap between the existing recycling solutions or bio-composites possibilities, and designers' current solutions. Not only do engineers have new materials and new product design solutions for eco-responsible products, but there are also now different tools available to help modeling and evaluating the solutions and the impact of the product life. Lastly, this definition of the end-of-life requirements points out the needs in terms of maintenance or parts fixing.

Unfortunately, the results of these rating tools depend on the available information. Most of the time, little information can be found about the product's life and end-of-life at design stage. Furthermore, in the case of the recycling processes under development, it is necessary to anticipate the potential of the technologies and their applications. Uncertainty in decision-making will therefore increase. Indeed in terms of end-of-life consequences, design decisions will undoubtedly appear 5–20 years later.

Before giving further explanation, it is important to remember that composite design is complex: it is necessary to define simultaneously (i) geometry and shape, (ii) constituents (matrix, fibers and their orientation), and (iii) manufacturing process. These three topics are linked and interdependent in the design and optimization process. For example, the choice of laminate limits shape possibilities, and depends on manufacturing capability. Consequently, as illustrated in Fig. 1, the eco-designer has to juggle with constraints from various sources, in addition to

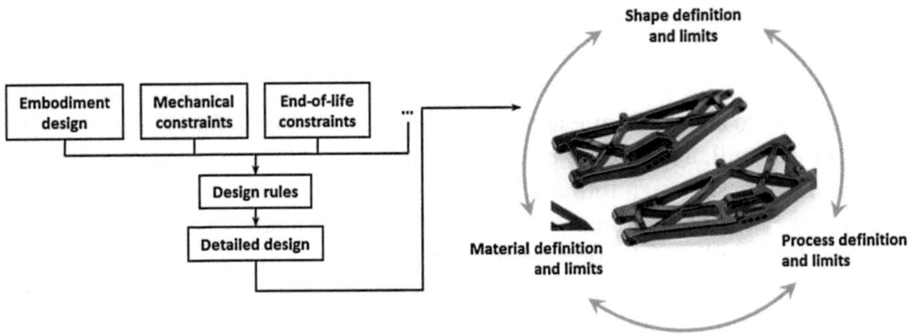

Fig. 1 Composite design constraints

internal constraints and relationships of the composite design. Up to now, these additional constraints have not been taken into consideration, except for the consequences of the new regulation (e.g. REACH) which focuses on manufacturing aspects.

4 Levels of Complexity for Integrated Design

Integrated design means addressing the recycling process development issue carefully, while at the same time proposing the possibility of including such evolutions in design methodology. We identified three main problems:

- the first is linked with recycling physics and scheduling to link and reach design methodologies;
- the second is dedicated to the uncertain and non-complete nature of the available information;
- the third problem has to do with competencies and skills needed for designing robust recovery problems.

4.1 Physics and Scheduling

This part deals with the knowledge of integrating the recycling process at the design stage.

These processes are for the most part under development, searching for breakthroughs, innovations and applications. The recycling rules that must be included in the design process are still under formalization, while designers must take decisions. Furthermore, decisions taken today will affect the product much later. Figure 2 shows the core elements that make up the skeleton of a product's end-of-life. Each phase of this process has its own limits and constraints to be

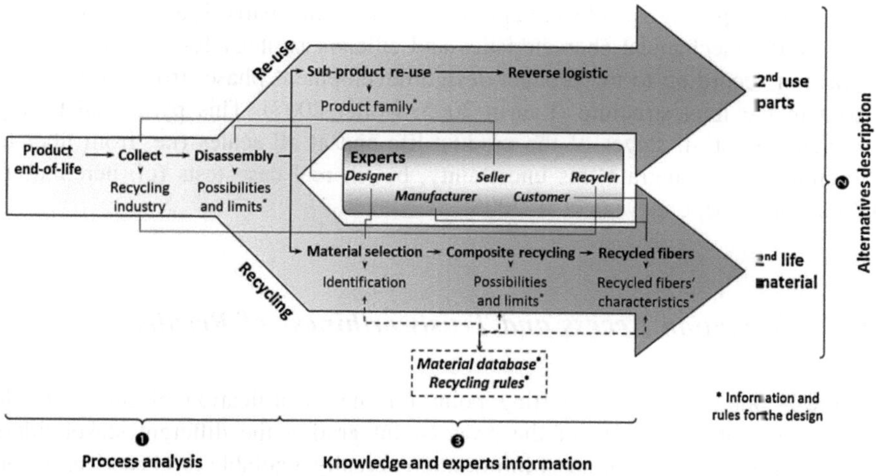

Fig. 2 Analysis of the recycling process, in the perspective of the product's design specifications and recycling process information

integrated or overshot. Different stakeholders are included in the loop and should point out the information, expected data and decision rules they apply to switch from one stage to another. The different time frames between the real recycling process and the parts' design phase, increases the integration difficulties. For existing and robust end-of-life paths, constraints and material re-processing and re-use are well known. For new recycling processes, robust validation can take time, yet designers have to take decisions now. An extreme example is nuclear plants; built 40 years ago, there are still no efficient end-of-life solutions as their dismantling is about to begin.

Research and development teams are banking on some kind of technological breakthrough to guide developments in recycling. Innovation is needed not only in the recycling phase, but also for all the key stages. It is important to improve disassembly techniques (see *design for assembly/disassembly* approaches) or selection efficiency (Boothroyd and Alting 1992; Aymonier et al. 2006). At the end of the process, it is essential to develop innovative and valuable uses to compete with virgin raw materials (for similar characteristics), or to find new opportunities, at the very early design of the recycling path.

As far as composites are concerned, new processes enable fibers to be recovered with very little distortion and fracture, compared to the initial reinforcement used in the original composite part; supercritical fluids can provide such opportunities (Loppinet-Serani et al. 2010; Kromm et al. 2003). However, the problem of misalignment and realignment of the recycled fibers still remains. Competences (knowledge and know-how) and fiber spinning and weaving skills have been integrated by recycling teams.

Other alternatives consist in reprocessing medium-sized flat rectangular pieces of pseudo-unidirectional (1D) or woven (2D) recovered carbon fabric. The innovation consists in proposing a patchwork approach for designing parts. Specific studies must identify the mechanical characteristics and efficient strategy for material characterization according to the product design development phase, from the recycled fibers to the final structure (Laurin 2005; Rollet 2007). This pyramidal testing problematic, at all stages of the product life and at all scales (i.e. from fiber to structure) must integrate this uncertainty, but in real case tests (cf. hereinafter) (Dennison 2010; Ladevèze et al. 2006).

4.2 Information Access and Trustworthiness of Results

Figure 2 can also be the starting point for the identification of the available information at each stage of the process. Integrating the different stakeholders helps identifying which kind of information is really available or needed, and who gets or requires it (Bernard et al. 2007).

In a classical product development V-cycle, two elements arise. Firstly, the kinds of data and their accuracy depend on the chosen end-of-life solution. As illustrated in Fig. 3 for a product re-use, the product life information might need to be certified and guaranteed for compliance. Thus, a material database will store all this information, structured for each recycling stage; then as an example, it will become available for the designer. Nevertheless in many cases, the kind of data required is known, but its real value is either unknown, or fuzzy, or ranges beyond the limits. As a result, design evaluation becomes uncertain. Therefore, end-of-life solutions are not fully defined at an early design stage, and these initial decisions will impact the environmental footprint.

In addition, as illustrated in the *Physics and timetable* part, an efficient end-of-life solution might not have been developed yet. Consequently, end-of-life

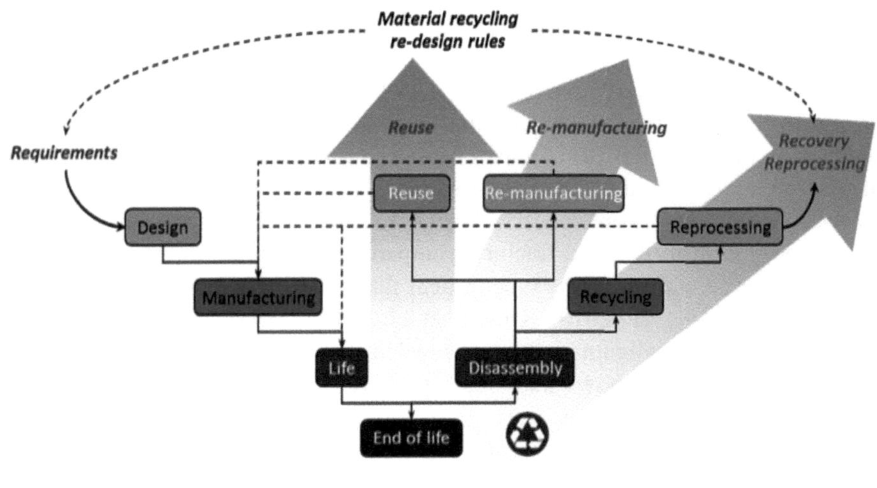

Fig. 3 V-cycle for information exchange

evaluations must be used with care, dividing the reliable results from the uncertain ones. Similar levels of information completeness can compare different solutions. Otherwise, the results should be taken as trends or qualitative comparisons.

So then, we firstly plan to map the design process cycle, with each key decision concerning life and end-of-life impact. These choices will need data, decision rules, etc. In addition, as previously shown in Fig. 2, all the information (requirements and constraints) will be captured in detail, in order to generate recycling rules. The connecting link between these two aspects, identified by using or generating the same data, will link designer to recycler.

4.3 Multidisciplinary Needs

The previous paragraph explained how we create the link between recyclers and designers. However in many cases, designers need data regarding a specific characteristic (e.g. maximum tensile stress) which recyclers are not able to give. Conversely, recyclers have to know about life damage, but designers can only inform on the use cases considered at design stage. Consequently, complementary information arises in this dual relationship. Material and product characterization is compulsory at different levels. People from material, chemical and mechanical fields will be able to provide a way of translating requests or requirements into real data. Indeed, many different characterizations must be carried out before and after the recycling process. For example, assembly consequences on the material end-of-life, and on the possible disassembly damage, must be identified. All these data guide the recycling process in order to minimize the variability incidence. In a second stage, the recycled material or re-processed semi-product has to be tested to

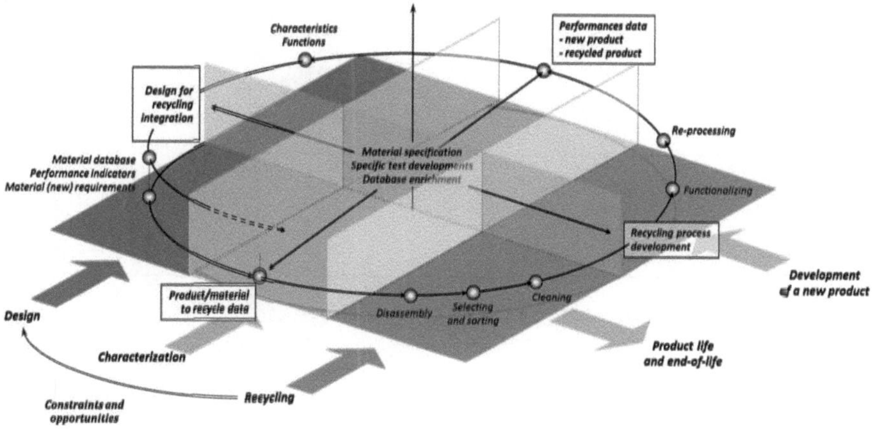

Fig. 4 Summary of the interrelation of the three skills

assess its quality and enrich design space for the designer. Figure 4 sums up the interaction between these three required skills.

From the characterization point of view, the key problem remains the adaptation of scale. The testing pyramid strategy (previously mentioned) helps to identify sufficient and necessary tests from the elementary sample level to the full system one (cf. Fig. 5). In most cases, specific tests must be developed in order to guarantee relevant and reliable results. This multi-level approach is also applied to the development of new recycling processes, and to their industrialization.

5 Composite Design for Recovery: First Lessons

The Recco project (in which we were involved from 2009 to 2011) sought industrial solutions to recover (i.e. recycle and re-process) wastes of carbon fiber based composites.

A matrix removal process has been developed for thermoset matrix based materials. It consists in the dissolution of the resin phase by water under supercritical conditions. At a pressure of about 200 bars and a temperature of 400 °C, water becomes a real solvent for the resin phase, which then can be easily and fully removed. After this, dry carbon reinforcement remains oriented according to the original composite sequence, with no major fiber degradation.

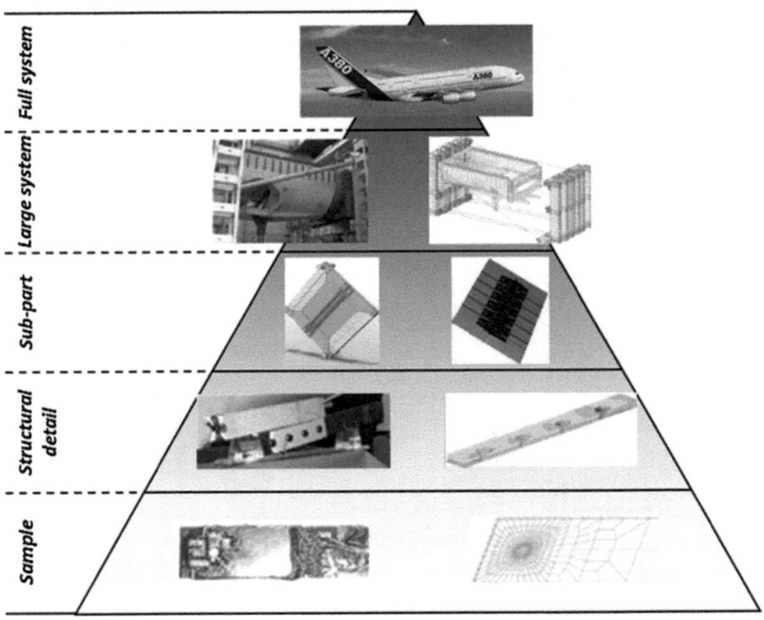

Fig. 5 Testing pyramid dimensions: from sample (micro scale) to structure tests (macro scale) (Laurin 2005)

The specificity of the solvolysis process is its maximal matrix removal, with a minimal water consumption; it also remains little polluting. Nevertheless, it is very sensitive to the presence of other non-organic material such as metal. For example, the solvolysis process oxidizes rivets, screws or inserts, and even parts of the process reactor in some cases. Then, it spreads small metallic particles in and on the carbon fibers, as illustrated in Fig. 6. Depending on the fibers orientation (unidirectional or woven), the solvolysis liquid flow concentrates this pollution on the outer edge of the fibers or where the tows cross.

This pollution has little influence on the dry fibers' mechanical strength, but it creates bridgings and local coatings that reduce the reprocessability (e g. spinning or re-weaving). Nevertheless, when imbuing with matrix polymer, this eases mechanical grip (by limiting fibers sliding and tows misalignment) and has little effect on mechanical properties at part scale. Regarding the physical characteristics, this pollution only modifies the behavior locally. But more tests must be carried out to analyze these solvolysis pollution impacts.

Besides, these analyses have provided further feedback. It has been noticed that cutting processes (needed for adjusting parts to the recycling reactor) were burning matrix and changing its properties. This carbon-based layer (which can reach up to 200 μm with hard cutting conditions) locally glues fibers together, as illustrated in

Fig. 6 SEM pictures of clean and polluted recycled carbon fibers: **a** Clean carbon fiber with small silicates, **b** Carbon fibers with slight nickel pollution, **c** Carbon fibers with considerable overallpollution from steel (Fe + Cr) and **d** Carbon fibers with edge pollution

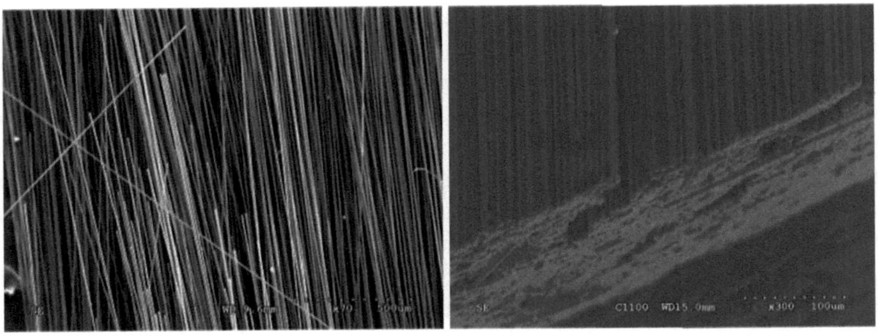

Fig. 7 SEM pictures of free (*left*) and glued carbon fibers at one edge, due to cutting process (*right*)

Fig. 7. This practical consequence can ease the handling of the recyclate, by reducing the shreds when spinning a long tow with short or medium-sized fibers. Then on the one side, fibers are kept aligned; but on the other, the spinning efficiency of the inter-fiber grip can also be reduced.

Up to now, we have improved our competence in textile manufacturing; this will provide a help for implementing the end of the recovery loop (particularly the remanufacturing stage). The pilot demonstrator is under development, but we are already able to transform composite parts into second generation tows, or patches of recycled carbon fabric.

Mechanical properties of this new reinforcement are nearly the same as virgin one. Nonetheless, fibers are of course shorter than new ones due to the recycling process; this constraint leads to reprocess new semi-products that would be more easy to use, in order to close the recycling loop.

6 Conclusions

We are now able to recycle carbon fiber/thermoset matrix composites, and more precisely to recover carbon fibers. This development highlights the fact that recycling possibilities and constraints should be included from the early design stages. Skills and competences (knowledge and know-how) in materials and mechanical characterization fields are required to provide information about parts to be recycled, and about materials or semi-products derived from this recycling.

From our point of view, designers, recyclers and characterizers' skills must be interwoven to achieve design under recycling constraints, and to develop robust recycling processes for second design perspectives. In order to facilitate discussion and communication within these three areas of expertise, we have to work on defining requirements for and from designer, but we must also take the recycler's point of view into account. Systematic formalization of requirements and semantic alignment between the different (yet not so distant) communities should be implemented.

An environmental assessment of the solvolysis recycling process still has to be carried out in order to compare material and energy consumptions, pollution, cost, etc. for a same quantity of new or recycled composite.

The increasing use of composites in industry has brought about the development of end-of-life solutions. Regulatory constraints and financial perspectives (cost-cutting) will give rise to design for recovery or eco-design approaches.

References

Aymonier C, Loppinet-Serani A, Reverón H, Garrabos Y, Cansell F (2006) Review of supercritical fluids in inorganic materials science. J Supercrit Fluids 38(2):242–251

Bernard A, Ammar-Khodja S, Perry N, Laroche F (2007) Virtual engineering based on knowledge integration. Virtual Phys Prototyping 2(3):137–154

Berry S (1996) Design for recycling. Eng Des 22(4):8–14

Boothroyd G, Alting L (1992) Design for assembly and disassembly. CIRP Ann-Manuf Technol 41(2):625–636

Bourmaud A, Baley C (2009) Rigidity analysis of polypropylene/vegetal fibre composites after recycling. Polym Degrad Stab J 92(6):1034–1045

Calcott P, Walls M (2005) Waste, recycling and design for environment: role of markets and policy instruments. Res Energy Econ J 27:287–305

Dennison A (2010) The test pyramid: a framework for consistent evaluation of RFID tags from design and manufacture to end use. www.idspackaging.com/Common/Paper/Paper_256.

Duflou JR, De Moor J, Verpoes I, Dewulf W (2009) Environmental impact analysis of composite use in car manufacturing. CIRP Ann-Manuf Technol 58:9–12

Feng LY (2010) The biomaterial for green composites. JEC Compos Mag Nat Fibres Environ 55:29–30

Ferro P, Amaral J (2006) Design for recycling in the automobile industry: new approaches and new tools. J Eng Des 17(5):447–462

Gaustad G, Olivetti E, Kirchain R (2010) Design for recycling: evaluation and efficient alloy modification. J Ind Ecol. doi:10.1111/j.1530-9290.2010.00229.x

Kriwet A, Zussman E, Seliger G (1995) Systematic integration of design-for-recycling into product design. Int J Prod Econ 38(1):15–22

Kromm FX, Lorriot T, Coutand B, Harry R, Quenisset JM (2003) Tensile and creep properties of ultra-high molecular weight PE fibres. Polym Test 22(4):463–470

Ladevèze P, Puel G, Romeuf T (2006) Lack of knowledge in structural model validation. Comput Methods Appl Mech Eng 195:4697–4710

Laurin F (2005) Approche multiéchelle des mécanismes de ruine progressive des matériaux stratifiés et analyse de la tenue de structures composites. PHD Thesis, Univ de Franche-Comté

Loppinet-Serani A, Aymonier C, Cansell F (2010) Supercritical water for environmental technologies. J Chem Technol Biotechnol 85(5):583–589. doi:10.1002/jctb.2323

Mantaux O, Aymonier C, Antal M (2009) Recycling of carbon fibre reinforced composite materials with supercritical water dissolution. In: Proceedings of the 16th Journées Nationales Composites

Mantaux O, Chibalon L, Lorriot T, Aurrekoetxea J, Puerto A, Arostegi A, Urrutibeascoa I (2004) Recycling study of end of life products made of ABS resin. J Mater Sci Technol 20(1):125–128

Mohamad M (2010) Natural fibres for the 3rd millenium. JEC Compos Mag Nat Fibres Environ 55:23–28

Perry N, Kromm FX, Mantaux O, Pilato A (2010a) Composite eco-design. In: IFIP AMPS 2010 conference, Como

Perry N, Mantaux O, Leray D, Lorriot T (2010b) Composite recycling: design for environment approach requirements. In: IDMME virtual concept 2010, Bordeaux

Pimenta S, Pinho ST (2011) Recycling carbon fibre reinforced polymers for structural applications: technology review and market outlook. Waste Manage 31–2:378–392

Rollet Y (2007) Vers une maîtrise des incertitudes en calculs des structures composites. PHD Thesis, ONERA—Ecole Polytechnique, Palaiseau

Seager TP, Theis TL (2004) A taxonomy of metrics for testing the industrial ecology hypotheses and application to design of freezer insulation. J Cleaner Prod 12:865–875

Vallet F, Millet D, Eynard B (2010) How ecodesign tools are really used. In: Requirements list for a context-related ecodesign tool. CIRP design conference proceeding, Nantes

Weager B (2010) High-performance biocomposites: novel aligned natural fibre reinforcements. JEC Compos Mag Nat Fibres Environ 55:31–35

A Design for Disassembly Approach to Analyze and Manage End-of-Life Options for Industrial Products in the Early Design Phase

Claudio Favi and Michele Germani

Abstract In modern society there has been an increase in consumption and discard of goods and products due to the high growth of the world population. Moreover, in the manufacturing field rapid technology cycles quickly render products obsolete and as a consequence consumers dispose of products more intensively. Product disassembly is becoming an important phase of the product lifecycle to consider from the environmental and economic point of view. It occurs to minimize the maintenance time and describe the End-of-Life (EoL) strategies, for example component reuse/recycling. These EoL closed-loop scenarios should be considered during the early phases of design process when decisions influence product architecture and in the product structure. In this context, the purpose of this chapter is to describe an approach to support the designer's evaluation of disassemblability by using the 3D CAD model structure and suitable key indices related to product features and environmental costs. A software system allows the product model to be analyzed and evaluates the product disassemblability degree. Experimental case studies facilitate the approach demonstration and highlights product environmental performance due to the application of the proposed approach.

List of Abbreviations

CAD	Computer aided design
DB	Database
DFD	Design for disassembly
DFE	Design for environment

C. Favi (✉) · M. Germani
Department of Industrial Engineering and Mathematical Sciences, Università Politecnica delle Marche, via Brecce Bianche n.1, 60131 Ancona, Italy
e-mail: c.favi@univpm.it

M. Germani
e-mail: m.germani@univpm.it

E. Henriques et al. (eds.), *Technology and Manufacturing Process Selection*, Springer Series in Advanced Manufacturing, DOI: 10.1007/978-1-4471-5544-7_15,

DFx	Design for x
DFPR	Design for product retirement
DFR	Design for recycling/reusing/remanufacturing
EOL	End of life
EU	European Union
ICT	Information and communication technology
LCA	Life cycle assessment
ND	New design
OD	Original design
PCB	Print circuit board
PLM	Product lifecycle management
PP	Polypropylene
SLCA	Simplified life cycle assessment
US	United States

1 Introduction

Industrial waste management is one of the most important aspects of the environmental problem which has to be faced in order to minimize the disposal of materials in landfills and to favor the recycling or reusing of nonrenewable substances.

Environmental consciousness is rapidly becoming a fundamental product design focus in a variety of industries. Increasing market competition, globalization and focus on environmental aspects are forcing industries to change their product development strategies from mass production to a more sustainable model. Furthermore, in recent years the EU and US have issued directives, such as the disposal of electronic and electrical products and equipment, and also the restrictions on the use of hazardous substances, which force manufacturers to respect environmental issues (European Parliament 2000, 2003a; US Environmental Protection Agency 2000). The accurate management of possible End-of-Life (EOL) scenarios is a fundamental eco-design strategy for companies which must often assume responsibility to "withdraw" the product at the end of its life (Rose and Ishii 1999). End-of-Life strategies, such as reusing, remanufacturing, primary and secondary recycling, incineration and disposal options, may be taken into account. It is very important to consider these aspects during the early design stage when designers can change different characteristics of products, such as materials or liaisons, with minimal impact on the manufacturing process and production costs (Rose 2001). Furthermore, managing the EOL strategies reduces the cost of product disposal and increases possible revenues for the companies. This is a great opportunity for designers, who can conceive and develop industrial products with an appropriate EOL scenario (Capelli et al. 2007).

A set of design methods and tools has been developed in recent years to support engineers for EOL evaluation and disassembly aspects (Hauschild et al. 2004). In particular, eco-design approaches such as design for disassembly (DFD) give a set of principles and rules used to guide designers in designing products which are easy to disassemble (Bogue 2007). DFD makes the de-manufacturing plan of components simple and efficient, and must be considered for product components with a high quality/value which can be separated from the others for reusing, recycling or remanufacturing. DFD supports designers in their choices, but it is difficult to manage and to integrate in the traditional design process. Furthermore, even if DFD rules are well-known, it is difficult to take them into account during the design phases, when time is limited and structural and functional evaluations dominate other design drivers. However, manufacturing companies need to consider DFD and EOL management as a fundamental prerequisite for the product design and development (Dewhurst 1993). In fact, separating components and subassemblies makes the consistency of EOL closed-loop scenarios and the management of products at EOL possible (Herrmann et al. 2008). In this way the future amount of waste can be reduced and the consumption of nonrenewable resources can be limited. The disassemblability analysis is mainly based on the time and cost assessment of disassembly operations and technologies. On the other hand, sustainability analysis can be done to evaluate the environmental performances of the product during lifecycle and also at final disposal. Life Cycle Assessment (LCA) is a standardized methodology which analyzes the EOL problems concerning the environmental impact (ISO 2006). LCA has a powerful method to evaluate the environmental sustainability of products and technologies, but only considers environmental items, and this aspect limits the analysis potential of the method.

The purpose of this chapter is to define a modern approach to evaluate environmental and economic aspects of product EOL in the early design stage. In particular, this approach permits the aforementioned obstacles to be overcome and provides a useful tool which supports the designers in the assessment and management of product EOL scenarios and technologies. The results of the EOL analysis are underlined by the calculation of six new indices. These EOL indices, based on the disassembly costs and environmental impacts, are proposed to evaluate and measure the feasibility of the different EOL scenarios and to stimulate the designers' interests in this field. The indices analysis focuses the attention of designers on disassemblability problems to favor the EOL closed loop lifecycle in the early phase of product development. The main purposes of the indices evaluation are to maximize product disassemblability during the design phase using the analysis of the environmental and economic feasibility of different EOL scenarios, and to encourage the closed loop lifecycle (3R model—Reusing, Recycling and Remanufacturing). In this way, industrial firms are able to design "green" products which can satisfy the recent standards and regulations, such as the waste management of hazardous substances and, in addition, can increase revenues derived from taking their products back at EOL. A tool was developed for the application of this approach to make an easy and automatic disassembly analysis and the analysis of EOL technologies can be adopted for the specific product parts.

2 Background Analysis and Context

The growth of industrial production together with the increasing consumption of goods and services ring warning bells for the environmental sustainability of the planet. The final disposal of goods and products is the most critical issue of this alarm and must be considered in order to reduce the amount of waste and to save large quantities of resources. Due to these reasons, managing EOL products has become a sector of rapidly growing interest for manufacturers and dismantling centers. The three main perspectives for the management and the retirement of EOL waste which drive the new generation of designers and engineers of industrial firms are:

- Respect of legislation/regulation
- Reduction of environmental impact
- Possibility to make a profit (Ramani et al. 2012).

Product design plays a critical role in obtaining such benefits. In particular, product design features such as product assembly architecture, product structure, material properties, product functionality and modularity can be developed considering the feasible products' EOL scenario. EOL management and their relative aspects (reverse logistics, disassembly, etc.) shall be considered in the early design stage to facilitate the product recovery efficiently (Kwak et al. 2009). The assessment of different alternatives gives the possibility to know the status of the product at EOL and develop discern reverse logistic systems.

2.1 EOL Scenarios for Industrial Products

Different alternative routes are possible for industrial products which arrive at EOL. Two main routes can be distinguished for EOL of industrial products:

1. EOL Open-Loop scenarios ("from Cradle to Grave")
2. EOL Closed-Loop scenarios ("from Cradle to Cradle" or "3R approach").

Figure 1 shows the whole product lifecycle and the two alternative EOL loops. The two loops can be distinguished based on the ecological aspects linked to product disposal and "credits" to the environment. In fact, the EOL closed-loop scenario gives a new "second life" to products, without losing the impact which the product or component has accumulated during its lifecycle (material extraction, manufacturing energy, etc.). These scenarios provide a negative value of environmental impacts and therefore a benefit to the ecosystem. On the other hand, the EOL open-loop scenarios give a positive value of environmental impact (landfill occupation, pollution, etc.) and therefore damage the ecosystem. The current legislations and industry perspectives are focusing their attention on the EOL closed-loop scenarios and encourage the Reusing, Remanufacturing and Recycling loops (Gehin et al. 2008).

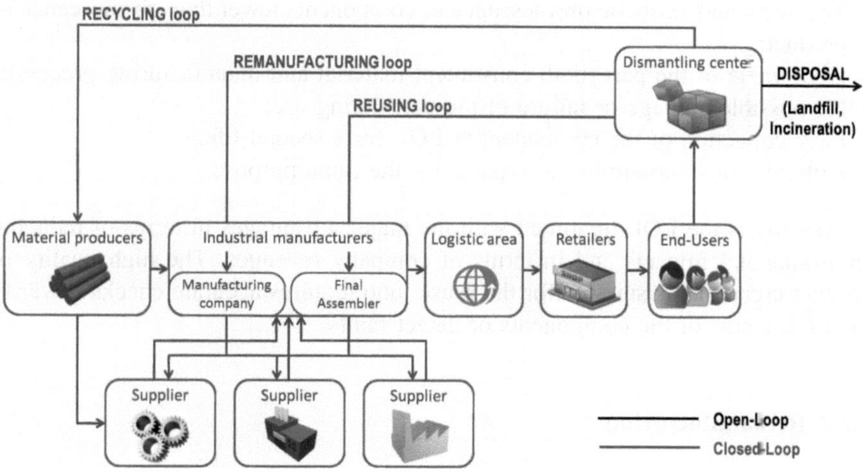

Fig. 1 Lifecycle assessment of industrial products with different eol scenarios

2.2 Closed-Loop Scenarios

EOL closed-loop scenarios involve waste processes which convert EOL products into remarketable products, components, or materials (Ramani et al. 2010). A list of possible treatments which can be classified in this loop are:

- Reusing
- Remanufacturing/Refurbishing
- Recycling.

The closed-loop scenarios enable industrial manufacturers to comply with legislation while gaining some economic advantage as well. As a result, more companies have become interested in EOL management, and successful cases have been reported by various industries, including Information and Communication Technology (ICT) and consumer electronics, household appliances, industry equipment, and automobiles.

2.2.1 Reusing

The Reusing process is the best chance for products or components to have a "second life". According to BS 8887-2 (BS 2009) the Reusing process can be defined as: *The process by which a product or its components are put back into use for the same purpose at EOL.* The Reusing loop can be an option if products or components satisfy specific conditions:

- Technical and aesthetic obsolescence of components lower than obsolescence of product;
- High value of the part (both constituent material and manufacturing process);
- No possible damage or failure of the part during use;
- Easy collection of the component at EOL for a second life;
- Difficulty or impossibility to replace for the same purpose.

Reusing is the EOL treatment with the major advantages in terms of reducing environmental impacts and in terms of company revenues. The high quality of product components suitable for the reuse loop requires accurate checks to verify the EOL status of the components or detect faults.

2.2.2 Remanufacturing

The remanufacturing process is currently a new business challenge for industrial firms and a new perspective to overcome the problems of the world's industrial crisis. According to BS 8887-2 (BS 2009) the Remanufacturing process can be defined as: *The process of returning a used product to at least its original performance with a warranty that is equivalent or better than that of the newly manufactured product*. The Remanufacturing loop can only be put into practice if particular characteristics are satisfied:

- Technical and aesthetic obsolescence of components lower than obsolescence of product;
- High value of the part (both constituent material and especially manufacturing process);
- The desired level of disassembly can be easily made for the component level;
- Easy collection of the component at EOL for a second life;
- High degree of modularity.

The remanufacturing loop has a highly potential perspective to reduce environmental impacts and to make profits for the industrial company. However, a significant amount of time and financial investment are required to set up the system and to train skilled engineers to design suitable products with specific characteristics.

2.2.3 Recycling

The Recycling process permits the constituent material of components to be retrieved to make new recycled material with the same characteristics of the virgin material. According to BS 8887-2 (BS 2009) the Recycling process can be defined as: *The processing of waste materials for their original purpose or for other purposes, excluding energy recovery*. Materials can only be recycled if:

- Components are made from materials which are recyclable and which can be collected easily at EOL for recycling;
- Components contain high value constituent materials (and in large enough quantities to enable cost effective extraction);
- Components or sub-assembly are constructed from a single or few separable materials;
- The desired level of disassembly can be easily made for the component level.

Recycling is currently the most popular activity to reduce environmental impact and retrieve revenues from industrial products at EOL. In particular this treatment permits a large reduction in the quantity of industrial waste. The drawbacks of this EOL treatment are generally the low quality of recycled materials and the possible contamination of the recycled materials caused by mixed waste.

2.3 Open-Loop Scenarios

In the overall described context the EOL Open-Loop scenarios are not considered as an alternative for the development of sustainable products. In fact, the open-loop scenarios provide an economical option for the management of EOL waste but have the highest value of environmental impacts (Rose and Ishii 1999). For these reasons the open-loop scenario will not be considered in depth in this chapter. However, for the sake of completeness, two different EOL open-loop scenarios are described in the next session: Incineration and Landfill.

2.3.1 Incineration (with Energy Recovery)

The Incineration process is a selective EOL treatment which greatly reduces the quantity/volume of industrial waste. Only certain industrial products which have primarily organic (containing carbon) components with little heavy metal content can be used in the Incineration scenarios (Ashby 2009). Plastics, for example, can be incinerated to produce heat and electric power but this option needs further investigation as an environmentally wise solution. Generally the Incineration process allows electrical and thermal energy from the combustion of organic waste to be recovered. According to BS 8887-2 (BS 2009) the Incineration process (with energy recovery) can be defined as: *The process of the combustion of organic waste materials to generate electric power (or combined heat and power).* No particular attention is necessary to recover or disassemble the product. Organic materials can be preliminary selected from metals or other materials by destructive de-manufacturing processes.

2.3.2 Landfill

Burying industrial product waste carries a very high environmental load (Rose and Ishii 1999). Furthermore, the landfill scenario is the most significant for land occupation and has high management costs without profits for the industrial company. For these reasons, landfill is a "banned" option concerning the existing communitarian regulations as a possible solution for the management of EOL products (European Parliament 2000, European Parliament 2003b). According to BS 8887-2 (BS 2009) the Landfill process can be defined as: *The process of disposing of waste by burial.*

Due to the aforementioned problems, landfill will not be more deeply analyzed in this context.

2.4 Environmental and Economic Parameters Affecting the Different EOL Scenarios

A list of economic and environmental items affecting EOL scenarios and treatments are proposed in this section. In most cases these aspects are not considered during the earlier design phases but are estimated when the product arrives at the dismantling center in order to select the specific disposal scenario. In most cases the management of EOL scenarios is based on the experience of the dismantlers (Bogue 2007). In fact, dismantlers know the real value of the product or constituent components very well and choose the best scenario to maximize profits using the cost/benefits analysis tool. In addition, dismantlers know the hazardousness of special materials or chemical substances.

In the case of closed-loop scenarios the common costs can be summarized as follows:

- Disassembly costs;
- Cleaning plant and operation costs;
- Reverse logistic costs (transportations sorting and plants);
- Remanufacturing/Refurbishing/Regeneration costs (only for the Remanufacturing process).

Other detailed costs will be evaluated case by case for each closed-loop scenario in the following parts (Das et al. 2000).

On the other hand, the company profits are based on the specific EOL scenarios which refer to the component. A list of benefits and revenues are reported below:

- Revenues due to the sales of reused/remanufactured/recycled parts or materials;
- Cost savings as virgin materials and production energy are not required;
- Energy savings due to the fact that the energy required for the recycling process is less than the energy required for material extraction (embodiment).

3 Eco-Design Methods and Tools to Conceive and Develop Sustainable Products Focused on EOL

A lot of research exists regarding the eco-design topic since environmental interest in product engineering has become so widespread. It is well known that although only 5–7 % of the entire product cost is related to early design, the decisions made during this stage weigh up to 70–80 % of the total product cost (Boothroyd et al. 2002). Correspondingly, it is possible to hypothesize the same for environmental impacts. That is, whether or not a product is relatively sustainable is largely determined during the early design stage (Sousa and Wallace 2006). Due to high levels of uncertainty regarding design details at the early design phase, new methods are essential to provide designers with a basis to determine the degree of sustainability of a given product. This requires the next generation of engineers to be trained in the context of sustainability, along with a global perspective, to solve problems of sustainability on multiple scales (Miheclic et al. 2008).

In the described context, the focus of the engineering activities moves from a static point of view of the product as an object to a global dynamic analysis of the product and its interaction with users and the environment. The use phase and especially the EOL phase are becoming interesting issues to be considered during the early design activities to perform environmental and economic product features (Dewulf et al. 2006).

3.1 Design for Environment

Environmental aspects are the main characteristics to be considered when designers conceive new industrial products focused on EOL. Approaches known as Design for the Environment (DFE) are defined as the systematic integration of environmental considerations into product and process design. They aim to minimize the costs and *adverse environmental impacts of products throughout their entire life cycles* (ISO 2002). Information requirements of engineering designers for eco-design have to be served in such a way that all lifecycle phases of the product are eco-friendly. DFE is one of target design methodologies (DFx—Design for x) that have been developed in the recent years. DFx mainly aims to improve the specific target it wants to maximize, but ignores all the other aspects of product lifecycle (Kuo et al. 2001). It is not easy to succeed in integrating different target design methodologies and it does not solve the highlighted problem. This is due to the difficult management of the design and decisional inconsistencies (Brissaud et al. 2007). In the last 15 years target design methodologies have had a connotation towards improving the environmental aspects of the product. DFE practices are oriented to develop "green" products and components while maintaining product, price, performance, and quality standard (Ramani et al. 2010). DFE suggests rules and guidelines for designers or engineers to include environmental considerations in the design phase. Another example is Design for Recycling/Reusing (DFR), which is aimed to improve environmental aspects of industrial products at the end of life (EOL) and encourage the recyclability of materials with simple design rules (Villalba et al. 2004). Therefore, the criteria and the procedures of DFx do not consider all the lifecycle phases of a product. Another important aspect which is lacking in the DFx methodology is the difficulty to identify simple parameters and objectives to compare the alternative design solutions.

Eco-design is closely connected to DFx as far as regulations and guidelines are concerned, but eco-design extends these concepts in all the lifecycle phases of the product as well as the use or EOL phase (Luttropp et al. 2006).

3.2 Life Cycle Assessment

Life Cycle Assessment (LCA) is the most popular approach which is able to analyze industrial products and activities in an analytical way. LCA is an environmental accounting and management approach which considers all aspects of resources used and environmental releases associated with an industrial system "from Cradle to Grave" (ISO 2006). Typically, the result of LCA analysis is a numerical value of environmental profiles distributed in different environmental categories (Goedkoop and Spriensma 2000). The most important environmental damage to be considered in the described context of industrial products are resource depletions, energy consumption and greenhouse emissions. Energy consumption and greenhouse emissions are the most important environmental parameters to evaluate (Senthil et al. 2003). It is also possible to evaluate other environmental parameters using the well-known LCA indicators called EI99. The EI99 indicators divide the damage to the ecosystem into three major categories:

- Human Health
- Ecosystem Quality
- Resources.

These indicators can be normalized and weighted in order to obtain a dimensionless value of environmental impact which considers the whole lifecycle (Goedkoop and Spriensma 2000). This indicator can be useful to compare alternatives in design choices.

Environmental load can be analyzed case by case. The whole product lifecycle is analyzed to describe the environmental impacts in different phases. LCA is a holistic view of environmental interactions which covers a range of activities, from the extraction of raw materials from the Earth to the production and distribution of energy through the use, reuse and final disposal of a product. LCA is a tool intended for relative comparison and not absolute evaluation, thereby helping decision makers to compare all major environmental impacts in the choice of alternative courses of action (Curran 1996). LCA analysis is both time and resource consuming due to the collection of the product data needed to perform it (Life Cycle Inventory phase-LCI). For this reason complete LCA can be carried out mainly to assess the environmental profile of an existing product, where manufacturing, use and dismantling can be accurately estimated.

Simplified Life Cycle Assessment (SLCA) approaches try to increase LCA usability in the early design stages of industrial products (Kaebernick et al. 2003). The main goal of SLCA is to reduce the complexity of several of the tasks involved, while maintaining the main features and soundness of a complete LCA. One projection for the development of such simplified methods has been the possibility to employ them during the early design stages of a product, when the data available is incomplete and lacking in details. SLCA adopts dedicated tools to estimate the environmental impacts of product alternatives and to predict environmental costs or burdens for manufacturers (Kaebernick et al. 2003). A large number of simplified LCA methods and tools have been developed in the last years to perform LCA analysis easily. In the engineering design phase it is not easy to perform an environmental consideration for the complete life cycle when not all product data is available and fixed. Therefore, it is important to evaluate simplified methods and to study what type of information they require and what kind of results they can produce. New emerging Computer Aided Design (CAD) tools are trying to support the integration of SLCA methods within the traditional flow of product design activities (Cerdan et al. 2009).

3.3 Design for Disassembly

Design for Disassembly is a particular target design methodology (DFx) which aims for the rapid separation of the components of an industrial product. The product disassemblability is relevant in two different phases of product lifetime: during the use phase for maintenance operations and during the EOL phase. Successful DFD entails the application of three critical disciplines:

- the selection and use of materials;
- the design of components and the product architecture;
- the selection and use of joints, connectors and fasteners (Bogue 2007).

Several authors have developed rules and guidelines to design products for easy disassembly. DFD methods involve all the phases of a product lifecycle design, from the conceptual design stage to detailed design stage. Dewhurst (1993) evaluates the depth of disassembly for particular components in a product to establish the effective cost convenience for disassembly operations. Zussman et al. (1994) employed a utility theory for assessment of EOL alternatives, which also takes into account the uncertainties in future economic and technical conditions. The authors considered three objectives in this evaluation:

- profit maximization through selection of the best EOL option;
- maximization of the number of reusable parts;
- minimization of the amount of landfill waste.

Relationships have been developed to evaluate disposal costs and recovery benefits (Zussman et al. 1994). Ishii et al. (1993) introduces the concept of "clump" for their analysis on Design For Product Retirement (DFPR). A "clump" is a collection of components or sub-assemblies which share common characteristics. The "clumps" in a product can be recycled and reprocessed without further disassembly. In the last decade the most important works on disassembly methodologies have focused on extrapolating data from 3D CAD models. In particular, many researchers have developed algorithms to find the best disassembly sequence for components in an industrial product. Their topics are mainly oriented to the selective disassembly of components due to the high value (Dini et al. 2001). Kara et al. (2005), for example, proposes an evaluation method to detect the possible paths for the disassembly of a specific component from the product. The achievement of the best disassembly sequence (e.g. the one that minimizes time) can be obtained through appropriate algorithms. The results of this analysis are the AND/OR diagrams, a very important tool to describe the different possible paths of disassembly. Information processing, as well as the implementation of these algorithms, requires long data elaborations.

Gungor and Gupta (1998) present a methodology to evaluate different disassembly strategies in order to choose the best one. The proposed methodology tries to minimize the direction of movement changes and tool changes during disassembly. Algorithms are introduced to improve the methodology and to find the geometrical constraints. The authors also introduce a method to handle uncertainty arising in disassembly due to defective parts and breakage during the disassembly process.

Srinivasan et al. (1997) analyzes the type of connection between components, the arrangement of components (product architecture), the direction of extraction and the first component to be disassembled to minimize time. A further step in this direction is the ability to recognize the type of mechanical liaison between components, thus to generate an optimum disassembly sequence directly from the CAD product model during the design stage. Algorithms have been developed to solve Disassembly Sequence Planning, i.e. the determination of the sequence for disassembling component parts using combinatorial structure models (Adenso-Díaz et al. 2007). These objectives can greatly reduce the time associated with disassembly operations and thus the costs. Another perspective in a virtual DFD method is called "virtual disassembly". This approach permits a "virtual" disassembly of components from a virtual product model through a virtual-reality-based software tool (Mo et al. 2002). Virtual Reality is a new technology which creates a real-time visual/audio/haptic experience with computer systems including hardware and software. It provides a potential way for disassembly simulation. This tool can support collaborative de-manufacturing (disassembly, service, recycling and disposal) between manufacturer/de-manufacturer, disposer and designer

(Mo et al. 2002). However, in spite of all the described methods to evaluate the best disassembly sequence and time, no additional information or guidelines are given to improve product disassemblability (Lambert 2001). Design suggestions can be available for designers during product development in order to conceive a correct product architecture, the right selection of materials and minimize disassembly time for particular components (Gungor and Gupta 2001).

A list of rules and guidelines can be adopted during design phases to increase product disassembly and to reduce de-manufacturing operations and time. These rules can be managed by expert engineers who know the product features and characteristics (Chan and Tong 2007). The use of these rules and guidelines in design departments permits a high level of selective disassembly to be reached with the scope to encourage the 3R approaches (Reusing, Recycling and Remanufacturing) (González and Adenso-Díaz 2005). A list of requirements are summarized and described in details. These guidelines can be classified and organized with the aim to suggest design improvements by the point of view of disassemblability. For example, specific guidelines related to material selection or liaison typology can be highlighted during design step to encourage closed-loop EOL scenarios.

4 Numerical Evaluation of Product Disassemblability: The Approach

The proposed approach is developed in the general context of industrial products inherent to mechatronic products such as large and small appliances, electric and electronic products, lighting systems and ICT tools. In particular, the target of the work is focused on the management and evaluation of EOL options for products and components in the early design phase with the aim to reduce environmental impacts and increase dismantler profits. New indices are developed to be used by designers and engineers to measure the disassemblability of products. In particular, these indices assess the economic feasibility of the different EOL scenarios and are combined with the LCA parameters to have a global vision of product lifecycle. A general framework of the software tool is proposed on the basis of this approach with the scope to help the designers in product disassembly evaluation. The tool is able to calculate the new metrics easily and suggest how to modify the product features to minimize both disassembly operations and time.

4.1 Disassembly Approach

The disassembly approach described below is oriented to increase the disassemblability of industrial products and inspire designers in the realization of EOL closed-loop lifecycle. The great challenge of this approach is to start with the product structure to reach the product sustainability by the analysis of suitable

EOL indices. The integration with the traditional design tools, such as the CAD tool, permits data for the disassemblability analysis to be shared and retrieved. The assessment of dedicated EOL indices combined with the disassembly path gives important suggestions to modify the original product design and to increase the environmental and economic performances of products at EOL. The environmental impacts, assessed by the SLCA tool, are closely related to the proposed indices with the scope to create a global view of the EOL aspects. The approach is made to increase the percentage of components which can have a closed-loop lifecycle, encouraging the Reusing, Recycling and Remanufacturing scenarios. A framework of the disassembly approach is proposed in Fig. 2.

The starting point of the approach is to collect the data derived from the 3D CAD system regarding the product structure, which refers to all the important features concerning the general architecture of product (sub-assemblies and components), Bill of Materials (BOM), production costs, etc. All this information can be used to define the liaison type and shape between components in the products and to retrieve the best disassembly sequence, necessary for the calculation of the EOL indices. The relationship among different components in the product is considered using a N $\times$ N "Precedence Matrix", where N is the total number of components. The elements in the diagonal are null. The rows of the matrix are filled using "1" for the components in the column which have disassembly precedence or "0" for the others. The disassembly phases are calculated by analyzing the precedence matrix. The first phase (Phase 0) contains the components which can be removed without any precedence. The components of phase "n" can only be removed if all the components of the phase "n − 1" have been disassembled. A detailed example of this process is reported in Fig. 3.

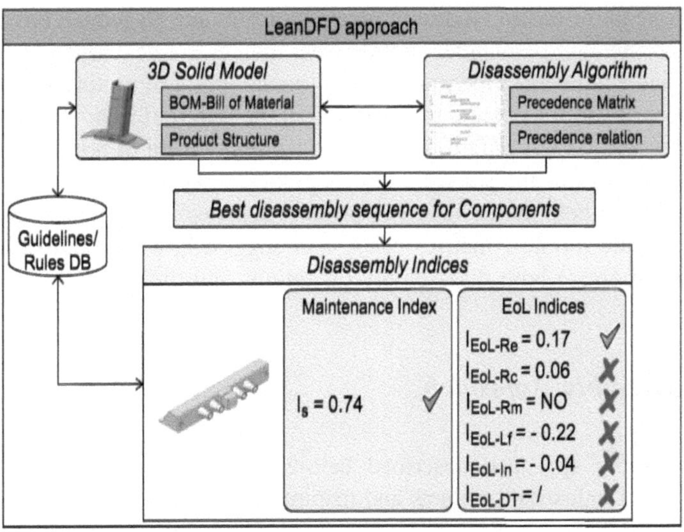

Fig. 2 Framework of disassembly approach

Fig. 3 DFD approach example to calculate disassembly precedence, feasible sequences and and/or diagram

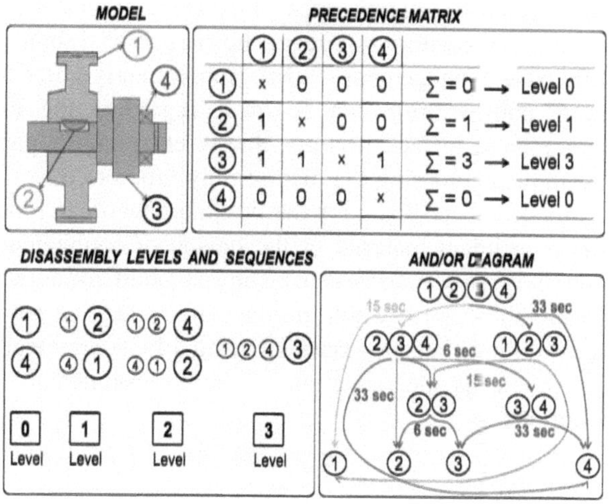

The first positive outcome of this disassembly analysis is the possibility to retrieve all the feasible disassembly sequences for a particular component or subassembly in the product (AND/OR diagram). The AND/OR diagram is the most important result of the disassemblability analysis. The AND/OR diagram is a powerful design tool to manage product disassembly and to recognize which disassembly sequence can be improved and how. In fact, the AND/OR diagram highlights the feasible disassembly sequences and the related time necessary to achieve each disassembly path. Using this diagram, it is possible to evaluate the best disassembly sequence for each product component or subassembly necessary for the selective disassembly. The criteria used for the selective disassembly of product components by the analysis of AND/OR diagram is the sequence which minimizes disassembly time.

The calculation of disassembly time for each possible disassembly sequences is fulfilled by the definition of the liaison type between two components at each "phase" of disassembly. Using this approach is it possible to clearly define a disassembly time for each liaison and so the best disassembly sequence would minimize the disassembly time (Fig. 3). Time is a common parameter used for the calculation of the indices proposed for selective disassembly. The indices also take into account cost and environmental and collateral aspects which effect the selection of a specific EOL scenario. The standard disassembly times of each liaison, referred to a specific condition, are stored in the relative repository (standard_time DB). Furthermore, the condition of each liaison in the moment of the de-manufacturing is considered with the introduction of several correction factors which multiply the specific times in order to obtain realistic times for the removal of each particular liaison.

Another important feature of the proposed method is to use this disassembly time to calculate the disassembly cost of manual operations, a key parameter in evaluating the EOL indices. Other parameters for the calculation of EOL indices, such as costs or properties of constituent material, can be easily retrieved by common repositories used in the industrial design departments such as Product Lifecycle Management (PLM). Subsequently, after the calculation of the indices, more consideration can be made regarding the EOL scenarios for a specific component. In fact, it is possible to encourage a particular EOL scenario by, for example, decreasing the disassembly time or cost of disassembly operations. This is important feedback for the designers who can modify the geometry, the shape, the constituent material or the liaison of components in the product. New alternative EOL scenarios can be investigated by designers to increase the product sustainability in a whole product lifecycle. The advantages in terms of environmental impact can be assessed using a LCA approach. The LCA indicators checks and confirms the reduction of the environmental damage for the new adopted solutions.

4.2 Disassembly Indices to Measure the Feasibility of Different EOL Scenarios

Six new indices are proposed as metrics to describe and measure the feasibility of the described EOL scenarios. These indices are able to support design choices for the selection of suitable materials for components in products, to design efficient joint methods and to encourage the closed loop lifecycle of the product. On the basis of these indices calculated for each component or sub-assembly, an analysis for product recyclability, reusability, and re-manufacturability can be made. Furthermore, it is possible to use these metrics in the design stage to inspire a specific EOL scenario, for example the recyclability of the product. The results of the proposed approach can be used, for example, to support the EU policy for the waste treatment of industrial products.

The main parameters introduced to define the indices are directly linked to economic aspects of product disassembly. The parameters introduced in these indices can be easily estimated in the design phase by the use of traditional design tools or by the management of product information. A list of these parameters is proposed below:

- C_p is the value of the part at EOL considering use deterioration (€). This terms can be assessed with dedicated lifecycle analysis or approx. as a percentage of the initial economic value of the part;
- C_c is the cost of cleaning operations for part regeneration (€). This value can be calculated taking into account the surface of the components to be cleaned [m^2] and an estimated cost per surface unit (€/m^2). This aspect includes costs of personnel, structures and tools;
- C_d is the disassembly cost of parts (€). It is related to disassembly time (T_d), the "Disassembly Correction Factor" (DCF), labor cost (L) and eventually the cost of special tools or equipment (C_t), according to the following equation (Eq. 1):

$$C_d = \left(\frac{T_d}{DCF} \times L\right) + C_t \tag{1}$$

- C_r is the cost of items/parts to be replaced with new ones (material production and manufacturing cost for the same new part) (€);
- m is the mass of the disassembled part (kg);
- R_f is the recycling factor (dimensionless);
- C_{Rc} is the economic value of recycled material (€/kg);
- E_{saving} is the difference between production energy necessary to obtain the virgin material and the recycling energy necessary for the material recycling treatment (MJ/kg);
- C_E is the energy cost (€/MJ);
- C_m is the production cost of virgin material (€);
- R_p is the market price of the new component produced starting from a component of the product retrieved by the "closed loop" (€). The second life of the component/sub-assembly generates a profit for the company which uses the remanufactured part in new product;
- C_{RL} is the cost of reverse logistics including the transport system from retailers to dismantling centers and from dismantling centers to different suppliers (€). This value can be calculated taking into account the volume of the components (m^3) and an estimated cost per unit of volume (€/m^3). This includes costs of personnel, structures and tools;
- C_{ReMan} is the cost of refurbishment for the specific components considering the inspection phase, the regeneration process or the remanufacturing operations to obtain the new part starting from the selected component (€);
- C_l is the cost for land-filling the part (€). This cost considers EOL treatments such as pollutant separation, product compaction, transport and others;
- P_c is the material heating value (MJ/kg);
- C_{dd} is the disassembly cost of parts in relation to destructive operations of disassembly (€).

The indices calculation permits different EOL scenarios to be analyzed for the different components under investigation and enables the entire product EOL to be managed. The proposed EOL indices are described below:

- *Reusing Index.* This index considers the opportunity for components to be reused in the same or other similar products. This first EOL index considers the components and not the materials. This EOL scenario is only possible if the component EOL life time is longer than the product life time (Eq. 2).

$$I_{EOL-Ru} = \frac{C_p - C_c - C_{RL} - C_d}{C_r} \tag{2}$$

- *Recycling Index.* This index compares the difference in terms of production costs for virgin materials and the cost of the recycling process. In particular, it takes into account the energy savings resulting from the Material recycling process. Another important constraint for the calculation of the recycling index is the quantity of material which can be recycled (recycling factor) in the current supply chain. This index establishes the real effective opportunity in terms of energy and cost reduction (Eq. 3).

$$I_{EOL-Rc} = \frac{(m \times R_f \times C_{Rc}) + (m \times E_{saving} \times C_E) - C_c - C_{RL} - C_d}{C_m} \tag{3}$$

- *Remanufacturing Index.* Equation (4) provides a numerical formula to calculate this index. The Remanufacturing Index ($I_{EOL\text{-}Rm}$) is based on different cost types and revenues involved in the "Remanufacturing loop".

$$I_{EOL-Rm} = \frac{(C_m \times m) + (m \times E_{saving} \times C_E) - C_{RL} - C_d - C_c - C_{ReMan}}{R_p} \tag{4}$$

- *Landfill Index.* Landfill does not generate any benefit from an economic point of view and therefore must never be considered as sustainable EOL treatment. In some cases, however, inert materials could be an alternative to recycling. Landfill treatment, by its nature, does not require careful disassembly and therefore the related costs are low. Only when all other possible EOL treatments are not profitable for the selected component can the landfill index be estimated (Eq. 5):

$$I_{EOL-Lf} = \frac{(-C_l)}{C_p} \tag{5}$$

- *Incineration Index.* Incineration is an opportunity for the EOL treatment of particular materials with a high heating value or for materials which cannot be easily recycled. This index establishes if particular combinations of materials can be directly incinerated for energy production (Eq. 6). In this case the components can be separated by destructive disassembly techniques and the disassembly operations can be made without particular attention. The time required for disassembly is therefore greatly reduced.

$$I_{EOL-Inc} = \frac{(m \times P_c \times C_E) - C_{dd}}{C_m} \tag{6}$$

- *Different Treatment Index.* This EOL treatment is necessary for particular types of materials which require further processes because they are considered potentially dangerous for the environment. This is the case, for example, of Printed Circuit Boards (PCBs), for recovering expensive materials (gold, palladium, etc.) or other hazardous materials, such as Ethernit or heavy metals which must be treated before being disposed of due to their high risks to human health. The index evaluation is a simple query for designers (Eq. 7).

$$I_{EOL-DT} = YES\_or\_NO \tag{7}$$

4.3 Disassemblability Tool Architecture and Features

A preliminary software tool architecture and integration with traditional design tools has been proposed in Fig. 2. The entire structure of the software and the data flow exchange is reported in Fig. 4.

The proposed tool can be built using four main modules:

1. *CAD importer*. Used to read both the product structure of assembly, which is the analyzed product, and geometrical entities for each occurrence. The aim is to convert this information into a proprietary data structure to define a specific geometrical kernel to use for geometry visualization and analysis. This is required to view the product and to analyze it extracting useful information for disassembly paths and liaison calculations. The importer provides functions to read directly from the common CAD systems. The definition of a data structure to store geometrical information allows the DFD engine to work independently from the CAD system used to model the analyzed product.
2. *DB reader*. Used to read data from the Liaisons_DB and the Costs_DB, to provide data for the DFD engine. The relational repositories contain the liaison definitions (Liaisons_DB), materials and relative processes (Materials_DB) which can be defined by designers or engineers. Liaisons are classified in a one level tree according to their class membership: for instance, screws, bolts and threaded rods are combined within the same group (threaded liaison) because they have similar properties. Each item of the Liaisons_DB has its own standard disassembly time which refers to a specific type of liaison under particular conditions. For example, the disassembly time of a screw is parameterized on its diameter, length, weight, head type, wear, working environment and material. The Liaison DB also contains data regarding the tools used to disassemble a

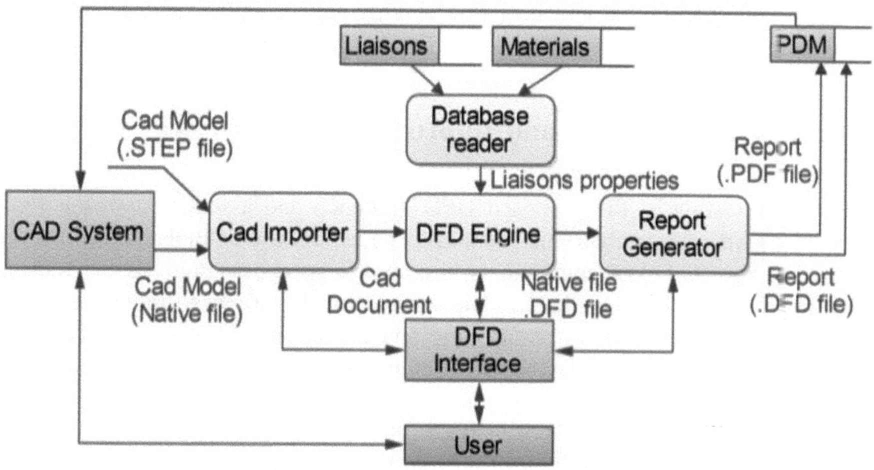

Fig. 4 DFD software structure using "data flow diagram" method

liaison and unitary costs. The Materials_DB contains data used to calculate EOL indices, such as material unitary cost, its transformation processes, recycling factor, economic value of recycled material, material heating value, etc.

3. *Report generator*. Used to define reports required by users to view EOL indices and graphs summarizing the project strengths concerning its disassemblability.
4. *Engine*. The main module of DFD system, providing the functions to define liaisons, precedence among components and to calculate disassembly paths and EOL indices. The engine's objective is to define a file containing information concerning EOL indices and product disassemblability, starting from a CAD document. Designers are driven toward the definition of disassemblability precedence and liaisons among components or sub-assemblies through a specific visual block.

The use of the tool described previously in this section leads to relevant advantages. With the described DFD application tool the designer of industrial products can easily consider all possible EOL scenarios as early on as in the design phase. The accurate calculation of disassembly time and costs as well as the calculation of the EOL indices can evaluate the percentage of components able to be recycled, remanufactured or reused. In this way the designer can modify some characteristics of the new product to improve the disassemblability, and as a consequence the sustainability of the product or of some important components. Another great advantage is the possibility of also considering a selective disassembly. The tool is able to calculate all the feasible disassembly sequences, so the designer can evaluate the disassembly times and costs of particular components or subassemblies, with the aim of favoring the maintenance of products. Selective disassembly is also very useful for products containing hazardous substances. In fact, for these particular products the current communitarian regulations impose manufacturing companies to take them back at the end of their life for the correct treatment of materials. For this reason, industries which are involved in these sectors have the interest to design components or subassemblies which can be disassembled quickly and at a very low cost.

5 Case Study Analysis and Improvements

Preliminary testing of the proposed DFD approach and assessment of the EOL metrics is given for mechatronic products. The estimation of EOL indices for the original design (O.D.) of components in a cooker hood case study is the starting point for the new redesign (N.D.). The analysis of the proposed indices together with the calculation of disassembly time and operation provide successful solutions for product redesign. Two product components are analyzed in detail as an example of the application of the method: a *Blower system* (with their support) and the *Electrical system*. The cooker hood disassemblability evaluation is made considering all phases of product lifecycle using the LCA approach. The LCA parameters together with the EOL indices give easily understandable metrics of the environmental and economic benefits of the proposed redesign. The designer's choices in the context of disassembly involve not only the product EOL but also the product use and material selection to manufacture the components. For these reasons a global point of view for the cooker hood is presented in Table 1.

The new design (N.D.) solution, adopted using the proposed approach, shows that there is a decrease of more than 30 % in the environmental parameters calculated (CO_2 production, Energy consumption and Resources exploitation), which consider all phases of the product lifecycle. Other advantages of the new adopted solutions for the hood product are the possibilities to recycle material (approx. 38 %), reuse components (17 %), and remanufacture components (approx. 7 %). Therefore, for the hood lifecycle, more than 62 % of product components can have a closed-loop lifecycle. There is an important, significant reduction in the percentage of EOL landfill waste with the new design of cooker hood compared to the old solution ($I_{EOL\text{-}Lf}$ 84.3 vs. 31.1 %).

5.1 Blower System Support Redesign

A first example of the proposed approach application is the *Blower system support* of the cooker hood. The blower module is characterized by a conveyor system for air, an electric motor and a rotor. The blower system is responsible for air movement. The Blower system is reinforced and assembled with the Blower system support. This support is a central module of the cooker hood and it is produced using materials (plastics and metals). The importance to reuse or remanufacture parts of the blower system such as the electric motors is a key feature added to the possibility to recycle the Blower System. For this reason a rapid separation system is implemented in the blower support to reduce the demanufacturing time and costs as depicted in Fig. 5.

Table 1 Comparison between the original design (O.D.) and the new design (N.D.) of a free-standing cooker hood

Free-standing cooker hood			
Parameters		O.D.	N.D.
Number of used materials		>20	<12
Component numbers		205	82
Environmental	Equivalent CO_2 (kg)	1222.5	725.7
	Energy cons. (MJ)	45e3	9.3e3
	Resources (EI99 Pt)	184	72
Cost	*LCC [€]*	953.8	667.2
End-of-life	$I_{EOL\text{-}Rc}$	6.2 %	38.5 %
	$I_{EOL\text{-}Ru}$	1.1 %	16.8 %
	$I_{EOL\text{-}Rm}$	3.2 %	7.2 %
	$I_{EOL\text{-}Inc}$	0.6 %	0.6 %
	$I_{EOL\text{-}Lf}$	84.3 %	31.1 %
	$I_{EOL\text{-}Dt}$	4.6 %	5.8 %

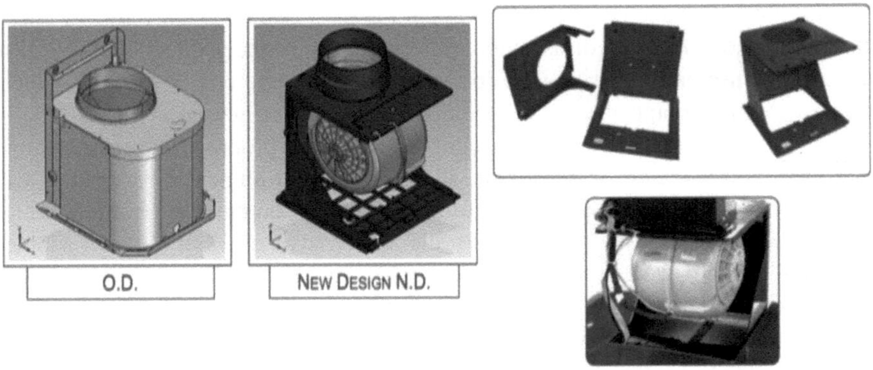

Fig. 5 Solution adopted in the new design configuration of blower system support and comparison with the original solution

The new adopted solution is characterized by the use of sliding guides to link the only two parts of the blower support, and the use of a unique type of plastic (PP-flame retardant). This improvement is realized by the assessment of the disassembly time, very critical for this particular sub-assembly. The application of desired rules and guidelines suggest this new assembly solution. The indices evaluation and the comparison with the previous solution are reported in Table 2. The indices value shows how recycling is the best EOL treatment for the new design (N.D.) of the Blower system support.

5.2 Electrical System Support Redesign

Similarly to the Blower, the necessity to improve the disassemblability of the Electrical System Support is also related to managing the PCB, Capacitor and Electro-mechanical Transformer linked to it. The original design solution is

Table 2 Evaluation results of EOL indices for the two components

	Blower S. S.		Electrical S. S.	
EOL indices	O. D.	N. D.	O. D.	N. D.
Material	Different materials	PP f.r.	Different materials	PP f.
Components	12	2	13	1
$I_{EOL\text{-}Rc}$	0 %	1.7 %	0 %	9.4 %
$I_{EOL\text{-}Ru}$	0 %	0 %	0 %	1.3 %
$I_{EOL\text{-}Rm}$	/	/	/	/
$I_{EOL\text{-}Inc}$	/	0 %	/	/
$I_{EOL\text{-}Lf}$	−64.2 %	−18.8 %	−12.7 %	−10.3 %
$I_{EOL\text{-}Dt}$	/	/	/	/

characterized by a galvanized steel support to which all the previously listed items are attached. All these electrical items are fixed using bolt and nut systems. The new design solution, which has been developed with the described approach, is an Electrical Box containing the entire Electrical System items (for example Capacitor, PCB system, Electromechanical transformer, etc.). Additionally, the Electrical Box is linked to the hood by cylindrical Snap-fits which allow the entire hood sub-assembly to be separated rapidly. The Electrical Cover and the Electrical Box are linked by the use of rectangular snap-fits. Another important feature of the new design solution (N. D.) of the Electrical System, and in particular for the Electrical box, is the selection of the same plastic material used for the other components of the hood (PP flame retardant). Figure 6 shows the new design solution adopted for the Electrical System module and the comparison with the original design.

The index results for the Electrical System Support are reported in Table 2. A comparison with the original design shows an increased value of the closed-loop EOL indices, i.e. Recycling and Reusing Index.

6 Results and Concluding Remarks

This research study is dedicated to support the phase of product analysis for disassemblability during the early design activities, when the 3D CAD model is defined and can be used to explore different solutions and make decisions. The definition of the precedence and liaisons between components of an assembly permits the disassembly time to be calculated. Considering the cost of labor and

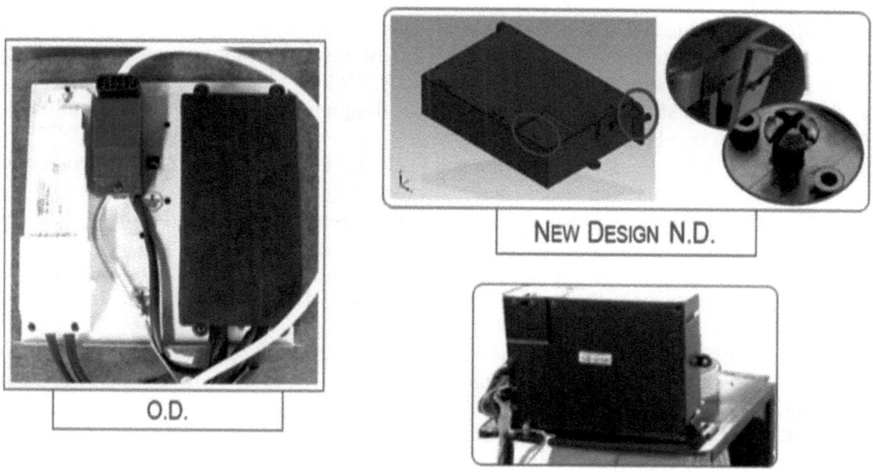

Fig. 6 Solution adopted in the new design configuration of electrical system and comparison with the original solution

the tools necessary in the disassembly process, it is possible to evaluate the disassembly cost of each component or subassembly. The great advantage of such an approach and the related software is given by the possibility to assess the product sustainability oriented to the EOL phases of the product lifecycle. The six presented indices allow designers of industrial products to evaluate the percentage of product components which have a closed-loop lifecycle. The shown case study gives an important result in terms of product EOL closed-loop, encouraging the Recycling, Reusing and Remanufacturing. Future efforts have the aim of studying different products to test the robustness of the formulated indices. It is also necessary to improve the DFD tool system in order to automatically extract information by reading the 3D model and adding software functionalities. In this way the system could provide designers with a powerful tool for immediate assessment of product disassembly and for the management of the different EOL scenarios.

Finally, it is important to underline how the described sustainability evaluation is possible during the design stage. This is absolutely fundamental in order to reduce the impact of the necessary design modifications in terms of manufacturing process and, in particular, production costs. This characteristic is essential to encourage manufacturing companies to consider the disassembly and in particular the EOL management in the design process as the other design drivers. As a consequence not only will the new products be optimized in terms of performance or costs, but they will also be developed according to the eco-design guidelines with the aim of reducing environmental impact and waste.

References

Adenso-Díaz B, García-Carbajal S, Lozano S (2007) An efficient GRASP algorithm for disassembly sequence planning. OR Spectr 29(3):535–549

Ashby MF (2009) Materials and the environment: eco-informed material choice. Elsevier, Oxford

Bogue R (2007) Design for disassembly: a critical twenty-first century discipline. Assembly Autom 27(4):285–289. doi:10.1108/01445150710827069

Boothroyd G, Dewhurst P, Knight W (2002) Product design for manufacture and assembly third edition. Taylor and Francis, New York

Brissaud D et al (2007) Product Eco-design and materials: current status and future prospected. 1st International seminar on society and materials

BS 8887-2 (2009) Design for manufacture, assembly, disassembly and end-of-life processing (MADE)

Capelli F, Delogu M, Pierini M, Schiavone F (2007) Design for disassembly: a methodology for identifying the optimal disassembly sequence. J Eng Des 18(6):563–575. doi:10.1080/09544820601013019

Cerdan C, Gazulla C, Raugei M, Martinez E, Palmer PF (2009) Proposal for new quantitative eco-design indicators: a first case study. J Clean Prod 17(18):1638–1643. doi:10.1016/j.jclepro.2009.07.010

Chan JWK, Tong TKL (2007) Multi-criteria material selections and end-of-Life product strategy: Grey relational analysis approach. Mater Des 28(5):1539–1546. doi:10.1016/j.matdes.2006.02.016

Curran MA (1996) Environmental life cycle assessment. McGraw-Hill, New York

Das SK, Yedlarajiah P, Narendra R (2000) An approach for estimating the EOL product disassembly effort and cost. Int J Prod Res 38(3):657–673

Dewhurst P (1993) Product design for manufacture: design for disassembly. Ind Eng 25:26–28

Dewulf W, Willems B, Duflou JR (2006) Estimating the environmental profile of early design concepts. In: Innovation in life cycle engineering and sustainable development, Part 3. Springer, Netherlands

Dini G, Failli F, Santochi M (2001) A disassembly planning software system for the optimization of recycling processes. Prod Plan Control Manag Oper 12(1):2–12. doi:10.1080/09537280150203924

European Parliament and Council (2003) Directive 2002/96/EC of 27 January 2003 on waste electrical and electronic equipment (WEEE)

European Parliament and Council (2003) Directive 2002/95/EC of 27 January 2003 on the restriction of the use of certain hazardous substances in electrical and electronic equipment (RoHS)

European Parliament and Council (2000) Directive 2000/53/EC of 18 September 2000 on end-of life vehicles

Gehin A, Zwolinski P, Brissaud D (2008) A tool to implement sustainable end-of-life strategies in the product development phase. J Clean Prod 16(5):566–576. doi:10.1016/j.jclepro.2007.02.012

Goedkoop M, Spriensma R (2000) The eco-indicator 99. A damage oriented method for life cycle impact assessment methodology. Report 17 April 2000, Second edition

González B, Adenso-Díaz B (2005) A bill of materials-based approach for EOL decision making in design for the environment. Int J Prod Res 43:2071–2099. doi:10.1080/00207540412331333423

Gungor A, Gupta SM (1998) Disassembly sequence planning for products with defective parts in product recovery. Comput Ind Eng 35(1–2):161–165

Gungor A, Gupta SM (2001) Disassembly sequence plan generation using a branch-and-bound algorithm. Int J Prod Res 39(3):481–509

Hauschild MZ, Jeswiet J, Alting L (2004) Design for environment—do we get the focus right?. CIRP Ann Manuf Technol 53(1):1–4. doi:10.1016/S0007-8506(07)60631-3

Herrmann C, Frad A, Luger T (2008) Integrating the end-of-life evaluation and planning in the product management process. Prog Ind Ecol 5(1/2):44–64

Ishii K, Eubanks CF, Marks M (1993) Evaluation methodology for post-manufacturing issues in life-cycle design. Concurr Eng Res Appl 1(1):61–68. doi:10.1177/1063293X9300100107

ISO 14040 (2006) Environmental management—life cycle assessment—principles and framework

ISO/TR 14062 (2002) Environmental management—integrating environmental aspects into product design and development

Kaebernick H, Sun M, Kara S (2003) Simplified life cycle assessment for the early design stages of industrial products. CIRP Ann Manuf Technol 52:25–28

Kara S, Pornprasitpol P, Kaebernick H (2005) A selective disassembly methodology for end-of-life products. Assembly Autom 25(2):124–134. doi:10.1108/01445150510590488

Kuo TC, Huang SH, Zhang HC (2001) Design for manufacture and design for X: concepts, applications, and perspectives. Comput Ind Eng 41:241–260

Kwak MJ, Hong YS, Cho NW (2009) Eco-architecture analysis for end-of-life decision making. Int J Prod Res 47(22):6233–6259

Lambert AJD (2001) Automatic determination of transition matrices in optimal disassembly sequence generation. Proceedings of the IEEE international symposium on assembly and task planning, pp 220–225. doi: 10.1109/ISATP.2001.928993

Luttropp C, Lagerstedt J (2006) Eco-design and the ten golden rules: generic advice for merging environmental aspects into product development. J Clean Prod 14:1396–1408. doi:10.1016/j.jclepro.2005.11.022

Miheclic JR, Paterson KG, Phillips LD, Zhang Q et al (2008) Educating engineers in the sustainable futures model with a global perspective. Taylor & Francis, London

Mo J, Zhang Q, Gadh R (2002) Virtual disassembly. Int J CAD/CAM 2(1):29–37

Ramani K, Ramanujan D, Bernstein WZ, Zhao F, Sutherland J, Handwerker C, Choi JK, Kim H, Thurston D (2010) Integrated sustainable life cycle design: a review. J Mech Des 132(9)

Rose CM (2001) Design for environment: a method for formulating product end-of-life strategies. Dissertation, Stanford University

Rose CM, Ishii K (1999) Product end-of-life strategy categorization design tool. J Electron Manuf 9(1):41–51. doi:10.1142/S0960313199000271

Senthil K, Ong SK, Nee AYC, Tan RBH (2003) A proposed tool to integrate environmental and economical assessments of product. Environ Impact Asses 23:51–72

Sousa I, Wallace D (2006) Product classification to support approximate life-cycle assessment of design concepts. Technol Forecast Soc Change 73:228–249

Srinivasan H, Shyamsundar N, Gadh R (1997) A framework for virtual disassembly analysis. J Intell Manuf 8:277–295. doi:10.1023/A:1018537611535

United States (U.S.) Environmental Protection Agency (2000). Solid waste and emergency response EPA 530-N-00-007

Villalba G, Segarra M et al (2004) Using the recyclability index of materials as a tool for design for disassembly. Ecol Econ 50:195–200. doi:10.1016/j.ecolecon.2004.03.026

Zussman E, Kriwet A, Seliger G (1994) Disassembly-oriented assessment methodology to support design for recycling. CIRP Ann Manuf Technol 43(1):9–14

Index

A

Aircraft
- industrialization sequence, 96
- life cycle, 81, 86, 87, 91, 92, 225
- materials and processes, 87
- production cost, 74, 77, 88, 95, 96
- structures, 83–85, 89

Architecture
- modular, 9, 10
- product, 1, 4, 6, 12, 297, 300, 307–308
- system, 1, 9, 13, 17, 253

Automotive
- Battery technology, 10, 11
- Industry, 10, 12, 59, 62

B

Balancing procedure, 169, 170

C

Carbon footprint, 143, 176, 187, 284

CO_2 emission, 119, 243, 281

Cost
- life cycle, 2, 59
- model, 30, 59–61, 64, 65, 67, 68, 70–73, 76, 116, 125
- performance, 61, 107
- second order, 59, 60
- traditional accounting, 114

D

Data base, 113, 208

Decision-making, 2, 43, 47, 115, 122, 126, 188, 194, 195, 199, 202, 217, 220, 224, 260, 261

Decision-making process, 63, 220

Demanufacturing, 299, 303, 308, 309, 311, 317

Design
- alternatives, 13, 119, 218, 223, 224, 229, 230, 232, 235, 262, 306
- evaluation, 262, 290, 297
- for environment, 218, 241, 305
- methodology, 288, 307
- phases, 21, 27, 29, 30, 203, 219, 299, 304, 309

Development process, 1, 4, 15, 16, 21, 26–30

DIN EN ISO 14040/44, 132

Disassembly, 265, 277, 286, 289, 291, 297, 299, 300, 302, 303, 307–316, 320

Dismantling, 245, 275, 282, 284, 289, 300, 304, 307, 313

Double helix, 21, 22, 27–31

E

Eco-design, 219, 257, 259, 261–263, 268, 274, 276, 277

Electromechanical systems, 257, 258, 264, 275, 277

End-of-life, 125, 196, 201, 223, 258, 261, 282, 283, 286, 290, 291, 298

Enegy
- consumption, 122, 134, 140, 141, 149, 150, 153–160, 163–165, 169, 171, 176, 178, 183, 184, 187, 196, 222, 231, 252
- flows, 115, 119–121, 126, 132, 137, 146
- management, 272

Environment
- performance, 105, 114, 139, 140, 217–220, 224, 227, 230, 231, 257, 258, 261, 297, 299

Equipment

E. Henriques et al. (eds.), *Technology and Manufacturing Process Selection*, Springer Series in Advanced Manufacturing, DOI: 10.1007/978-1-4471-5544-7,

manufacturing, 59, 125, 154, 155, 157, 159–165, 169
Evaluation
life cycle-oriented framework, 143, 193, 245, 246
of materials, 107, 126, 144
of process technologies, 193, 194, 197, 203, 206, 212
of product technologies, 193, 194, 197, 203, 206, 212
Procedure model for, 147, 199
Expert exchange, 252
Expert system, 186–188

F
Feed
animal, 176, 179
compound processing, 177
Flexibility
Framework for architecture, 7, 9, 10, 12, 13, 17
importance of, 4
in a project, 1, 8, 9, 13
in architecture, 4, 10, 12
in automotive industry, 10
rule, 14
valuation of, 7, 8, 14
Forecasting, 2, 3, 9, 16, 50, 123, 125, 126, 160
Forging, 60, 62, 72–74
Fuzzy rules, 187, 188

G
GaBi, 135, 145, 148

I
Incremental
improvement, 11, 45–47
innovation, 45
Industrial case study, 26, 31
Information system, 114, 127, 177, 202, 241, 245, 249
Injection moulding, 217, 220, 229, 231, 235
Input-output flow analysis, 123, 261
Investment appraisal, 116, 124, 125, 202, 204
ISO 9001, 39, 44–46, 57, 91

L
Legacy, 22, 25–31
Life cycle
assessment, 116, 124, 126, 132, 133, 135, 140, 141, 144, 147, 149, 150, 176, 218, 219, 241
cost, 59, 88, 91, 125, 204, 217
design, 218, 219, 241
engineering, 203, 218
integrated analysis, 232
model, 33, 37, 41, 46, 56, 126
phases, 37, 125, 170, 220
profits, 201, 202
project, 37, 45, 149
revenues, 194, 204, 205
six sigma, 34
technical system model, 33, 35, 37–39

M
Machine-tool, 123, 157, 160, 195
Manufacturing
economic activity within, 21, 22
process, 6, 59–61, 67, 77, 82, 83, 91, 95, 131
technology, 62, 68, 82, 133
Marketing innovation, 47
Material
agricultural raw, 177
aluminum, 32, 75
carbon fiber, 282
carbon fiber reinforced plastics (CFRP), 207
composite, 87, 281
efficiency, 105–107, 116, 206
flow cost accounting, 106, 206
lightweight, 273
lightweight metallic alloys, 72
loops, 115, 116
properties, 59, 60, 64, 66, 68, 69, 77, 198, 283, 300
selection, 60–62, 69, 77, 78
substitution, 59, 60, 63, 65, 67, 69, 71, 77, 78
three-dimensional contoured thermoplastic sandwich structures (t3S), 207, 208, 210, 211
Material flow cost accounting (MFCA)
cost categories, 109
flow cost matrix, 110
flow cost model, 108, 110, 116
flow quantity model, 108, 110, 126
flow structure model, 108
implementation, 23, 25, 30, 106, 107, 114, 127, 128
plan, 106, 122, 124, 125

procedure, 108
Modularity, 6, 203, 259, 268, 273, 300
Monte-carlo simulation, 8, 9, 15

N
Net present value, 8, 125, 202, 204

O
Online measuring devices, 186, 188
Organizational innovation, 47

P
Performance criteria, 201, 207
Performance measures
economical, 123, 207
technical, 123, 207
Peripheral systems, 159, 160, 163
Planning, 1, 33, 39, 42, 45, 49, 54, 93, 112, 153, 155–157, 160, 164, 165, 182, 196, 199, 204, 261
Planning method, 112, 153, 156, 164
Process
base models, 193–195, 197, 203, 206, 210, 212
chains, 106, 114–116, 122, 125, 128
control, 68, 88, 177, 182, 185–189
innovation, 10, 35, 47, 90, 129
model, 60
Product
design, 4, 5
development, 1, 2, 4, 15, 22, 25
engineering process, 96, 154, 155, 165, 171
innovation, 10, 11
lifecycle management, 241, 245
model, 195, 243, 297, 308
sustainability, 299, 305, 309, 312, 316, 320
Production
customer specific, 185
system, 21–28
Progressive die stamping, 61, 67

Q
Quality management system, 44, 48

R
Radical innovation, 45
Reconfiguration, 21
Recycling, 76, 83, 87, 116–118, 134, 141, 147, 155, 196, 202, 208, 223, 228, 243, 244, 273, 281–289, 291, 293, 294, 297, 299, 300, 302, 308, 309, 313, 316, 320
Resource efficiency, 21, 22, 112, 122, 127, 147
Routing, 55, 96, 99–102

S
Six sigma
design for (DFSS), 34, 35, 54
DMA(DV)C roadmap, 38
DMAIC roadmap, 35, 38, 46, 49–51
IDOV roadmap, 38, 45, 55
life cycle, 33, 34, 37, 47, 57
project life cycle model, 33, 37
Squeeze casting, 60, 62, 75
Stage-gate, 24, 26, 30
Substantial innovation, 38, 47
System design, 6, 26, 27, 29, 38

T
Technology chains, 131–134, 136, 139, 144, 145, 147
Tool alternatives, 224, 232

U
Uncertainties, 4–7, 9
Uncertainty
strategies to address, 6

V
Value, 1, 3, 4, 7, 9, 10, 23, 24

MIX
Papier aus verantwortungsvollen Quellen
Paper from responsible sources
FSC® C105338

If you have any concerns about our products,
you can contact us on
ProductSafety@springernature.com

In case Publisher is established outside the EU,
the EU authorized representative is:
Springer Nature Customer Service Center GmbH
Europaplatz 3, 69115 Heidelberg, Germany

Printed by Libri Plureos GmbH
in Hamburg, Germany